Turbulent Fluid Flow

Turbulent Fluid Flow

Peter S. Bernard
University of Maryland
USA

WILEY

This edition first published 2019

Registered Offices
John Wiley & Sons, Inc., 111 River Street, Hoboken, NJ 07030, USA
John Wiley & Sons Ltd, The Atrium, Southern Gate, Chichester, West Sussex, PO19 8SQ, UK

Editorial Office
The Atrium, Southern Gate, Chichester, West Sussex, PO19 8SQ, UK

For details of our global editorial offices, customer services, and more information about Wiley products visit us at www.wiley.com.

Wiley also publishes its books in a variety of electronic formats and by print-on-demand. Some content that appears in standard print versions of this book may not be available in other formats.

Library of Congress Cataloging-in-Publication Data applied for
9781119106227

Cover Design: Wiley
Cover Images: © Willyam Bradberry/Shutterstock

Set in 10/12pt WarnockPro by SPi Global, Chennai, India

Printed in the UK

To my children, Jennifer, Alexander, and Rebecca

Contents

Preface

During the time period in which this book was written, there have been significant advances in computation and experimental techniques that suggest that a new era in the long history of turbulent flow study is commencing: one in which answers to many long-held questions are being found and one in which the capabilities for turbulent flow simulation are beginning to become viable for engineering design and research. As an example, some canonical flows are now being simulated and measured at what appears to be the asymptotic small viscosity (high Reynolds number) range which has previously only been the realm of theoretcal speculation. Moreover, simulations of the flow past geometrically complex shapes such as automobiles that utilize billions of gridpoints are now being attempted. At the same time, as significant as the recent advances in simulation and measurement have been, there remains a great distance between current capabiliites in predicting turbulent flows and a future time when the engineering of real-world turbulent flows can be done with high accuracy and efficiency as a routine matter. This means that a very large part of turbulent flow prediction today relies on traditional modeling and coarse simulation techniques that have been under continuous development for the last five decades.

In view of where the field of turbulence is today, an introductory book such as this one must give an accounting of both the range of new developments in the field of turbulence as well as a description of modern modeling and simulation techniques together with the many significant and fundamental results that support, motivate, and justify them. With these goals in mind, this book is meant to be a readible presentation of the subject that covers many fundamental and new results with some detail, although it avoids the depth of discussion that can be found in the many articles that are cited as part of the development. In many cases, the latter offer the preferred means of getting to a more comprehensive understanding of specific topics that cannot be reasonably included in a volume such as this. This book will be successful if it prepares students to pursue any number of specialized directions within turbulent flow research or enhances the knowledge and capabilities of engineers who are engaged in predicting turbulent flows via commercial software.

The book is bracketed by chapters in which ground vehicle aerodynamics is used as a means of focussing discussion on the nature of the turbulent flow problem. This includes, in the first chapter, what needs to be determined and what kind of phenomena is to be expected and, in the last chapter, how successful current methodologies are in achieving predictions of turbulent flows. In between these chapters, the line of development considers in turn the fundamental processes whose analysis has led to a variety of models that are incorporated in predictive schemes today. An essential part of these developments is considering what is being discovered in recent times. Specifically, what are the main issues being investigated today that may one day improve prediction techniques and give definitive insights into turbulent flow physics.

This book is the outgrowth of many years teaching graduate courses at the University of Maryland devoted to the theory, physics, and prediction of turbulent flow. I am indebted to the many students along the way who prodded me to keep the subject interesting and whose curiosity helped fuel my own interest. A number of problems are included after many of the chapters in the book. Some of these ask for details omitted in the line of development. Others suggest calculations that can be made via relatively uncomplicated codes (e.g., using MATLAB) that model and simulate turbulence or allow for the analysis of its properties from data sets that are readily available on the internet. Many related problems can be easily devised from extension of the problems given in the book.

I would like to express my appreciation to Neil Ashton, Marc Buffat, Jonathan Morrison, Arsensio Oliva, Richard Owens, Ulrich Rist, Philipp Schlatter, Eric Serre, Kidambi Sreenivas, Makoto Tsubokura, and James Wallace for supplying original figures, artwork, and photographs. I am indebted to Bruce Berger, Pat Collins, and Martin Erinin for reading some chapters from the book and providing useful commentary.

About the Companion Website

This book is accompanied by a companion website:

www.wiley.com/go/Bernard/Turbulent_Fluid_Flow

The website includes:

- Solutions manual

Scan this QR code to visit the companion website.

1
Introduction

1.1 What is Turbulent Flow?

Turbulent flow is ubiquitous in nature and technology. From breaking waves on a beach, to vortical eddies in the atmosphere that shake an airplane in flight, to flow across the hull of a submarine and separating into the ocean, to the flow disturbing the landing of a helicopter on the flight deck of a ship, fluid flows in a turbulent state. Why turbulent flow is so common is equivalent to the question of why flows often depart from a laminar state to become turbulent. The answer to this question can explain the ubiquitousness of turbulent flow and simultaneously provide a useful definition of what is meant by turbulent flow.

It is a fact that in many fluid flows there are sources of perturbation that are persistent and inevitable. These can arise from minute imperfections on boundary surfaces or slight variability in the incoming flow field. It is also conceivable [1, 2] that perturbations of a molecular origin can occur, as in the kinds of organized molecular behavior that leads to Brownian motion. How a fluid flow reacts to the presence of perturbations lies at the heart of whether or not the fluid motion is turbulent or will become turbulent. When perturbations to the velocity field appear in an otherwise laminar flow, the action of viscous forces is to diffuse the local momentum excess associated with the disturbance. Depending on the strength of the perturbation and the effectiveness of the viscous smoothing, the flow disturbances may grow, leading to the appearance of three-dimensional (3D), non-steady, disorganized motion that is referred to as turbulence or else be damped leading to the maintenance of laminar flow. The balance in this case is between the inertia of fluid particles in motion and the viscous forces acting on them to regularize the local flow field.

As will be considered more formally in the next chapter, the Reynolds number $R_e = \mathcal{U}\mathcal{L}/\nu$ where $\mathcal{U}$ and $\mathcal{L}$ are, respectively, characteristic velocity and length scales, and ν is the kinematic viscosity, is a measure of the ratio of inertia to viscous forces and so has a large role to play in characterizing whether fluid flow is turbulent or not. For small values of the Reynolds number, internal viscous forces dominate and the flow tends to be laminar, or laminarize if it is not initially so. For high Reynolds numbers viscous smoothing is insufficient to prevent the growth of instabilities, with turbulent flow being the result.

The transition of laminar flow to turbulence can follow a number of different routes depending on the magnitude and nature of the perturbations that are present [3]. For slight perturbations a linear instability is triggered that in a boundary layer,

Turbulent Fluid Flow, First Edition. Peter S. Bernard.

Companion website: www.wiley.com/go/Bernard/Turbulent_Fluid_Flow

for example, would be manifest as a pattern of streamwise disturbances known as Tollmein–Schlichting waves. As they develop downstream vortical structures appear in the flow whose breakdown and interactions signal the appearance of turbulent flow. If the initial perturbation to a laminar flow is sufficiently large, then *bypass* transition may occur for which there is a rapid development of the vortical structures leading to turbulence. Some aspects of the vortical structures occurring in transition and turbulent flow are discussed in Chapter 8.

Once initiated, turbulent flow persists unless there is a change in external conditions that could remove or reduce the mechanisms leading to instability. For example, flow acceleration in some circumstances has been observed to relaminarize the flow in boundary layers [4]. In many situations, such as pipe and channel flows, and boundary and mixing layers that will be discussed in this book, the flows may be seen to evolve from an upstream laminar state, through transition to the fully turbulent state downstream.

The subject of turbulence is primarily concerned with describing and predicting the quantitative and qualitative properties of fluid flow specifically in turbulent flow regions. The study of the conditions that might lead to the appearance of turbulent flow is the main interest of *stability theory*. Some overlap between these fields can be expected since the conditions leading to flow breakdown in transition may have some role in the fully turbulent region as well [5]. Moreover, turbulent flow may contain structural features in the form of vortices that also populate the transitional flow region [6]. Studying such aspects of transition can have benefit for the study of turbulent flow as well. For those flows where laminar, transitional, and turbulent flow exist simultaneously, as in a developing boundary layer, it can be important to predict the extent and properties of each separate region.

It will be seen subsequently that there are numerous ways that turbulent fluid motion differs from laminar motion, and this has important consequences for such aspects of flow analysis as predicting the forces on bodies or the diffusion of contaminants within the flow field. Moreover, the special properties of turbulent motion make it much more difficult than laminar flow to either solve for the fluid velocity field via numerical simulations or measure it in physical experiments. It will be seen that limitations of a variety of sorts shadow the analysis of turbulent flow so that in the study of any particular flow decisions often have to be made as to what methods ought to be brought to bear in studying the flow and how and to what extent they should be deployed. The combination of an important need to predict turbulent flow behavior and the fact that analysis and measurement techniques are not without limitations leads to the need for fluids engineers to acquire some degree of proficiency in understanding the advantages and disadvantages of available techniques and how best to interpret what they say about the flow field.

1.2 Examples of Turbulent Flow

Considering some specific examples where turbulent flow is present and important can help make clear the wide range of situations that are encompassed in the study of turbulence. Among the important categories of phenomena involving turbulent motion are those associated with the flow adjacent to solid surfaces. In the near-wall field lies the origin of the viscous and pressure forces that affect the motion of bodies. Turbulence

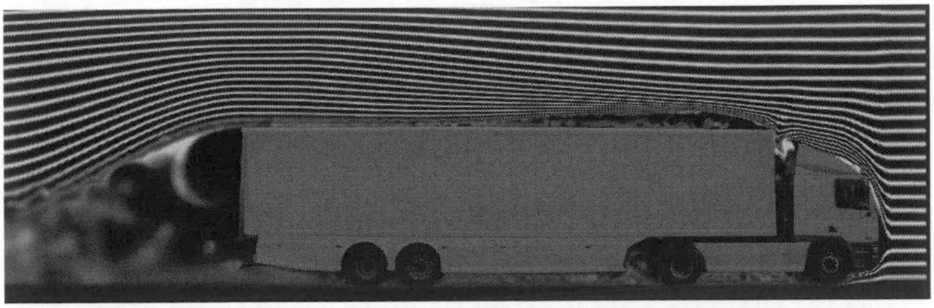

Figure 1.1 Artist's rendering of smoke visualized flow past a tractor trailer. The main areas affected by turbulence include boundary layers on the trailer surface, the tractor-trailer gap, under the chassis, and the large area of turbulence behind the vehicle. Courtesy of Don-Bur (Bodies and Trailers) Ltd.

produced in boundary layers is a common occurrence that has significant consequences for the strength and nature of the boundary forces. Turbulent flow is an integral part of the movement of automobiles and trucks, as seen in Figure 1.1 where knowledge from wind tunnel testing and simulations has been used to recreate the kind of smoke pattern to be expected in the flow around semi-tractor trailers. Turbulence develops over the trailer as boundary layers that separate into a turbulent wake that has a large influence on the overall drag force. Complicated turbulent eddying appears in the gap between the cab and trailer, and in the underbody with considerable consequences for drag and stability.

Turbulence produced on the wings and fuselage of an aircraft must be taken into account in the design process. For a wing at high angle of attack, as seen in Figure 1.2, turbulent vortices filling the wake and shedding from the edges of the wing are inextricably

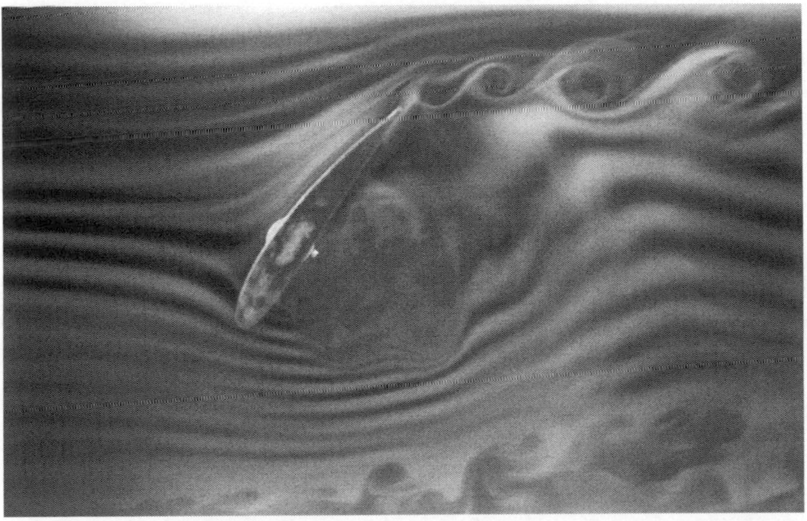

Figure 1.2 Fog wind tunnel visualization of a NACA 4412 airfoil at a low-speed flow (Re = 20,000). Turbulence fills the massively separated flow on the back of the airfoil and vortices roll up in the trailing edge wake. Image by Georgepehli, Smart Blade GmbH.

tied to the dynamics of the airplane by their effect on the pressure field. Similarly, turbulent flow is an important part of producing drag and lift forces affecting the hulls and keels of boats and such objects as runners, sky divers, skiers, ice skaters, and projectiles including baseballs, footballs, and golf balls.

Turbulent flow next to walls is also an essential part of the forces in internal flows such as channels and pipes. For low Reynolds numbers, the flow in a pipe will be laminar, that is Poiseulle flow, but beyond a transition Reynolds number depending on such factors as the smoothness of the boundaries and incoming flow, the motion in the pipe will be turbulent. In a smooth-walled pipe transition occurs at a Reynolds number based on mean flow velocity and diameter of approximately 2000 [7], a condition that is often exceeded in engineering applications. Under some circumstances, such as the presence of blockages, turbulent flow occurs in the human lung and heart. Turbulence is an essential aspect of the combustion process in engines, and the flow through heat exchangers, turbines, and numerous other devices.

Oftentimes, turbulent flows occur away from the immediate effects of solid boundaries in what are known as *free shear flows*. In the case of mixing layers two streams of differing velocity come together, leading to the development, amplification, and merging of vortical structures that have arisen via Kelvin–Helmholtz instability. An example of this is illustrated in Figure 1.3 where clouds mark the presence of a mixing layer formed from the presence of high wind shear aloft in the atmosphere.

Jet flows under many circumstances develop a strong turbulent field promoted by instability at their interface with surrounding fluid leading to the development and breakdown of vortical structures. This is seen in the jet shown in Figure 1.4 and in the jet propelled by buoyancy forces that is formed from the venting of hot air into the atmosphere from a smokestack, as shown in Figure 1.5. In these images turbulent vortical structures are seen to develop at the base of the jet where it departs the orifice and grow in size downstream until the entire exiting volume of fluid is in turbulent motion.

Covering a large range of scales, turbulent motion is often found extending long distances downstream in the wakes of flows past objects, as is seen in the wind farm shown in Figure 1.6. The performance of downstream wind turbines is significantly

Figure 1.3 Clouds forming over Mount Duval, Australia are a visual indicator of vortices formed within a turbulent mixing layer produced by strong wind shear. Photograph by GRAHAMUK at Wikimedia.

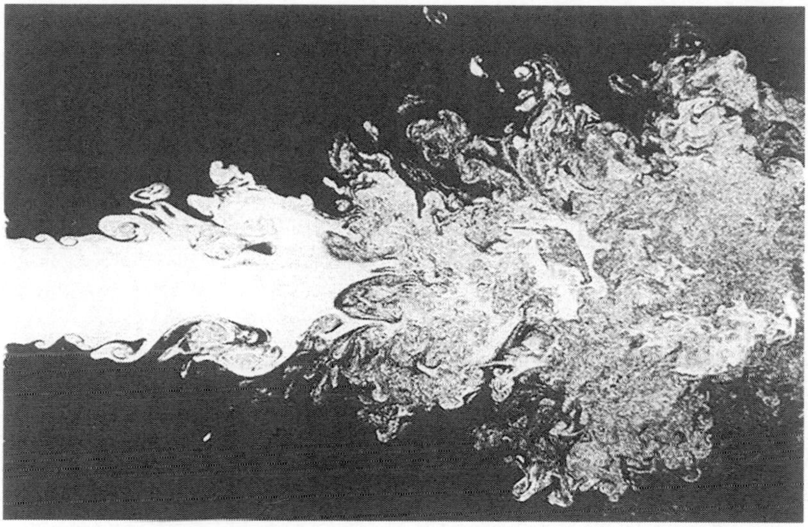

Figure 1.4 Transition to turbulence in a jet. Courtesy of J.-L. Balint and L. Ong.

Figure 1.5 A smoke plume from the Dunbar Cement Works chimney. cc-by-sa/2.0 - ©Walter Baxter - geograph.org.uk/p/3765299.

altered by being exposed to the long turbulent wakes of upstream wind turbines. Another example of turbulent wake flows is found in the von Kármán vortex streets shown in Figure 1.7 that form as wind blows past some appropriately shaped and sized Canary Islands. Breakdown of the wake vortices into turbulence is the end stage of the upstream perturbation to the oncoming flow field.

In many applications turbulence per se represents just one aspect of the essential physics which might also include sound and shock waves, combustion, chemical reactions, natural convection, two-phase gas–liquid flows as well as flows with particulates, sedimentation and slurries, and electromagnetic phenomena as in plasmas. The presence of turbulence in such circumstances can have a profound influence on

Figure 1.6 A picture of the Horn Rev offshore wind farm in Denmark. Turbulence appearing behind the wind turbines is marked by moisture condensation from the atmosphere. Photo by Christian Steiness. Original image link: http://i.imgur.com/qruVcnu.jpg.

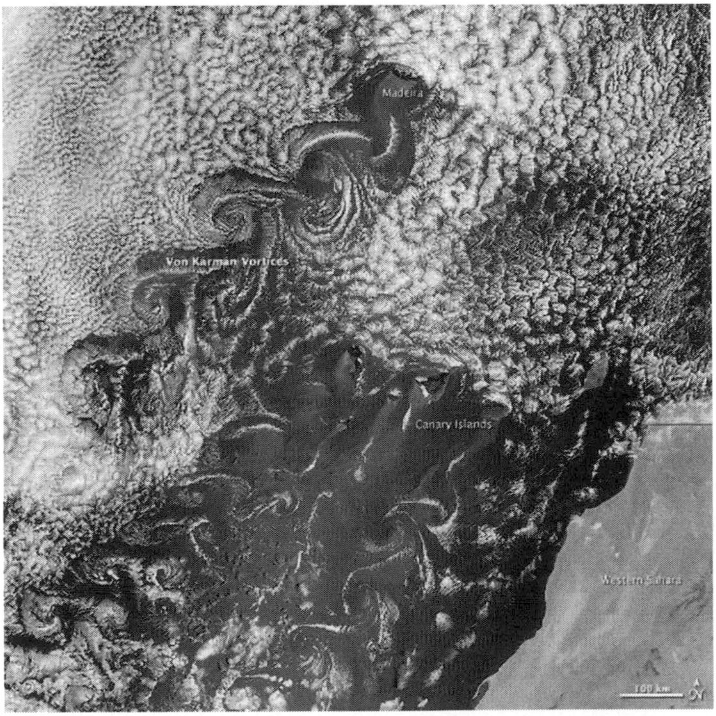

Figure 1.7 Clouds mark the presence of turbulent von Kármán vortex streets forming in the wake of the Canary Islands. NASA image by Jeff Schmaltz, LANCE/EOSDIS Rapid Response.

phenomena associated with the additional physics. In many cases there is a two-way coupling in which turbulence both affects and is affected by the presence of other physical phenomena. For example, liquid–gas flow in pipes can take on a number of different regimes depending on the turbulent interaction between phases.

For the many flows where turbulent motion is present it is natural to wonder how its presence might affect the way in which flows are studied. The next section gives some insight into this question by considering in some detail the kinds of turbulent flow phenomena that are present in a particular application, specifically the flow past a semi-tractor trailer truck. Such flows are of great practical interest and harbor a variety of phenomena that are shared by many other common flows of interest. After discussing what one would like to know about turbulence in such a practical setting, we then describe what sorts of methodologies might be applied to acquire the desired information. Presenting some of the essential knowledge associated with each of these approaches toward studying turbulent flow phenomena then forms the main goal of this book.

1.3 The Goals of a Turbulent Flow Study

The flow past a semi-tractor trailer truck is complex and includes a variety of phenomena that are likely to be worthy of investigation and analysis to fluids engineers. For example, traveling at 65 mph a typical truck might use approximately 30% of its fuel consumption to overcome aerodynamic drag [8]. Multiplied over the entire population of trucks traveling on any given day, modifications to trucks that reduces drag can provide a very significant environmental and economic gain. Other aspects of truck flows that are subjects of investigation include stability to cross winds, noise generation, splash and spray of road water, and fouling of the windshield from airborne particulates. In all these aspects, turbulent flow plays an integral part and it is important to have a capacity for predicting and understanding its action.

One relatively straightforward aspect of a truck flow consists of the boundary layers that form on the trailers' flat surfaces. For example, on windless days or if the truck is moving head-on directly into the wind, boundary layers will develop down the length of the upper surface and sides of the trailer. As illustrated in Figure 1.1, the boundary layer on the top surface is turbulent and develops in response to external conditions associated with the free stream velocity and the near-wall pressure field associated with the truck movement. The initial development of the boundary layer is strongly influenced by the upstream flow conditions including buffeting from fluid separating from the cab and gap as well as complex geometrical features of the of the leading edge of the trailer. In a laboratory setting with laminar upstream flow conditions, transition generally occurs when the Reynolds number $Re_x = x\mathcal{U}/\nu \approx 400{,}000$. For the circumstances similar to that of a truck traveling at 105 kph (65 mph) in air with $\nu = 1.48 \times 10^{-5}$ m^2/s, the downstream distance in a zero-pressure gradient boundary layer where transition occurs is $x \approx 20$ cm (8 in). Thus, even in the best of circumstances when a laminar boundary layer is able to form, the non-turbulent part of the boundary layer will be of little consequence to a typical trailer with length up to 16 m (53 ft). For the flow to be laminar over the truck, it would have to travel at 1 mph. Boundary layers in turbulent flow have been studied extensively and a considerable amount is known about their statistics and properties. Much of this will be considered in subsequent chapters.

At the rear of the truck the boundary layers separate off into a wake region that forms behind the blunt shape of the trailer, as seen in Figure 1.1. The flow here is unlike the sleek aerodynamic wake behind a wing and is not amenable to simple analysis. Vortical eddies produced in the upstream boundary layers join with strong vortical recirculation regions that develop as an outgrowth of separation of the turbulent boundary layer off the rear edge. Out of this complicated non-steady flow pattern a pressure field is established whose amplitude fails to balance the high pressure on the front surfaces of the cab and trailer. The result is a significant net contribution to the drag force. The problem of predicting drag on trucks is thus in part closely associated with predicting the details of the non-steady wake flow. This generally means resolving the time history of vortical features in the wake in the course of simulating the flow field.

The truck cab presents itself to the oncoming flow as a geometrically complex bluff body, with the flow over the front surfaces having the character of a potential flow. The wake flow behind the cab is similar to the rear of the trailer and can be expected to produce turbulent upstream conditions for the trailer flow. Depending on whether fairing is placed to cover the gap between cab and trailer, there may be a strong cavity flow in the region between them. The pressure field on the rear of the cab and front of the trailer will affect the overall drag prediction.

The direction and magnitude of the drag on a truck depend strongly on the relative strengths and directions of the ambient wind in relation to the velocity of the truck. In effect, the relative wind as seen by an observer traveling with the truck is the essential quantity needed to determine the drag. The relative velocity is liable to point in any direction. When it is not head-on there is potential for recirculating flows to form along the sides of the trailer, including significant separation. Such phenomena are associated with large transients in the strength and orientation of side forces that can affect the stability of trucks, particularly when they pass each other.

Specialized aspects of the truck flow that may also be of interest include the noise producing flow around the side mirrors, the flow past the moving tires, the cooling effect of air traveling through the engine compartment, and flow in the irregular underbody region of varying cross-sectional area. Non-steadiness of many aspects of the flow imply that capturing the movement and history of the turbulent field is essential for reproducing the physics of such flows. Finally, it is certain that the flow within the engine is mostly turbulent, compressible, involves combustion, two phases, atomizing liquid jets and large gradients in temperature, and other flow properties.

It is evident from this discussion that ideally one would like to have accurate knowledge of the time-dependent velocity field and pressure for the entire flow past a truck. Out of such data a means is provided for obtaining the complete transient forces and moments acting on the truck. In particular, the velocity field enables computation of the surface shear stress that combines with the pressure to create the force field on the body of the truck. From this the drag and moments acting on the truck can be computed to help determine projected fuel efficiency and stability. Knowing the entire flow also allows for prediction of such practical features as the paths of airborne particles that may interact with the windshield, the calculation of noise generated by the truck, and the dispersion of exhaust gases and particulates into the wake. A capability of simulating the turbulent engine flow can be important for achieving additional gains in minimizing pollution and maximizing engine efficiency. Current capabilities for acquiring the desired data fall short of the ideal and result in the need to make decisions as to what kind of information can be obtained and at what accuracy.

1.4 Overview of the Methodologies Available to Predict Turbulence

Assuming that one is interested in studying a particular turbulent flow, such as the truck flow described in the previous section, decisions have to be made as to what route one should take to best analyze the flow field. As it happens, it is often not obvious what the best strategy should be, and, in fact, there are a variety of different approaches including physical experiments, computation, and modeling that can be pursued with no one of them dependably giving a complete and satisfactory answer to questions about all flow fields. Here, as a preamble to what will be the main preoccupation of the book, we consider the advantages and disadvantages of various strategies for predicting turbulent truck flow as an example of what might occur in the general case.

1.4.1 Direct Numerical Simulation

Since knowledge of the complete transient flow field past a truck is tantamount to knowing all the forces acting everywhere on the truck at all times, any method that can provide this data would be an ideal engineering tool. The potential for acquiring such data numerically depends on the capabilities of schemes for solving the Navier–Stokes equations on a computational mesh surrounding the truck. If successful, such a numerical solution would supply the surface shear stress and pressure field so that forces and moments could be computed as well as other much-needed information. With an accurate numerical approach physical experiments are unnecessary and parametric studies are relatively easy to accomplish.

The kind of numerical calculation that has just been described is known as a direct numerical simulation (DNS) in the sense that the simulation is performed directly without the use of simplifying steps. Computer storage and speed became sufficiently large in the late 1980s to enable the first DNS of wall-bounded turbulent flows [9]. For the most part early simulations of shear flows were limited to the study of fully developed channel flow at Reynolds numbers just beyond the transition region. In subsequent years there have been dramatic advances in all aspects of computer technology and as a result great improvements in geometrical complexity and Reynolds numbers in the turbulent flows that can be well simulated.

For a DNS to be appropriate for truck flows it is necessary that the grid resolution in space and time be sufficiently fine to capture the smallest spatial or transient variations in the velocity field. As will be seen in later chapters, turbulent flow contains dissipative processes depending on viscosity that occur over very small distances in comparison to the scale of the overall flow field. The need to resolve such motions creates the need for large meshes that drive up the cost of simulations beyond the point where they are practical using modern supercomputers.

However, while DNS of industrial flows may be precluded because of affordability issues, nonetheless DNS is feasible for a variety of laboratory flows and thus has been and continues to be a great source of knowledge about the physics and modeling of turbulent flow. From such insights come better means of predicting and understanding more complex flows such as those with trucks. DNS solutions can also give insights into how best to develop predictive schemes that compensate for reduced grid resolution by incorporating turbulence models. This is the field of turbulence modeling that represents an alternative to DNS and physical experiments.

1.4.2 Experimental Methods

Since DNS is not feasible for studying complex turbulent flows such as that past a truck, it is necessary to consider alternative approaches. Performing physical experiments in such cases has strong appeal, since real flows at high Reynolds number can be used as the subject of the study. Indeed, experimental methods are used today in a significant capacity in fluids engineering and it is interesting to consider what such studies might offer.

While the ultimate goal of an experimental study of truck flow might be to measure the flow past a full-sized truck under typical road conditions, it is rarely if ever possible to achieve such measurements in practice. For a start, facilities such as wind tunnels that are large enough to accommodate trucks [10] are rare. More typical is the use of scale-model trucks in wind-tunnel studies [11]. In this case there is likely to be some difficulty in establishing complete dynamical similarity between the model and the prototype, for example in maintaining the same Reynolds number. It is also the case that the drag on a truck is normally experienced as it moves over a stationary surface. In wind tunnel studies, the truck is stationary and so to mimic true road conditions a moving floor should be used. Such a capability is not available in all facilities and drag measurements in such instances will almost certainly be affected.

Measurements in wind tunnels are also likely to reflect blockage effects in which the presence of the model reduces the cross-sectional flow area, leading to a distortion of the flow exterior to the boundary layers forming on the truck surfaces. It is also the case that not all wind tunnels can accommodate the full range of external boundary conditions of interest. For example, imposing side flows requires turning the model, which can present difficulties in many tunnels unless the models are at a sufficiently small scale. It can be expected that capturing transient effects associated with stability will also be difficult to achieve in a wind tunnel experiment.

Independent of the particular flow configuration that is to be studied, decisions have to be made concerning what quantities will be measured and where in the flow field they will be obtained. Thus, sampling of the 3D velocity field around an object such as a truck by measurement techniques is not practical beyond a relatively small number of discrete points. In fact, there is no easy way to capture experimentally the full range of velocity and pressure data that could be obtained from a DNS if one were feasible. To measure the force distribution over the truck surface, flow measurements in its vicinity must be made but this is only doable on a small subset of the total surface, for example by outfitting the model with a finite number of pressure taps.

One special strength of physical measurements is that they allow determination of the integrated effect of the forces on the truck. For example, the drag on a truck or a truck model can be measured through the use of a force balance that records the entire net force and moments on the object to which it is connected. Generally, such data represent a time-integrated mean force, which is all that is necessary to obtain the drag. Other integrated methodologies for determining drag consist of taking measurements of the average velocity over a 2D plane in the wake of the body. This can be done by sweeping out the area via a rake of probes. The connection of such data to drag arises from theoretical considerations tying the momentum deficit in the wake to the drag force [12].

Flow visualization methods are also useful in studying turbulent flows and in particular the flow past ground vehicles. In this case tufts placed at convenient locations on the surface of the vehicle can give an indication of where separation occurs. Similarly, oil covering the surface of the truck can be used to get a qualitative idea of the location of structural flow features on the surfaces such as separation lines. Another possibility is to introduce smoke or other marker particles to visualize structural aspects of the flow around the truck, as is illustrated in Figure 1.1.

The engineering of turbulent flow may often involve parametric studies as in determining the effect of geometric modifications on drag forces, noise generation, or other flow properties. For an experimental program such tests can require costly fabrication of multiple forms of the model vehicle. For example, for clay models they may need to be resculpted many times to test different shapes. For this reason, as well as those discussed above, it is evident that while experimental measurement programs can provide much needed information they are likely to be done in concert with alternative analyses to get the full range of data that is required in engineering studies. Taken together with the limited applicability of DNS it becomes necessary to consider a variety of approximate means of analyzing turbulent flow that collectively form a wide range of turbulence models. Such approaches allow for numerical analyses of flows aided by insights taken from physical experiments and DNS.

1.4.3 Turbulence Modeling

Turbulence modeling in many guises attempts to fill the gap between the requirements of turbulent flow prediction and the limitations of DNS and experimental methods in satisfying them. As we will see, turbulence modeling most often comes with the penalty of lost accuracy while nonetheless still supplying information about flows that is useful in a practical sense. In other words, by not expecting an exact solution, but only an approximate one, more room is created for developing alternative methodologies for analyzing turbulent flows. Turbulence modeling attempts to use theoretical and empirical results about the behavior of turbulent flow to develop equations whose solutions to varying degrees approximate the actual flow fields.

The most widespread methodology for modeling turbulent flow is based on solving the Reynolds-averaged Navier–Stokes (RANS) equations that govern the dynamics of the mean velocity field. In this case the averaging is done to the equations themselves prior to solving them, in contrast to DNS where averages are obtained only after the Navier–Stokes equations are solved. Of increasing popularity in recent years as computing resources have grown in size and speed are large eddy simulations (LES) that provide approximations to turbulent flow fields that are simplified by not modeling the detailed motion of small-scale phenomena. Such effects appear instead in the guise of models of one sort or another. In a more recent trend, a number of hybrid methods have been developed that attempt to combine the best aspects of LES and RANS modeling into one self-contained approach. Exactly how the modeling is done in RANS, LES, and hybrid LES/RANS and whether or not it can be justified consumes a large part of this book.

It is difficult to predict a priori how well a particular turbulence model will succeed when applied in complex flow situations such as that associated with a truck. Whether or not such calculations are accurate depends on the particular statistic that is being computed and what the criterion for "accuracy" might happen to be. For example, for many popular models the drag might be computed reasonably well, say to within 5% of measured values, yet errors in the underlying pressure distribution may nonetheless be quite significant. In fact, offsetting errors in the pressure prediction computed over different portions of a body may allow for the net pressure force to fall near measured values. How one reacts to this situation depends on the goals of the computational study. Inaccuracies in modeling turbulent flow at the rear of a truck may mean that the wake vortex structure is incorrect, but this may not be of concern if designing the shape of upstream features such as the side mirror is the focus of the study.

Decisions as to whether to pursue RANS vs. LES depend partly on the availability of computer resources. A RANS calculation for the truck flow can be a formidable computation because of the size and complexity of the geometry of the flow. It is likely, however, to be much less expensive than an equivalent LES computation, but whether or not there is value in pursuing LES might depend on the likelihood and importance of achieving better results and at what cost.

Affecting the choice of RANS or LES is also whether or not transient information is needed, as it would be in predicting shed vortices linked to sound generation or investigating stability in the face of sudden side winds. Generally, RANS is less adept at providing transient data than LES so the latter may be the only option. Conversely, for parametric studies requiring many similar calculations, the fact that RANS solutions can be obtained in less time than LES may be a deciding factor in the decision to employ RANS.

Within the RANS and LES approaches, there are numerous modeling choices to consider. Some of the more commonly found methodologies will be described in the following, though this is not to say that other choices in the literature might provide some improvements for specific applications. Indeed, it is the pursuit of better performance that has caused the fracturing of what were originally a small number of modeling approaches into the great number there are today.

1.5 The Plan for this Book

The next chapter considers the kinds of quantities that one might want to determine or measure in a turbulent flow. This is followed by a presentation of the basic equations of motion that can be used as a framework for later discussions of the physics of turbulence. Starting with Chapter 4 and continuing through Chapter 8, we consider in turn the major aspects of the turbulent flow physics whose modeling is the goal of the various methodologies that have been developed for predicting turbulent motion. It will be seen that the most commonly employed turbulence models, whether RANS or LES, are based upon a number of basic notions about the physical nature of turbulent flow that in one way or another can be expressed mathematically to yield a set of equations amenable to numerical solution. The discussion combines empirical and theoretical evidence for the way in which turbulent flow behaves with consideration of how this understanding of the physics finds its way into models.

After having discussed the relationship between the major aspects of turbulence physics and its modeling, Chapter 9 considers the complete RANS models that are built from these analyses. Then Chapter 10 considers LES and hybrid LES/RANS models. Chapter 11 is concerned with various aspects of free turbulent flows, a category of applications that is not usually in the direct line of development of turbulence models. Finally, Chapter 12 considers some aspects of how a variety of RANS and LES models, many of which are discussed in earlier chapters, fare in the prediction of ground vehicle flows, including trucks.

References

1 Tsuge, S. (1974) Approach to the origin of turbulence on the basis of two-point kinetic theory. *Phys. Fluids*, 17, 22–33.

2 Gallis, M.A., Bitter, N.P., Koehler, T.P., Torczynski, J.R., Plimpton, S.J., and Papadakis, G. (2017) Molecular-level simulations of turbulence and its decay. *Phys. Rev. Letters*, 118, 064 501.

3 Kachanov, Y.S. (1994) Physical mechanisms of laminar-boundary-layer transition. *Ann. Rev. Fluid Mech.*, 26, 411–482.

4 Narasimha, R. and Sreenivasan, K.R. (1979) Relaminarization of fluid flows. *Adv. Appl. Mech.*, 19, 221–309.

5 Park, G.I., Wallace, J.M., Wu, X., and Moin, P. (2012) Boundary layer turbulence in transitional and developed states. *Phys. Fluids*, 24, 035 105.

6 Bernard, P.S. (2013) Vortex dynamics in transitional and turbulent boundary layers. *AIAA J.*, 51, 1828–1842.

7 Eckhardt, B., Schneider, T.M., Hof, B., and Westerweel, J. (2007) Turbulence transition in pipe flow. *Ann. Rev. Fluid Mech.*, 39, 447–468.

8 Choi, H., Lee, J., and Park, H. (2014) Aerodynamics of heavy vehicles. *Ann. Rev. Fluid Mech.*, 46, 441–468.

9 Kim, J., Moin, P., and Moser, R. (1987) Turbulence statistics in fully developed channel flow at low Reynolds number. *J. Fluid Mech.*, 177, 133–166.

10 Davies, A. (2017) Climb inside the world's largest wind tunnel, *Tech. Rep. February 26*, www.wired.com.

11 Lacey, J. (2016) Wind tunnel concepts for testing heavy trucks, *Tech. Rep. 2016-8144*, SAE Technical Paper.

12 Onorato, M., Costelli, A., and Garrone, A. (1984) Drag measurement through wake analysis, *Tech. Rep. 840302*, SAE Technical Paper.

2

Describing Turbulence

2.1 Navier–Stokes Equation and Reynolds Number

The spatial and temporal development of the velocity field in incompressible turbulent flow may be assumed to obey the mass conservation and Navier–Stokes equations [1] given by

$$\frac{\partial U_i}{\partial x_i} = 0 \tag{2.1}$$

and

$$\rho\left(\frac{\partial U_i}{\partial t} + U_j \frac{\partial U_i}{\partial x_j}\right) = \rho g_i - \frac{\partial P}{\partial x_i} + \mu \frac{\partial^2 U_i}{\partial x_j^2}, \tag{2.2}$$

respectively, where the velocity field in three dimensions $\mathbf{U}(\mathbf{x}, t)$ at position $\mathbf{x} = (x_1, x_2, x_3)$ at time t has components (U_1, U_2, U_3). $P(\mathbf{x}, t)$ is the pressure field, $\rho(\mathbf{x}, t)$ is the density, and μ is the dynamic viscosity. Note that, on occasion, the position coordinates will be denoted as x, y, z and the velocity components as U, V, W. In all cases, unless indicated otherwise, repeated indices are to be summed.

Equations (2.1) and (2.2) are written in standard index notation corresponding to rectangular Cartesian coordinates. An alternative expression for these and other equations, which has the advantage of being applicable to all coordinate systems as well as providing a simpler view of the physics in some circumstances, is in terms of direct vector and tensor notation (see [1] for an accounting of the essential ideas). Written this way, Eqs. (2.1) and (2.2) are, respectively

$$\nabla \cdot \mathbf{U} = 0 \tag{2.3}$$

and

$$\rho\left(\frac{\partial \mathbf{U}}{\partial t} + (\nabla \mathbf{U})\mathbf{U}\right) = \rho \mathbf{g} - \nabla P + \mu \nabla^2 \mathbf{U}, \tag{2.4}$$

where $\nabla \mathbf{U}$ is the velocity gradient tensor with components $(\nabla \mathbf{U})_{ij} = \partial U_i / \partial x_j$. Note that for any tensor T and vector $\mathbf{v}$, $T\mathbf{v}$ is the vector with components $(T\mathbf{v})_i = T_{ij} v_j$. Thus, $((\nabla \mathbf{U})\mathbf{U})_i = (\partial U_i / \partial x_j) U_j$, as appears in Eq. (2.2). The term $\nabla^2 \mathbf{U}$ is the vector Laplacian defined as $\nabla^2 \mathbf{U} = \nabla \cdot \nabla \mathbf{U}$. In this, $\nabla \cdot \nabla \mathbf{U}$ refers to the divergence of the tensor $\nabla \mathbf{U}$ and, by definition, for any tensor T, $(\nabla \cdot T)_i$ is the vector $\partial T_{ij} / \partial x_j$ so that $(\nabla^2 \mathbf{U})_i = \partial^2 U_i / \partial x_j^2$ as also appears in Eq. (2.2). From these definitions it may be seen that Eqs. (2.2) and (2.4)

Turbulent Fluid Flow, First Edition. Peter S. Bernard.

Companion website: www.wiley.com/go/Bernard/Turbulent_Fluid_Flow

are identical. Some limited use of direct notation will be used in what follows, particularly where it allows for simplified notation.

Equation (2.4) shows that the momentum balance in a moving fluid is essentially between the inertia represented by the non-linear term $(\nabla\mathbf{U})\mathbf{U}$ and the viscous forces $\mu\nabla^2\mathbf{U}$. The pressure force tends to adapt to whichever of the other terms is dominant and it need not be considered explicitly when comparing the relative magnitude of forces. The viscous force derives from momentum diffusion at the molecular level. For non-dense gases this is by molecular collisions while for liquids it is due to a local tug between molecules in close proximity. By its nature, diffusion smoothes away perturbations as they arise in the flow. Conversely, the inertial term, by its non-linear structure, contains a mechanism by which perturbations self-reinforce so that they are enhanced in magnitude and extent. Depending on a number of factors, including the strength of the perturbations and the speed of the flow, the conflicting tendencies between inertia and diffusion lead to either the maintenance of initially laminar conditions or the breakdown of the flow to turbulence.

To gauge the relative importance of the inertial and viscous terms it is helpful to consider the magnitude of their ratio. In particular, introducing length and velocity scales, $\mathcal{L}$ and $\mathcal{U}$, respectively, that are appropriate to a particular flow, the force terms can be scaled to then reveal their relative magnitude. Thus, introducing the scalings $\mathbf{x}^* = \mathbf{x}/\mathcal{L}, \mathbf{U}^* = \mathbf{U}/\mathcal{U}$ and applying them to the ratio of the inertial to viscous terms in Eq. (2.4) leads to

$$\frac{\mu|\nabla^2\mathbf{U}|}{\rho|(\nabla\mathbf{U})\mathbf{U}|} = \frac{1}{R_e}\frac{|\nabla^{*^2}\mathbf{U}^*|}{|(\nabla^*\mathbf{U}^*)\mathbf{U}^*|}, \tag{2.5}$$

where ∇^* refers to differentiation in the $\mathbf{x}^*$ coordinates. Assuming that the choice of scales is appropriate to the fluid flow of interest, the terms in the scaled coordinates may be assumed to be $O(1)$, that is, have a magnitude in the neighborhood of 1, say between 1/10 and 10. According to Eq. (2.5) the magnitude of the ratio on the left-hand side of the equation is determined by the Reynolds number

$$R_e = \frac{\mathcal{U}\mathcal{L}}{\nu} \tag{2.6}$$

appearing on the scaled right-hand side of the equation. Thus, for small R_e, viscous smoothing dominates while for higher Reynolds numbers inertia effects dominate. For high Reynolds numbers the growth in perturbations cannot be stopped by viscous action and flows tend to be turbulent, containing a seemingly random pattern of motion.

2.2 What Needs to be Measured and Computed

Since the opportunity of obtaining the transient 3D velocity field in turbulent flows of practical interest is rarely available, it becomes necessary to develop a variety of alternative quantities that can be computed or measured and have some use for answering questions about turbulent flow and its consequences. As suggested previously, averaged fields often represent a practical means for describing turbulence and a variety of averaged quantities are commonly employed in this regard. Measured and computed flow fields can also be leveraged in different ways to acquire insights into the structural features of turbulent flow as, for example, by determining spectra or correlations between quantities at different physical locations. Besides the quantities considered here, later chapters will introduce additional, more specialized quantities obtainable from flow measurements and computation.

2.2.1 Averaging

There are several different ways in which averaging can be applied to turbulent flows. For the most part the choice of which kind of averaging procedure to follow is dictated by the constraints or opportunities presented in particular experimental or numerical circumstances. For example, in wind tunnel experiments, the simplest and most practical way to obtain mean statistics is often to average a turbulent field in time. In this case the average velocity is calculated from

$$\overline{\mathbf{U}}(\mathbf{x},t) = \frac{1}{T}\int_{t-T/2}^{t+T/2} \mathbf{U}(\mathbf{x},s)ds, \tag{2.7}$$

where T is a sufficiently long time interval so that the left-hand side of the equation is not random and $\mathbf{U}(\mathbf{x},s)$ is available from continuously sampling the flow velocity at position $\mathbf{x}$ over the interval from $t-T/2$ to $t+T/2$. Alternatively, if the velocity is only known at N discrete times t_i, $i = 1, 2, \ldots, N$ then

$$\overline{\mathbf{U}}(\mathbf{x},t) = \frac{1}{N}\sum_{i=1}^{N} \mathbf{U}(\mathbf{x},t_i), \tag{2.8}$$

represents another way of getting the time average. Once again, N should be large enough so that the average is a deterministic quantity. In general, a time average is most legitimate for *stationary* flows where the mean statistics do not change in time so that, for example, $\overline{U}(\mathbf{x})$ is independent of t. In this case, the existence of a suitable T that provides converged statistics is guaranteed. For non-stationary flows it may not be possible to find a value of T for which the left-hand side of Eq. (2.7) is deterministic and yet not affected by the mean drift of the random field. Also, the identity

$$\overline{\frac{\partial U}{\partial t}} = \frac{(U(t+T/2) - U(t-T/2))}{T}, \tag{2.9}$$

obtained by applying Eq. (2.7) to $\partial U/\partial t$, suggests that the average of $\partial U/\partial t$ will always be random unless T can be made very large, as in stationary turbulence.

A more natural approach toward averaging in transient flows is that of *ensemble averaging* in which measurements are made at the same point and time for many identical experiments. In the case of engine cylinder flow, for example, one can imagine that each four-stroke cycle of the engine may be considered to be one "experiment," and so averaging in this case requires obtaining data at the same point in the cycle for many repetitions of the flow field. For N experiments the average in this case consists of

$$\overline{\mathbf{U}}(\mathbf{x},t) = \frac{1}{N}\sum_{j=1}^{N} \mathbf{U}^j(\mathbf{x},t), \tag{2.10}$$

where $\mathbf{U}^j$ is the velocity measured in the jth experiment.

Ensemble averaging is particularly convenient for theoretical work since it readily commutes with the time and space derivatives occurring in the equations of motion. For example, for ensemble averaging

$$\overline{\frac{\partial \mathbf{U}}{\partial t}}(\mathbf{x},t) = \frac{\partial \overline{\mathbf{U}}}{\partial t}(\mathbf{x},t), \qquad \overline{\nabla \mathbf{U}}(\mathbf{x},t) = \nabla \overline{\mathbf{U}}(\mathbf{x},t), \tag{2.11}$$

where all terms are non-random. Ensemble averaging of the Navier–Stokes equation yields an exact equation governing the mean velocity in a general non-steady and non-uniform flow. To justify using time averaging in this instance the flow would have to be stationary.

In some situations, particularly in numerical work, it is convenient to use spatial averaging to get flow statistics. For a given volume of fluid, $\mathcal{V}$, this is defined in an obvious way via

$$\overline{\mathbf{U}}(\mathbf{x},t) = \frac{1}{\mathcal{V}} \int_{\mathcal{V}} \mathbf{U}(\mathbf{x}',t) d\mathbf{x}', \tag{2.12}$$

where $\mathcal{V}$ is generally taken to be centered around the point $\mathbf{x}$. Volume averaging is most appropriate if the turbulent flow is *homogeneous* over the averaging region in the sense that mean statistics do not vary over this domain. In some cases, particularly flows with geometrical symmetries, averaging may be taken along particular lines or within surfaces in the flow field over which the velocity field has a uniform mean behavior. For example, Eq. (2.12) is appropriate to use in a channel flow with $\mathcal{V}$ taken to be planes parallel to the bounding surfaces since mean statistics do not vary over such regions. Averaging over planes in this case is equivalent to averaging over many experiments and provides a convenient means of getting rapidly converged statistics. Similar to the case of time averaging, spatial averaging is not generally appropriate for flows with non-uniform turbulence properties, that is, *non-homogeneous turbulence.* In particular, the average of terms containing spatial derivatives in the directions over which averaging is implemented cannot be computed accurately. In the example of a channel flow spatial averaging over planes parallel to the boundaries is legitimate because the non-homogeneity of the mean flow in this instance is in the direction normal to the averaging planes.

In some contexts it is useful to generalize the idea of averaging to encompass only special subsets of the full ensemble of possible outcomes of the flow field. These are known as *conditional averages,* where a criterion is set up that selects only certain events from the flow field and then an average is made over these. Such ideas are particularly useful for exposing structural features of turbulence that otherwise might be unseen in the full random signal. A somewhat related but different philosophy underlies the averaging carried out as the basis of LES methods, known as *filtering,* that will be discussed in Chapter 10. In this case the averaging process weighs different scales of the motion more or less than others (e.g., the influence of large scales versus small scales) so that this bias is reflected in the averaged fields that result. Since the averaging is not complete in this case, the resulting statistics retain some aspects of the randomness of the underlying turbulent field.

It will be taken for granted in this book that ensemble, time, and spatial averages are equivalent when circumstances allow more than one of the techniques to be applied, though this conclusion lacks formal mathematical proof. Subsequent discussions of a theoretical nature often will be based on ensemble averaging while many of the experimental results quoted will have been determined by time averaging. Other specialized averages will be noted as appropriate in later chapters.

2.2.2 One-Point Statistics

In addition to the mean velocity field, averages may be sought for the pressure and vorticity $\boldsymbol{\Omega} \equiv \nabla \times \mathbf{U}$ as well as density and other quantities such as temperature and species concentrations in flows with complex physics. These fields feed into the common RANS approach of solving the averaged equations for the mean quantities through the incorporation of models.

The part of the turbulent velocity field that differs from the local mean at a given point $\mathbf{x}$ at a given time t is referred to as the *fluctuating velocity* and is defined via

$$\mathbf{u} = \mathbf{U} - \overline{\mathbf{U}}. \tag{2.13}$$

In the absence of turbulence $\mathbf{u} = 0$ the fluctuating field is of intrinsic importance in the presence of turbulent flow. The fluctuating velocity field allows one to view the turbulent velocity field as the sum of a deterministic mean field and a random component according to

$$\mathbf{U} = \overline{\mathbf{U}} + \mathbf{u} \tag{2.14}$$

where

$$\overline{\mathbf{u}} = 0. \tag{2.15}$$

This result is a consequence of the necessary condition that for a well-defined average

$$\overline{\overline{\mathbf{U}}} = 0, \tag{2.16}$$

since $\overline{\mathbf{U}}$ is deterministic and hence equal to its average. In all cases Eq. (2.16) holds for ensemble averaging and it will also be true for time averaging in stationary turbulence or volume averaging in homogeneous turbulence. It does not necessarily hold for time averaging in non-stationary turbulence or spatial averaging in non-homogeneous turbulence, so in these cases Eq. (2.15) may be violated.

Statistics formed from the quadratic products of the components of the fluctuating velocity field are useful in characterizing several main aspects of turbulent motion. In the first place, the strength of the turbulent field is a direct reflection of the magnitudes of the variances of the velocity components, $\overline{u^2}, \overline{v^2}, \overline{w^2}$. Taken together these form the turbulent kinetic energy per unit mass,

$$K \equiv (\overline{u^2} + \overline{v^2} + \overline{w^2})/2 = \overline{u_i u_i}/2 = \overline{\mathbf{u} \cdot \mathbf{u}}/2. \tag{2.17}$$

The variances of the fluctuating velocity components and K provide a succinct way of deciding whether or not turbulence is present in a flow and how strong it might be. For example, in a wind tunnel where U is to be measured the ratio

$$\frac{\sqrt{\overline{u^2}}}{\overline{U}} \tag{2.18}$$

determined in the entrance region of the tunnel gives an idea of the extent to which the incoming stream might be a source of perturbation to an experiment. If this ratio is small, say 0.001, it means that the upstream turbulence is but one tenth of a percent of the mean flow speed and can be considered to be a slight perturbation. Higher percentages of the ratio may reach a range where disturbances in the free stream cause rapid transition to turbulence in downstream flows of interest.

The relative values of the variances $\overline{u^2}, \overline{v^2}, \overline{w^2}$ give some indication as to whether or not the local turbulent field has a bias toward one direction or not. The exact absence of any such bias means that the turbulent field is *isotropic*, as will be considered in Chapters 4 and 5. The equality of the separate velocity variances is one consequence of isotropy, among others. One region where anisotropy usually prevails is next to solid boundaries where the component of velocity normal to the surface is more strongly suppressed by the presence of the boundary than the components parallel to the surface. Thus, next to a wall situated at $y = 0$ one can expect that $\overline{v^2} < \overline{u^2}, \overline{w^2}$, as will be seen below in examples.

The full set of quadratic products of the velocity fluctuation vector is encompassed in the velocity covariance tensor

$$R = \overline{\mathbf{u} \otimes \mathbf{u}} \tag{2.19}$$

formed from the tensor product of $\mathbf{u}$ with itself and having components $(\overline{\mathbf{u} \otimes \mathbf{u}})_{ij} = \overline{u_i u_j}$. Note that for any vectors $\mathbf{a}$, $\mathbf{b}$, and $\mathbf{c}$, $(\mathbf{a} \otimes \mathbf{b})$ is the tensor with the property that $(\mathbf{a} \otimes \mathbf{b})\mathbf{c} = (\mathbf{b} \cdot \mathbf{c})\mathbf{a}$. In index notation Eq. (2.19) may be written as

$$R_{ij} \equiv \overline{u_i u_j}, \tag{2.20}$$

associated with the fluctuating part of a velocity field $\mathbf{U}$. The next chapter will show that R in the form

$$\sigma_t = -\rho R, \tag{2.21}$$

and known as the Reynolds stress tensor,[1] plays a major role in the analysis of turbulent fluid motion. Besides containing the diagonal components for which $i = j$ and whose meaning was discussed previously, R contains off-diagonal components such as $R_{12} = \overline{u_1 u_2} = \overline{uv}$ which will later be seen to be the central focus of attempts at modeling turbulent flows.

The significance of correlations such as $\overline{u_1 u_2}$, which can be written suggestively as

$$\rho \overline{u_1 u_2} = \overline{(\rho U_1) u_2} \tag{2.22}$$

using the definition of u_1 and Eq. (2.15), arises from its interpretation as representing the flux of momentum ρU_1 per unit area per second in the x_2 direction caused by the u_2 velocity fluctuation. In fact, as will be considered in detail in Chapter 6, any correlation of the form $\overline{u_i \Phi}$ where Φ is equal to the volume concentration of a transportable quantity in the fluid can be interpreted as representing the flux of Φ per unit area-sec across a surface oriented in the x_i direction driven by turbulent fluctuations. Since turbulent transport of momentum and other quantities is one of the primary physical processes of turbulent flow, there is great interest in understanding the behavior of correlations such as R in flows of practical interest. A significant part of our subsequent discussion will be involved with transport correlations in one way or another.

Associated with the components of the velocity field are probability density functions (*pdfs*) that provide, in principle, considerably more information about the random

---

1 For convenience, R_{ij} will also be referred to as the Reynolds stress tensor.

turbulent field than is contained in its mean and covariances. There is also sometimes interest in higher order statistics such as the skewness

$$S = \frac{\overline{u_1^3}}{\overline{u_1^2}^{3/2}}, \tag{2.23}$$

which gives an indication of bias or asymmetry in how the fluctuating velocity field is distributed between plus and minus values. In the same vein, the flatness factor or *kurtosis*, defined via

$$F = \frac{\overline{u_1^4}}{\overline{u_1^2}^{2}}, \tag{2.24}$$

is a measure of how much the velocity fluctuations are congregated at large and small values. In Chapter 5 it will be seen that the skewness and flatness factor belonging to velocity derivatives, such as $\partial u_1/\partial x_1$, play a major role in the equation governing the decay of isotropic turbulence.

The probability density function itself, if computed or measured, say for the velocity component u_1, provides information about the range of values u_1 takes and the likelihood that it assumes specific values in its range. In fact, if the probability density function of u_1 is $p(u)$, then by definition $p(u)du$ is the probability that u_1 takes on values between u and $u + du$. Clearly,

$$\int_{-\infty}^{\infty} p(u)du = 1 \tag{2.25}$$

since u must take on some value for each experiment. It also follows that

$$\overline{u_1} = \int_{-\infty}^{\infty} up(u)du = 0 \tag{2.26}$$

and

$$\overline{u_1^2} = \int_{-\infty}^{\infty} u^2 p(u)du \tag{2.27}$$

and so forth, for higher order moments.

Probability density functions can also be written for multiple variables, in which case they are referred to as *joint* probability density functions. For example, for velocity components u_1 and u_2, $p(u, v)dudv$ is the probability that both $u \leq u_1 \leq u + du$ and $v \leq u_2 \leq v + dv$ are satisfied by a realization (u_1, u_2) of the velocity field. A notable use of the joint pdf is in analyzing off-diagonal terms in the Reynolds stress tensor, as in

$$\overline{u_1 u_2} = \int\int uv\, p(u, v)dudv, \tag{2.28}$$

a relation which is useful in discerning the origin and nature of turbulent momentum transport.

2.2.3 Two-Point Correlations

As alluded to previously, there is a structural side to turbulent motion that is useful to measure and investigate, and cannot be explored via one-point statistics. Turbulent flow contains localized vortical features known loosely as "eddies" for whom the local, surrounding velocity field evolves as a somewhat coherent entity. Such flow events can best be understood through measures that take into account their physical and temporal extent. For example, a vortex passing by a point may be associated with a particular velocity pattern in its neighborhood. One route toward acquiring information about coherent events is via measurements made at two or more different spatial points at the same or different times. Such statistics can be analyzed theoretically and can be measured in physical experiments and in DNS.

Among the most important quantities involving the velocity field at different points are the two-point velocity correlation tensor defined by

$$\mathcal{R}(\mathbf{x},\mathbf{y},t) \equiv \overline{\mathbf{u}(\mathbf{x},t) \otimes \mathbf{u}(\mathbf{y},t)}, \tag{2.29}$$

or in index notation

$$\mathcal{R}_{ij}(\mathbf{x},\mathbf{y},t) \equiv \overline{u_i(\mathbf{x},t)u_j(\mathbf{y},t)}, \tag{2.30}$$

and the two-point triple velocity correlation tensor

$$\mathcal{S}(\mathbf{x},\mathbf{y},t) \equiv \overline{\mathbf{u}(\mathbf{x},t) \otimes \mathbf{u}(\mathbf{x},t) \otimes \mathbf{u}(\mathbf{y},t)}, \tag{2.31}$$

or in index notation

$$\mathcal{S}_{ij,k}(\mathbf{x},\mathbf{y},t) \equiv \overline{u_i(\mathbf{x},t)u_j(\mathbf{x},t)u_k(\mathbf{y},t)}, \tag{2.32}$$

where the indexing on the left-hand side of Eq. (2.32) emphasizes the fact that the k component is at a different location than the i and j components. When $\mathbf{x} = \mathbf{y}$,

$$\mathcal{R}(\mathbf{x},\mathbf{x},t) = \overline{\mathbf{u}(\mathbf{x},t) \otimes \mathbf{u}(\mathbf{x},t)} = R(\mathbf{x},t), \tag{2.33}$$

which is the Reynolds stress tensor Eq. (2.19). Setting $\mathbf{r} = \mathbf{y} - \mathbf{x}$ as the position vector connecting $\mathbf{x}$ with $\mathbf{y}$, an alternative way of indicating the two-point velocity correlation tensor is as $\mathcal{R}(\mathbf{x}, \mathbf{x} + \mathbf{r}, t)$, and if it so happens, as it does in homogeneous turbulence, that this correlation is the same for all $\mathbf{x}$ then it can be written as

$$\mathcal{R}(\mathbf{r},t) \equiv \overline{\mathbf{u}(\mathbf{x},t) \otimes \mathbf{u}(\mathbf{x}+\mathbf{r},t)}. \tag{2.34}$$

At $\mathbf{r} = 0$, $\mathcal{R}(\mathbf{r}, t)$ is identical to Eq. (2.19) while at sufficiently large $\mathbf{r}$ it can be expected that $\mathcal{R}(\mathbf{r}, t)$ will be at or close to zero since there is unlikely to be a flow mechanism by which correlation between fluctuating velocity components can persist over large distances. For example, when $\mathbf{r}$ is greater than the size of the largest eddies in the flow, there is little reason why a correlation should exist. In a similar way to $\mathcal{R}(\mathbf{r}, t)$ one can define the triple velocity correlation tensor for homogeneous turbulence at two points separated by $\mathbf{r}$ as

$$\mathcal{S}(\mathbf{r},t) \equiv \overline{\mathbf{u}(\mathbf{x},t) \otimes \mathbf{u}(\mathbf{x},t) \otimes \mathbf{u}(\mathbf{x}+\mathbf{r},t)}. \tag{2.35}$$

Equations (2.34) and (2.35) play central roles in the analysis of isotropic turbulent flows, as will be seen in Chapters 4 and 5. For convenience the notation in Eqs. (2.34) and (2.35) will be adopted even when considering such correlations in non-homogeneous turbulent flows, so that in such instances the $\mathbf{x}$ dependence is implied.

Of special importance in a number of contexts are the two-point longitudinal and transverse correlation coefficients $f(r)$ and $g(r)$, respectively, that correspond to the particular choices of $\mathcal{R}_{ij}(\mathbf{r}, t)$ in the formulas

$$\overline{u^2} f(r) \equiv \mathcal{R}_{11}(r\mathbf{e}_1) \tag{2.36}$$

and

$$\overline{v^2} g(r) \equiv \mathcal{R}_{22}(r\mathbf{e}_1). \tag{2.37}$$

Though not indicated explicitly, a time dependence is implied if the turbulent flow is non-stationary. The physical meaning of $f(r)$ and $g(r)$ is illustrated in Figure 2.1. In fact, Eq. (2.36) can be simplified to

$$f(r) = \overline{u(x)u(x+r)}/\overline{u^2}(x) \tag{2.38}$$

while Eq. (2.37) is equivalent to

$$g(r) = \overline{v(x)v(x+r)}/\overline{v^2}(x). \tag{2.39}$$

From its definition it is clear that $f(0) = 1$ and expanding $f(r)$ in a Taylor series around $r = 0$ gives

$$f(r) = 1 + r\frac{df}{dr}(0) + \frac{r^2}{2!}\frac{d^2f}{dr^2}(0) + \dots . \tag{2.40}$$

The identity

$$\overline{u^2}\frac{df}{dr}(r) = \overline{u(x)\frac{\partial}{\partial r}[u(x+r)]} = \overline{u(x)\frac{\partial u}{\partial x}(x+r)} \tag{2.41}$$

follows from Eq. (2.38) and when $r = 0$ this reduces to

$$\overline{u^2}\frac{df}{dr}(0) = \frac{1}{2}\frac{\partial \overline{u^2}}{\partial x}. \tag{2.42}$$

In homogeneous flow for which $\overline{u^2}$ is constant, and keeping just quadratic terms, Eq. (2.40) simplifies to

$$f(r) = 1 + \frac{r^2}{2}\frac{d^2f}{dr^2}(0), \tag{2.43}$$

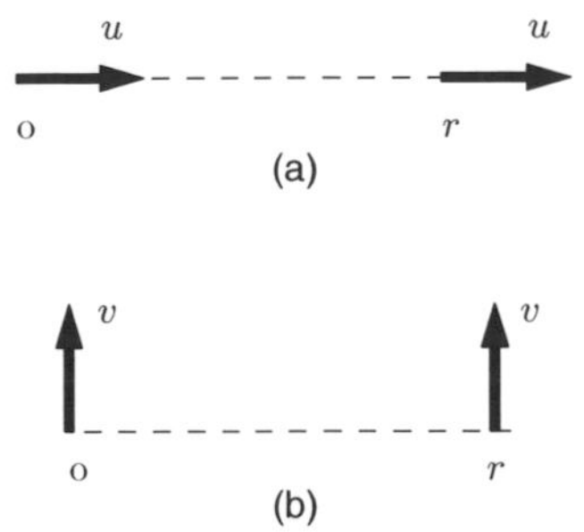

Figure 2.1 (a) Longitudinal and (b) transverse velocities appearing in the definitions of $f(r)$ and $g(r)$, respectively.

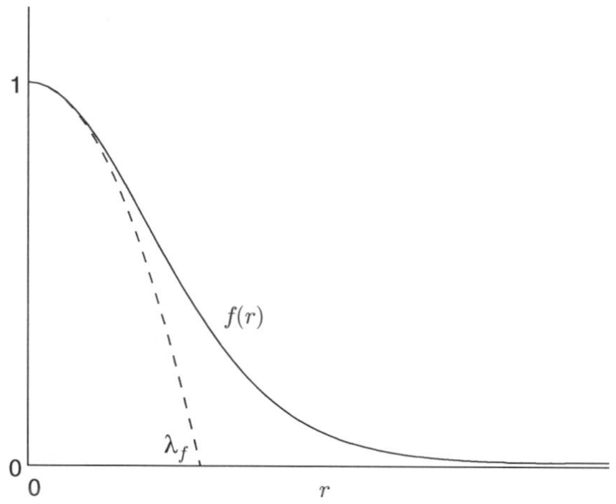

Figure 2.2 Definition of the microscale λ_f.

where the expression on the right-hand side is a parabolic approximation to $f(r)$, as shown in Figure 2.2. The distance from $r = 0$ at which the parabola intersects zero establishes a length scale, λ_f, that is known as the Taylor *microscale.* By definition

$$0 = 1 + \frac{\lambda_f^2}{2}\frac{d^2f}{dr^2}(0) \tag{2.44}$$

from which it is found that

$$\lambda_f = \sqrt{\frac{-2}{\frac{d^2f}{dr^2}(0)}}. \tag{2.45}$$

A similar definition and calculation can be done for $g(r)$, leading to the transverse microscale

$$\lambda_g = \sqrt{\frac{-2}{\frac{d^2g}{dr^2}(0)}}. \tag{2.46}$$

Microscales provide a useful way of describing the size of the flow features where viscous effects are important, as will be shown later.

While λ_f and λ_g are representative of the smaller scales in turbulent flow, a measure of the largest scales is given by integral scales such as

$$\Lambda_f \equiv \int_0^\infty f(r)dr. \tag{2.47}$$

The integral in (2.47) exists because $f(r)$ can be expected to have bounded support. Since the largest turbulent eddies establish the extent of the correlation in $f(r)$, it is to be expected that Λ_f is a measure of the largest scales of turbulent motion in a flow field. Ordinarily, events at the scale of Λ_f are very energetic, since they are likely to be the direct result of the stirring mechanism that is responsible for turbulence generation.

Quantities such as $f(r)$ and $g(r)$ can be difficult to measure in physical experiments so that they are more likely to be the subject of DNS studies or theoretical analyses. A correlation coefficient that has some similarities with $f(r)$ and yet is more readily computed in physical experiments is the time auto-correlation coefficient defined as

$$\mathcal{R}_E(\tau) = \overline{u(t)u(t+\tau)}/\overline{u^2}, \tag{2.48}$$

where $u(t)$ is a time sequence of velocities measured at a fixed point. This kind of data can be readily obtained from a simple velocity probe placed at a fixed position in a flow field. By the same kind of argument as was used in showing that $df/dr(0) = 0$ in homogeneous flow, it may be shown that $d\mathcal{R}_E/d\tau(0) = 0$ in stationary turbulent flow, and in this case a time microscale can be defined via

$$\tau_E = \sqrt{\frac{-2}{\frac{d^2\mathcal{R}_E}{d\tau^2}(0)}}. \tag{2.49}$$

In fact, $\mathcal{R}_E$ and τ_E are sometimes used as a basis for estimating $f(r)$ and λ_f, as will be shown in Chapter 4.

2.2.4 Spatial Spectra

Another way to reveal information about the kinds of motions that are present in turbulent flow is by expanding the velocity field in a Fourier series. In a typical situation velocity data may be known on a finite grid of points such as occurs when numerically studying periodic flow in a cubic box of dimension L. In this case a discrete Fourier transform can be applied to the velocity data using, for example, the command *fft* in MATLAB, and from this information about the energy spectrum can be obtained.

For turbulent flow in a cubic box with periodic boundary conditions the Fourier series decomposition of the velocity field consists of

$$\mathbf{u}(\mathbf{x}, t) = \sum_{\mathbf{k}} \hat{\mathbf{u}}(\mathbf{k}, t) e^{\imath \mathbf{k}\cdot\mathbf{x}} \tag{2.50}$$

where the wave number vector

$$\mathbf{k} = 2\pi\mathbf{n}/L, \tag{2.51}$$

$\imath = \sqrt{-1}$, and $\mathbf{n}$ denotes the set of integer triples (n_1, n_2, n_3) with each n_i running from $-\infty \to +\infty$. The countable set of vectors $\hat{\mathbf{u}}(\mathbf{k}, t)$ are the Fourier coefficients of $\mathbf{u}(\mathbf{x}, t)$ and may be obtained from the inverse transform

$$\hat{\mathbf{u}}(\mathbf{k}, t) = \frac{1}{L^3} \int_{\mathcal{V}_L} \mathbf{u}(\mathbf{x}, t) e^{-\imath \mathbf{k}\cdot\mathbf{x}} d\mathbf{x}, \tag{2.52}$$

where the integration domain, $\mathcal{V}_L$, is a cube with sides of length L. Knowledge of the complete set of Fourier coefficients associated with $\mathbf{u}$ is equivalent to knowledge of $\mathbf{u}(\mathbf{x}, t)$ itself and vice versa. The basis functions $e^{\imath \mathbf{k}\cdot\mathbf{x}}$ in Eq. (2.52) are mutually orthogonal under the inner product

$$\langle f, g \rangle \equiv \int_{\mathcal{V}_L} f(\mathbf{x}) g^*(\mathbf{x}) d\mathbf{x} \tag{2.53}$$

where $g^*(\mathbf{x})$ is the complex conjugate of the function $g(\mathbf{x})$. This means that

$$\langle e^{\imath \mathbf{k}\cdot\mathbf{x}}, e^{\imath \mathbf{l}\cdot\mathbf{x}} \rangle = \int_{\mathcal{V}_L} e^{\imath(\mathbf{k}-\mathbf{l})\cdot\mathbf{x}} d\mathbf{x} = 0, \qquad \mathbf{k} \neq \mathbf{l}. \tag{2.54}$$

The Fourier coefficients in the expansion in Eq. (2.50) give information about the degree to which specific wave-like contributions $e^{\imath\mathbf{k}\cdot\mathbf{x}}$ to the velocity field are important. For high wave numbers the basis functions account for small-scale motions, while for small **k** large-scale motions are represented.

Since $\mathbf{u}(\mathbf{x}, t)$ is real, the series in Eq. (2.50) must equal its complex conjugate, $\mathbf{u}^*(\mathbf{x}, t)$, so that

$$\mathbf{u}(\mathbf{x}, t) = \sum_{\mathbf{k}} \hat{\mathbf{u}}^*(\mathbf{k}, t)e^{-\imath\mathbf{k}\cdot\mathbf{x}} = \sum_{\mathbf{k}} \hat{\mathbf{u}}^*(-\mathbf{k}, t)e^{\imath\mathbf{k}\cdot\mathbf{x}}, \tag{2.55}$$

where the last equality comes about by changing **k** to $-\mathbf{k}$ in the summation. Setting Eq. (2.50) equal to the last expression in Eq. (2.55) and grouping terms according to **k** gives

$$\sum_{\mathbf{k}} [\hat{\mathbf{u}}^*(-\mathbf{k}, t) - \hat{\mathbf{u}}(\mathbf{k}, t)]e^{\imath\mathbf{k}\cdot\mathbf{x}} = 0. \tag{2.56}$$

Since Fourier modes are orthogonal it follows that the term in brackets must be zero; hence the useful identity

$$\hat{\mathbf{u}}^*(-\mathbf{k}, t) = \hat{\mathbf{u}}(\mathbf{k}, t), \tag{2.57}$$

holds.

The Fourier representation of the velocity in Eq. (2.50) can be made the basis for a spectral decomposition of the turbulent kinetic energy. First note from Eq. (2.52) and Eq. (2.57) that

$$\overline{\hat{\mathbf{u}}(\mathbf{k}, t) \otimes \hat{\mathbf{u}}^*(\mathbf{l}, t)} = \frac{1}{L^6} \int_{\mathcal{V}_L} \int_{\mathcal{V}_L} \overline{\mathbf{u}(\mathbf{x}, t) \otimes \mathbf{u}(\mathbf{y}, t)} e^{-\imath(\mathbf{k}\cdot\mathbf{x}-\mathbf{l}\cdot\mathbf{y})} d\mathbf{x}\, d\mathbf{y}. \tag{2.58}$$

For the flow in a periodic box that is being considered here, it may be assumed that the turbulence is homogeneous, so that $R(\mathbf{x}, \mathbf{x} + \mathbf{r})$ satisfies Eq. (2.34). In this case, setting $\mathbf{y} = \mathbf{x} + \mathbf{r}$ in Eq. (2.58) and transforming the **y** integration to **r** integration gives

$$\overline{\hat{\mathbf{u}}(\mathbf{k}, t) \otimes \hat{\mathbf{u}}^*(\mathbf{l}, t)} = \frac{1}{L^6} \int_{\mathcal{V}_L} d\mathbf{x}\, e^{-\imath(\mathbf{k}-\mathbf{l})\cdot\mathbf{x}} \int_{\mathcal{V}_L} d\mathbf{r}\mathcal{R}(\mathbf{r}, t)e^{\imath\mathbf{l}\cdot\mathbf{r}}. \tag{2.59}$$

By orthogonality, the **x** integration in this relation is only non-zero when $\mathbf{k} - \mathbf{l} = 0$, in which case Eq. (2.59) yields

$$\overline{\hat{\mathbf{u}}(\mathbf{k}, t) \otimes \hat{\mathbf{u}}^*(\mathbf{l}, t)} = \delta_{\mathbf{k}-\mathbf{l}\,0} \frac{1}{L^3} \int_{\mathcal{V}_L} \mathcal{R}(\mathbf{r}, t)e^{\imath\mathbf{l}\cdot\mathbf{r}} d\mathbf{r}. \tag{2.60}$$

Equation (2.60) shows that when $\mathbf{k} \neq \mathbf{l}$ so that $\mathbf{k} - \mathbf{l} \neq 0$, it follows that

$$\overline{\hat{\mathbf{u}}(\mathbf{k}, t) \otimes \hat{\mathbf{u}}^*(\mathbf{l}, t)} = 0 \qquad \mathbf{k} \neq \mathbf{l}. \tag{2.61}$$

Setting $\mathbf{l} = \mathbf{k}$ in Eq. (2.60) gives

$$\overline{\hat{\mathbf{u}}(\mathbf{k}, t) \otimes \hat{\mathbf{u}}^*(\mathbf{k}, t)} = \frac{1}{L^3} \int_{\mathcal{V}_L} \mathcal{R}(\mathbf{r}, t)e^{\imath\mathbf{k}\cdot\mathbf{r}} d\mathbf{r}. \tag{2.62}$$

Equation (2.62) can be regarded as expressing the Fourier coefficients in a transform of $\mathcal{R}(\mathbf{r}, t)$, so that its inverse relation gives the Fourier series representation of $\mathcal{R}(\mathbf{r}, t)$ as

$$\mathcal{R}(\mathbf{r}, t) = \sum_{\mathbf{k}} \overline{\hat{\mathbf{u}}(\mathbf{k}, t) \otimes \hat{\mathbf{u}}^*(\mathbf{k}, t)} e^{-\imath\mathbf{k}\cdot\mathbf{r}}. \tag{2.63}$$

In index notation this is

$$\mathcal{R}_{ij}(\mathbf{r},t) = \sum_{\mathbf{k}} \overline{\hat{u}_i(\mathbf{k},t)\hat{u}_j^*(\mathbf{k},t)}e^{-\imath\mathbf{k}\cdot\mathbf{r}}. \tag{2.64}$$

Setting $\mathbf{r} = 0$, taking the trace of Eq. (2.64) so that $i = j$, and noting that for any complex number z, $|z|^2 = zz^*$, a decomposition of the turbulent kinetic energy is derived in the form

$$K = \frac{1}{2}\sum_{\mathbf{k}} \overline{|\hat{\mathbf{u}}(\mathbf{k},t)|^2}, \tag{2.65}$$

showing how K is made up of contributions from different wave numbers. Defining $E_{\mathbf{k}} = \frac{1}{2}\overline{|\hat{\mathbf{u}}(\mathbf{k},t)|^2}$, then Eq. (2.65) gives

$$K = \sum_{\mathbf{k}} E_{\mathbf{k}} \tag{2.66}$$

where the quantities $E_{\mathbf{k}}$ form the 3D spectrum of K. To the extent that the terms on the right-hand side of Eq. (2.66) can be measured or computed this gives information about the relative importance of motions at different scales in contributing to the turbulent kinetic energy.

The decomposition in Eq. (2.66) can be made more useful by gathering terms whose wave vectors $\mathbf{k}$ have magnitude falling within shells $k\,dk < |\mathbf{k}| < (k+1)\,dk$, $k = 0, 1, 2, \ldots$ of thickness dk. In this case Eq. (2.66) can be rearranged to give

$$K = \sum_{k=0}^{\infty} E_k dk \tag{2.67}$$

where

$$E_k \equiv \frac{1}{2\,dk} \sum_{k\,dk<|\mathbf{k}|<(k+1)\,dk} \overline{|\hat{\mathbf{u}}(\mathbf{k},t)|^2}. \tag{2.68}$$

The discrete energy decomposition in Eq. (2.67) can be evaluated from gridded data whether from computation or measurements (see Problems 2.5 and 2.6).

In theoretical considerations of turbulent flow the energy decomposition is often considered in infinite space, which is equivalent to having $L \to \infty$ in the analysis of periodic flow in a box. In this case it is convenient to develop the spatial relations based on the 3D Fourier transform of $\mathcal{R}(\mathbf{r},t)$ given by

$$\mathcal{E}(\mathbf{k},t) \equiv (2\pi)^{-3} \int_{\Re^3} e^{\imath\mathbf{r}\cdot\mathbf{k}}\mathcal{R}(\mathbf{r},t)d\mathbf{r}, \tag{2.69}$$

that is the continuous analogue of Eq. (2.62) in which $\mathbf{k}$ is now the continuous wave-number vector and $\mathcal{E}(\mathbf{k},t)$, having components $\mathcal{E}_{ij}(\mathbf{k},t)$, is referred to as the energy spectrum tensor. Equation (2.69) is well defined since the two-point velocity correlation tensor $\mathcal{R}_{ij}(\mathbf{r})$ may be assumed to be bounded and non-zero on a finite region. Corresponding to Eq. (2.69) is the inverse transform

$$\mathcal{R}(\mathbf{r},t) = \int_{\Re^3} e^{-\imath\mathbf{r}\cdot\mathbf{k}}\mathcal{E}(\mathbf{k},t)d\mathbf{k}, \tag{2.70}$$

where $d\mathbf{k} = dk_1 dk_2 dk_3$ is a differential volume in wave-number space. Though not indicated explicitly, $\mathcal{R}(\mathbf{r},t)$ and $\mathcal{E}(\mathbf{k},t)$ can be determined at any $\mathbf{x}$ position in a flow so there is an implied **x** dependence in these quantities. Equations (2.69) and (2.70), in effect, provide a means of decomposing turbulence correlations into contributions from a continuous range of scales as represented by Fourier components $e^{\imath \mathbf{r}\cdot\mathbf{k}}$.

Taking the trace of Eq. (2.70), dividing by 2 and setting $\mathbf{r} = 0$ gives the spectral decomposition of the turbulent kinetic energy:

$$K(t) = \frac{1}{2}\int_{\Re^3} \mathrm{tr}\mathcal{E}(\mathbf{k}, \mathrm{t})\mathrm{d}\mathbf{k}, \tag{2.71}$$

where

$$\frac{1}{2}\mathrm{tr}\mathcal{E} = \frac{1}{2}\mathcal{E}_{ii}(\mathbf{k}, \mathrm{t}) \tag{2.72}$$

is the density of energy in wave number space. As was done in deriving Eq. (2.67), it is useful to collect the energy – which is typically scattered throughout **k** space – onto shells of fixed distance $k = |\mathbf{k}|$ from the origin. This is done by writing (2.71) in terms of spherical coordinates in wave number space so that

$$K(t) = \int_0^\infty \left[\frac{1}{2}\int_{|\mathbf{k}|=k} \mathrm{tr}\mathcal{E}(\mathbf{k}, \mathrm{t})\mathrm{d}\Omega\right] dk, \tag{2.73}$$

where $d\Omega$ is an element of solid angle and $d\mathbf{k} = d\Omega dk$. The term in square brackets in Eq. (2.73) is defined as the *energy density function* or *energy spectrum*:

$$E(k, t) \equiv \frac{1}{2}\int_{|\mathbf{k}|=k} \mathrm{tr}\mathcal{E}(\mathbf{k}, \mathrm{t})\mathrm{d}\Omega \tag{2.74}$$

yielding

$$K(t) = \int_0^\infty E(k, t)dk, \tag{2.75}$$

as a continuous analogue of Eq. (2.67). In this formula, $E(k, t)$ shows how the kinetic energy is distributed among the different scales of the flow. The nature of $E(k, t)$ in turbulent flows has long played an important role in the theoretical analysis of turbulence, and will be considered at length in later chapters.

2.2.5 Time Spectra

Complementing the spatial spectra that were defined in the previous section are time spectra formed from a Fourier transform of the time correlation function $R_E(\tau)$ defined in Eq. (2.48). In this case $\hat{\mathcal{R}}_E(\omega')$ is defined as the Fourier transform of $\mathcal{R}_E(\tau)$, where the frequency $\omega' = 2\pi\omega$ is in radians/second, and ω is the frequency with units of hertz, that is, cycles per second. Thus

$$\hat{\mathcal{R}}_E(\omega') = \int_{-\infty}^{\infty} e^{-\imath\tau\omega'}\mathcal{R}_E(\tau)d\tau, \tag{2.76}$$

which is complemented by the inverse transform

$$\mathcal{R}_E(\tau) = \frac{1}{2\pi}\int_{-\infty}^{\infty} e^{\imath\tau\omega'}\hat{\mathcal{R}}_E(\omega')d\omega'. \tag{2.77}$$

Evaluating (2.77) at $\tau = 0$ and defining the 1D energy spectrum as

$$\hat{E}_{11}(\omega) = 2\overline{u^2}\hat{\mathcal{R}}_E(2\pi\omega) \tag{2.78}$$

gives, after a change of variables in Eq. (2.77),

$$\overline{u^2} = \frac{1}{2}\int_{-\infty}^{\infty} \hat{E}_{11}(\omega)d\omega. \tag{2.79}$$

If the turbulent velocity field is stationary so that

$$\overline{u(t)u(t+\tau)} = \overline{u(t-\tau)u(t)} \tag{2.80}$$

then

$$R_E(\tau) = R_E(-\tau). \tag{2.81}$$

In this case Eq. (2.76) implies that

$$\hat{\mathcal{R}}_E(\omega') = 2\int_0^{\infty} \cos\tau\omega' \mathcal{R}_E(\tau)d\tau \tag{2.82}$$

and then that

$$\hat{\mathcal{R}}_E(-\omega') = \hat{\mathcal{R}}_E(\omega') \tag{2.83}$$

so that Eq. (2.77) gives

$$\mathcal{R}_E(\tau) = \frac{1}{\pi}\int_0^{\infty} \cos\tau\omega' \hat{\mathcal{R}}_E(\omega')d\omega'. \tag{2.84}$$

Thus $\mathcal{R}_E(\tau)$ and $\hat{\mathcal{R}}_E(\omega)$ through Eqs. (2.82) and (2.84) form a Fourier cosine transform pair in stationary turbulence. It follows also from Eqs. (2.78) and (2.83) that

$$\overline{u^2} = \int_0^{\infty} \hat{E}_{11}(\omega)d\omega. \tag{2.85}$$

The fact that $\mathcal{R}_E(\tau)$ is easily evaluated from data from a single probe means that it is a relatively simple matter to determine the 1D spectrum, $\hat{E}_{11}(\omega)$. In contrast, it is often relatively difficult to measure velocity data throughout a specified flow region or along a line so that obtaining spectra such as $E(k,t)$ is not easy. In Section 4.4.2 it will be shown that in at least some circumstances it is possible to approximately link the temporal and spatial variations in u and from this a connection between $\hat{E}_{11}(\omega)$ and $E(k,t)$ can be developed. In this way information from a continuous measurement of velocity at a point can give insight into how energy is distributed among spatial scales in a turbulent flow.

Reference

1 Bernard, P.S. (2015) *Fluid Dynamics*, Cambridge University Press, Cambridge.

Problems

2.1 Demonstrate why a time average taken in a non-stationary flow might not have a fluctuating velocity with zero mean.

2.2 Compute the form of λ_f in a general shear flow when $df/dr(0) \neq 0$. Are there any potential problems with this definition?

2.3 Evaluate $f(r), g(r), \lambda_f, \lambda_g$ for a point in a turbulent channel flow. Use data available on the internet, such as the channel flow data at http://turbulence.pha.jhu.edu/. Let the separation in the correlations be in the spanwise and streamwise directions and see if the orientation of the spatial separation in $f(r)$ and $g(r)$ affects the results.

2.4 Consider the function

$$u(x) = 5\sin(2\pi 3x/L) + .9\cos(2\pi 7x/L) - 1.5\sin(2\pi 8x/L) + 2\cos(2\pi 11x/L) \tag{2.86}$$

defined and periodic on the interval $0 \leq x \leq L$ and take $L = 2$. Let $u_i = u(x_i)$ be the function evaluated on the grid of N points $x_i = \Delta x(i-1), i = 1, 2, \dots, N$ where $\Delta x = L/N$. (Note that $x_{N+1} = L$). Plot the function. Find the spectral content of $u(x)$ by taking its discrete Fourier transform (e.g., using the command *fft* in MATLAB) and make a stem plot of the absolute value of the result. Show that the trigonometric terms in the given function show up as non-zero modes in the stem plot. Also, make note of the *aliasing* in which these modes also show up at the high wave number end of the spectrum.

2.5 Obtain turbulent flow velocity data u, v, w on a grid covering a finite region of 3D space. Compute the 3D spectrum given in Eq. (2.66) using the data. In MATLAB, for example, the commands $fftn$ followed by $fftswitch$ can be used to obtain the spectral decomposition $E_{\mathbf{k}}$. Now compute and plot E_k using Eq. (2.68).

2.6 Develop the analogue of Eqs. (2.67) and (2.68) that applies to turbulent velocity data in one space dimension. Evaluate the spectrum using actual velocity data available on the internet such as the channel flow data at http://turbulence.pha.jhu.edu/.

2.7 Why is Eq. (2.80) valid in stationary turbulence?

2.8 For stationary turbulence derive Eqs. (2.82), (2.84), and (2.85).

3

Overview of Turbulent Flow Physics and Equations

Turbulent fluid motion is nominally governed by the laws of mass, momentum, and energy conservation so that in the best of circumstances, as, for example, when it is feasible to perform a DNS, the solution is obtained by numerically solving the conservation equations. In the RANS approach the governing equations are averaged prior to solving them and when the laws of motions are cast in this form, a number of fundamental flow processes can be delineated that form the starting point for modeling and theoretical analysis. In this chapter the most widely applied of the averaged fluid equations are considered, including a discussion of the physical ideas associated with them that are important for understanding turbulent motion. Later chapters will add to these concepts besides considering the flow processes in greater depth.

3.1 The Reynolds Averaged Navier–Stokes Equation

Consider the Navier–Stokes equation expressed as

$$\rho\left(\frac{\partial \mathbf{U}}{\partial t} + (\nabla \mathbf{U})\mathbf{U}\right) = \rho \mathbf{g} + \nabla \cdot \sigma, \tag{3.1}$$

where for a Newtonian fluid the stress tensor

$$\sigma = -PI + d, \tag{3.2}$$

the deviatoric stress tensor

$$d = 2\mu e_s, \tag{3.3}$$

and

$$e_s = \frac{1}{2}(\nabla \mathbf{U} + (\nabla \mathbf{U})^t) \tag{3.4}$$

is the rate of strain tensor [1]. Here, $(\nabla \mathbf{U})^t$ denotes the transpose of the velocity gradient tensor and μ is the viscosity. Substituting Eqs. (3.2)–(3.4) into (3.1) recovers Eq. (2.4).

Before taking an average of Eq. (3.1) it is helpful to rewrite the advection term in flux form, which for incompressible flow means employing the identity

$$(\nabla \mathbf{U})\mathbf{U} = \nabla \cdot (\mathbf{U} \otimes \mathbf{U}) \tag{3.5}$$

Turbulent Fluid Flow, First Edition. Peter S. Bernard.

Companion website: www.wiley.com/go/Bernard/Turbulent_Fluid_Flow

where the term on the right-hand side is the divergence of the tensor $\mathbf{U} \otimes \mathbf{U}$. Using Eq. (3.5), (3.1) becomes

$$\rho\left(\frac{\partial \mathbf{U}}{\partial t} + \nabla \cdot (\mathbf{U} \otimes \mathbf{U})\right) = \rho \mathbf{g} + \nabla \cdot \sigma. \tag{3.6}$$

Prior to averaging Eq. (3.6) note the identity

$$\overline{\mathbf{U} \otimes \mathbf{U}} = \overline{\mathbf{U}} \otimes \overline{\mathbf{U}} + \overline{\mathbf{u} \otimes \mathbf{u}} \tag{3.7}$$

that follows after substituting Eq. (2.14) into the left-hand side of this relation and taking advantage of Eq. (2.15) and the linearity of the tensor product. The last term in Eq. (3.7) is the velocity covariance tensor previously defined in Eq. (2.19) that is related to the Reynolds stress tensor Eq. (2.21). Averaging Eq. (3.6) and applying Eqs. (3.7), (2.19), and (2.21) gives

$$\rho\left(\frac{\partial \overline{\mathbf{U}}}{\partial t} + (\nabla \overline{\mathbf{U}})\overline{\mathbf{U}}\right) = \rho \mathbf{g} + \nabla \cdot (\overline{\sigma} + \sigma_t), \tag{3.8}$$

or written in the conventional form using index notation, where $P = \overline{P} + p$,

$$\rho\left(\frac{\partial \overline{U}_i}{\partial t} + \overline{U}_j \frac{\partial \overline{U}_i}{\partial x_j}\right) = -\frac{\partial \overline{P}}{\partial x_i} + \rho g_i + \frac{\partial}{\partial x_j}\left(\mu \frac{\partial \overline{U}_i}{\partial x_j} - \rho R_{ij}\right). \tag{3.9}$$

Accompanying Eqs. (3.8) and (3.9) for incompressible flow is the averaged mass conservation equation derived from Eq. (2.3)

$$\nabla \cdot \overline{\mathbf{U}} = 0. \tag{3.10}$$

Subtracting Eq. (3.10) from Eq. (2.3) shows that $\nabla \cdot \mathbf{u} = 0$ so that the velocity fluctuation field is incompressible in its own right.

It is instructive to compare Eqs. (3.1) and (3.8). The former, governing the velocity field of an arbitrary fluid flow, is dependent only on the stresses contained in σ. In contrast, Eq. (3.8) describes the physics of a fictitious flow field that everywhere moves at the local mean velocity. The motion of this imaginary "fluid" depends on two effective stress fields: the mean of the true physical stress tensor (i.e., $\overline{\sigma}$) and the Reynolds stress tensor which, as noted previously, has a physical interpretation as momentum transport across surfaces. An average of Eq. (3.2) using (3.3) and (3.4) gives

$$\overline{\sigma} = -\overline{P}I + \mu(\nabla\overline{\mathbf{U}} + (\nabla\overline{\mathbf{U}})^t), \tag{3.11}$$

showing that the contribution that $\overline{\sigma}$ makes to $\overline{\mathbf{U}}$ as determined by Eq. (3.8) is not unlike the contribution that σ makes to $\mathbf{U}$ as determined by Eq. (3.6). In contrast to $\overline{\sigma}$ the Reynolds stress tensor σ_t brings a set of physics to the determination of the mean velocity field that has no parallel in the Navier–Stokes equation itself. Finding an accurate constitutive law governing the Reynolds stress tensor so that Eq. (3.8) can be solved for $\overline{\mathbf{U}}$ has proven to be exceedingly difficult. To date no fully satisfactory relation has been developed. The problem in this case is in translating knowledge about the transport process into robust formulas that enable its prediction in general circumstances. We consider in detail some aspects of the transport physics in Chapter 6 and efforts at modeling the Reynolds stress tensor in Chapter 9.

3.2 Turbulent Kinetic Energy Equation

The energy in a unit volume of fluid resides in both the kinetic energy $\frac{1}{2}\rho\mathbf{U}\cdot\mathbf{U} = \frac{1}{2}\rho|\mathbf{U}|^2$ associated with the bulk motion of the fluid and in the internal energy, ρe, where the internal energy/mass, e can generally be assumed to satisfy $e = cT$, where T is the temperature and c is a specific heat. For liquids $c = c_p$ appropriate to a constant pressure process, while for gases $c = c_v$ for a process at constant specific volume. The division of energy between kinetic and internal energies can be viewed as a split between energy associated with the mean drift of the molecules, which is the velocity $\mathbf{U}$ in the Navier–Stokes equation, and the energy associated with the translational, rotational, and vibrational motions that molecules have relative to the mean drift.

The division of total energy between kinetic and internal energies has a parallel in turbulent flow in the division of the total mean kinetic energy per unit mass given by $\overline{|\mathbf{U}|^2}/2$ into a sum of the kinetic energy of the mean field defined by

$$\overline{K} \equiv \frac{1}{2}|\overline{\mathbf{U}}|^2 \tag{3.12}$$

and the kinetic energy of the fluctuating field K that was defined in Eq. (2.17). In fact, using Eq. (2.14) and the linearity of the tensor product, it follows that the total mean kinetic energy

$$\frac{1}{2}\overline{|\mathbf{U}|^2} = \overline{K} + K, \tag{3.13}$$

which is the sum of the energy of the fictitious fluid traveling at the mean speed and the energy in the fluctuating motion as it varies above and below the velocity of the mean field.

Equations for $\overline{K}$ and K can be derived that have some similarities with the general equations for $|\mathbf{U}|^2/2$ and e, so it is useful to first review some salient points that can be gleaned from the conservation equations for the latter quantities. Neglecting the effect of gravity, the law governing the kinetic energy of a fluid follows by taking a dot product of the Navier–Stokes equation (3.6) with the velocity giving

$$\rho\left(\frac{\partial|\mathbf{U}|^2/2}{\partial t} + \mathbf{U}\cdot\nabla(|\mathbf{U}|^2/2)\right) = \mathbf{U}\cdot(\nabla\cdot\sigma), \tag{3.14}$$

where the identity

$$\mathbf{U}\cdot(\nabla\cdot(\mathbf{U}\otimes\mathbf{U})) = \mathbf{U}\cdot\nabla(|\mathbf{U}|^2/2) \tag{3.15}$$

has been used, a relation that is easily proven using index notation. The first term on the right-hand side of Eq. (3.14) is the total work/volume done by surface forces in changing kinetic energy [1]. It reflects differences in the stress field from one side of a material fluid element to the other that cause its acceleration and hence changes in kinetic energy.

The internal energy is governed by the first law of thermodynamics [1] that can be expressed in the form

$$\rho\left(\frac{\partial e}{\partial t} + \mathbf{U}\cdot\nabla e\right) = \sigma : (\nabla\mathbf{U}) + \nabla\cdot(k_e\nabla T), \tag{3.16}$$

where for tensors T and S, $T : S \equiv T_{ij}S_{ij}$ is the tensor inner product. The last term in Eq. (3.16) reflects changes in internal energy resulting from molecular heat conduction, with k_e being the thermal conductivity. The first term on the right-hand side of Eq.

(3.16), known as the deformation work, accounts for the production of internal energy in incompressible flow by the deformation of fluid elements by frictional surface forces. The appearance of $\nabla \mathbf{U}$ in the expression reflects the fact that it is local differences in the velocity field that result in the deformation of fluid elements. The deformation work term for incompressible flow is always non-negative since it can be shown using Eqs. (3.2)–(3.4) that

$$\sigma : \nabla \mathbf{U} = 2\mu e_s : e_s \geq 0. \tag{3.17}$$

The sum of the surface work terms in (3.14) and (3.16) gives the total work/volume by surface forces as

$$\nabla \cdot (\sigma \mathbf{U}) = \sigma : (\nabla \mathbf{U}) + \mathbf{U} \cdot (\nabla \cdot \sigma), \tag{3.18}$$

an expression which appears in the equation governing the sum of kinetic and internal energy derived by adding Eqs. (3.14) and (3.16), namely,

$$\rho \left(\frac{\partial (e + |\mathbf{U}|^2/2)}{\partial t} + \mathbf{U} \cdot \nabla (e + |\mathbf{U}|^2/2) \right) = \nabla \cdot (\sigma \mathbf{U}) + \nabla \cdot (k_e \nabla T). \tag{3.19}$$

Keeping these results in mind, now consider the equation for the mean of the kinetic energy in turbulent flow that is derived by averaging Eq. (3.14) and has the form

$$\rho \left(\frac{\partial \overline{|\mathbf{U}|^2}/2}{\partial t} + \overline{\mathbf{U} \cdot \nabla (|\mathbf{U}|^2/2)} \right) = \overline{\mathbf{U} \cdot (\nabla \cdot \sigma)}. \tag{3.20}$$

The second term on the left-hand side can be turned into a more transparent form by noting that

$$\begin{aligned} \overline{\mathbf{U} \cdot \nabla (|\mathbf{U}|^2/2)} &= \nabla \cdot (\overline{\mathbf{U}|\mathbf{U}|^2}/2) \\ &= \nabla \cdot (\overline{\mathbf{U}}\,\overline{|\mathbf{U}|^2}/2) + \nabla \cdot (\overline{\mathbf{u}|\mathbf{U}|^2}/2) \\ &= \overline{\mathbf{U}} \cdot \nabla (\overline{K} + K) + \nabla \cdot (\overline{\mathbf{u}|\mathbf{U}|^2}/2). \end{aligned} \tag{3.21}$$

Furthermore, the last term in Eq. (3.21) can be written as

$$\nabla \cdot (\overline{\mathbf{u}|\mathbf{U}|^2}/2) = -\nabla \cdot (\sigma_t \overline{\mathbf{U}})/\rho + \nabla \cdot (\overline{\mathbf{u}|\mathbf{u}|^2}/2), \tag{3.22}$$

after substituting Eq. (2.14) for $\mathbf{U}$ in the expression $\mathbf{U} \cdot \mathbf{U} = |\mathbf{U}|^2$. Assembling these results gives

$$\begin{aligned} \rho \left(\frac{\partial (\overline{K} + K)}{\partial t} + \overline{\mathbf{U}} \cdot \nabla (\overline{K} + K) \right) &= \overline{\mathbf{U}} \cdot (\nabla \cdot \overline{\sigma}) + \overline{\mathbf{u} \cdot (\nabla \cdot \sigma')} \\ &\quad + \nabla \cdot (\sigma_t \overline{\mathbf{U}}) - \rho \nabla \cdot \overline{\mathbf{u}(|\mathbf{u}|)/2}, \end{aligned} \tag{3.23}$$

where the decomposition $\sigma = \overline{\sigma} + \sigma'$ has been made.

The first two terms on the right-hand side of Eq. (3.23) collectively account for the average work accomplished by σ in changing the kinetic energy. The third term on the right-hand side of Eq. (3.23) represents the total work done by the Reynolds stress in altering the sum of kinetic energy of the mean and fluctuating fields and may be viewed analogously to the situation in Eq. (3.19), where the total work of σ affects the sum of kinetic and internal energies. The last term in Eq. (3.23) may be interpreted as the gradient of the turbulent flux of turbulent kinetic energy.

To complete the analogy between kinetic and internal energies in an arbitrary fluid flow on the one hand and $\overline{K}$ and K in turbulent flow on the other, consider the separate equations for $\overline{K}$ and K. The former may be derived by taking a dot product of (3.8) with $\overline{\mathbf{U}}$ yielding

$$\rho\left(\frac{\partial \overline{K}}{\partial t} + \overline{\mathbf{U}} \cdot \nabla \overline{K}\right) = \overline{\mathbf{U}} \cdot (\nabla \cdot \overline{\sigma}) + \overline{\mathbf{U}} \cdot (\nabla \cdot \sigma_t). \tag{3.24}$$

Subtracting this from (3.23) gives the K equation as

$$\rho\left(\frac{\partial K}{\partial t} + \overline{\mathbf{U}} \cdot \nabla K\right) = \overline{\mathbf{u} \cdot (\nabla \cdot \sigma')} + \nabla \overline{\mathbf{U}} : \sigma_t - \rho \nabla (\overline{\mathbf{u}|\mathbf{u}^2|}/2). \tag{3.25}$$

Equations (3.24) and (3.25) show that the total work done by the Reynolds stress, namely $\nabla \cdot (\sigma_t \overline{\mathbf{U}})$, that appears in Eq. (3.23), divides into a term affecting the mean kinetic energy in Eq. (3.24) and a term affecting turbulent kinetic energy in (3.25), in much the same way that the total work by σ was partitioned between Eqs. (3.14) and (3.16). Following the same kind of interpretation as before we see that the work of σ_t in Eq. (3.24) is such as to accelerate the mean field, while that in Eq.(3.25) has the appearance of deformation work that will extract energy from the mean field to place it in the fluctuating field.

If the analogy between $\overline{K}$ and K on the one hand and the kinetic and internal energies on the other were perfect, then it should be that $\nabla \overline{\mathbf{U}} : \sigma_t \geq 0$. In fact, the positivity of this term is not guaranteed in turbulent flow so that energy can flow from both the mean field to the fluctuating field and vice versa in some circumstances. The latter phenomenon is known as *backscatter* and may be associated with the potentiality for small vortices to combine to form larger ones. This brings energy to the large scales that become part of the mean flow field.

The fluctuating work term $\overline{\mathbf{u} \cdot (\nabla \cdot \sigma')}$ in Eq. (3.25) plays a significant role in affecting K. This is made evident by reworking it into a more useful form by the following steps. Define

$$d' = 2\mu e_s' \tag{3.26}$$

where the fluctuating rate of strain is

$$e_s' = \frac{1}{2}(\nabla \mathbf{u} + (\nabla \mathbf{u})^t) \tag{3.27}$$

and note that

$$\sigma' = -pI + d'. \tag{3.28}$$

In this case, for incompressible flow and using the symmetry of d', it is readily derived using index notation that

$$\overline{\mathbf{u} \cdot (\nabla \cdot \sigma')} = -\nabla \cdot \overline{p\mathbf{u}} + \nabla \cdot \overline{d'\mathbf{u}} - \rho \epsilon_T \tag{3.29}$$

where, using Eq. (3.26),

$$\epsilon_T \equiv \frac{1}{\rho} \overline{\nabla \mathbf{u} : d'} = 2\nu \overline{e_s' : e_s'} \tag{3.30}$$

is the average dissipation rate per unit mass of the kinetic energy caused by internal friction in the fluctuating field. A further manipulation gives

$$\epsilon_T = \epsilon + \nu \overline{\nabla \mathbf{u} : (\nabla \mathbf{u})^t}, \tag{3.31}$$

where

$$\epsilon \equiv \nu\overline{\nabla\mathbf{u} : \nabla\mathbf{u}} = \nu\overline{|\nabla\mathbf{u}|^2}. \tag{3.32}$$

In homogeneous isotropic turbulence the second term in Eq. (3.31) is zero according to the identity

$$\overline{\nabla\mathbf{u} : (\nabla\mathbf{u})^t} = \nabla\cdot(\nabla\cdot\overline{\mathbf{u}\otimes\mathbf{u}}), \tag{3.33}$$

which holds for incompressible flow and is also readily proven using index notation. Consequently,

$$\epsilon_T = \epsilon \tag{3.34}$$

holds for homogeneous turbulence and ϵ is known as the *isotropic dissipation rate.*

Considering the second term on the right-hand side of Eq. (3.29) it may be shown that

$$\nabla\cdot\overline{d'\mathbf{u}} = \mu\nabla^2K + \mu\overline{\nabla\mathbf{u} : (\nabla\mathbf{u})^t} \tag{3.35}$$

for incompressible flow. Using this together with Eq. (3.31) in Eq. (3.29), and noting that the term $\overline{\nabla\mathbf{u} : (\nabla\mathbf{u})^t}$ cancels, it is derived that

$$\overline{\mathbf{u}\cdot(\nabla\cdot\sigma')} = -\nabla\cdot\overline{p\mathbf{u}} + \mu\nabla^2K - \rho\epsilon. \tag{3.36}$$

Incorporating Eq. (3.36) into (3.25) yields the standard form of the K equation

$$\frac{\partial K}{\partial t} + \overline{\mathbf{U}}\cdot\nabla K = \frac{1}{\rho}\nabla\overline{\mathbf{U}} : \sigma_t - \epsilon - \frac{1}{\rho}\nabla\cdot\overline{p\mathbf{u}} + \nu\nabla^2K - \nabla\cdot\overline{\mathbf{u}(|\mathbf{u}|^2/2)}. \tag{3.37}$$

The terms on the right-hand side account for the total work of the Reynolds stresses in producing K from the mean field, isotropic dissipation rate, total work by the fluctuating pressure field (often referred to as the *pressure transport* term), viscous diffusion, and turbulent kinetic energy transport, respectively. Since, according to Eq. (3.32), $\epsilon \geq 0$, it is clear that the effect of this term in Eq. (3.37) is to act as a sink of turbulent kinetic energy caused by the action of frictional viscous forces.

Equation (3.37) offers some insight into the kinds of physical processes that are important in determining the distribution of turbulent kinetic energy in a flow field. In this regard, it is helpful to consider an integral form of this equation. Thus, using the identity $\overline{\mathbf{U}}\cdot\nabla K = \nabla\cdot(\overline{\mathbf{U}}K)$, Eq. (3.37) can be written as

$$\frac{\partial K}{\partial t} = \nabla\cdot(-\overline{p\mathbf{u}}/\rho - \overline{\mathbf{U}}K + \nu\nabla K - \overline{\mathbf{u}(|\mathbf{u}|^2/2)}) + \frac{1}{\rho}\nabla\overline{\mathbf{U}} : \sigma_t - \epsilon. \tag{3.38}$$

Integrating this relation over a fixed control volume $\mathcal{V}$ with boundary $\mathcal{A}$, outward normal $\mathbf{n}$ and applying the divergence theorem

$$\int_{\mathcal{V}} \nabla\cdot\mathbf{v}dV = \int_{\mathcal{A}} \mathbf{n}\cdot\mathbf{v}\,dA \tag{3.39}$$

that holds for an arbitrary vector field $\mathbf{v}(\mathbf{x})$ yields

$$\begin{aligned}\frac{\partial}{\partial t}\int_{\mathcal{V}} KdV &= \int_{\mathcal{A}} \mathbf{n}\cdot(-\overline{p\mathbf{u}}/\rho - \overline{\mathbf{U}}K + \nu\nabla K - \overline{\mathbf{u}(|\mathbf{u}|^2/2)})dA \\ &+ \int_{\mathcal{V}} \frac{1}{\rho}\nabla\overline{\mathbf{U}} : \sigma_t dV - \int_{\mathcal{V}} \epsilon dV.\end{aligned} \tag{3.40}$$

The first integral on the right-hand side represents the net amount of flow work acting on the control surface and the amount of viscous and turbulent diffusion of K through the control volume surface. The four terms in the integral in this expression are zero at solid boundaries, so their collective contribution to the K balance is only to redistribute pre-existing K throughout the flow field, and not to cause its creation or destruction. In fact, the only potential source of kinetic energy that can contribute to the balance lies in the second term on the right-hand side, with the last term representing exclusively a loss of energy.

These considerations show that in Eq. (3.37) the sole source of K comes from the first term on the right-hand side, representing the deformation work performed by σ_t. This same term may also represent a loss of K in some circumstances. A continual loss of K is provided by the $-\epsilon$ term. The remaining terms act to redistribute the energy that already exists. The model chosen for σ_t is thus clearly of great importance in establishing the K field. Later it will be seen that standard models for σ_t tend to result in the deformation work term being strictly positive, thus confirming its role as a production term in the K equation. On the other hand, since it can also be negative in regions of backscatter, the potential for serious modeling errors is present if such models are used.

Despite the potential pitfalls in accounting for the production term in Eq. (3.38) it is, in fact, the dissipation term ϵ that generally poses the greatest difficulty in arriving at a fair accounting of the physics of the kinetic energy field. Chapter 5 considers the decay of isotropic turbulence in isolation from other effects with a view to how its influence in the K equation can be modeled. Such knowledge is limited, however, so that the practical determination of ϵ is fraught with difficulty. Some insight into why this is so is made evident in the next section, where an exact equation governing the distribution of ϵ is derived.

3.3 ϵ Equation

The importance of the dissipation rate, ϵ, in counterbalancing production of turbulent kinetic energy is evident from Eq. (3.38). Solutions to K require estimates of ϵ which lead to the common practice of solving a modeled form of its own exact equation simultaneously with the K equation. To derive the ϵ equation it is convenient to begin with the Navier–Stokes equation written in index notation. Take a derivative of Eq. (2.2) with respect to x_j resulting in an equation for $\partial U_i/\partial x_j$. Multiply this by $2\nu\ \partial u_i/\partial x_j$ and average, yielding an equation for ϵ. In particular, the term $\partial U_i/\partial t$ gives, after these operations,

$$2\nu\overline{\frac{\partial u_i}{\partial x_j}\frac{\partial^2 U_i}{\partial x_j \partial t}} = 2\nu\overline{\frac{\partial u_i}{\partial x_j}\frac{\partial^2 u_i}{\partial x_j \partial t}} = \frac{\partial \epsilon}{\partial t} \tag{3.41}$$

and similarly for the remaining terms. How best to organize the many terms that result from this process is not entirely self-evident. Nonetheless, the traditional approach is to write the ϵ equation in the form:

$$\frac{D\epsilon}{Dt} = P_\epsilon^1 + P_\epsilon^2 + P_\epsilon^3 + P_\epsilon^4 + \Pi_\epsilon + T_\epsilon + D_\epsilon - \Upsilon_\epsilon, \tag{3.42}$$

where

$$P_\epsilon^1 = -\epsilon_{ij}^c \frac{\partial \overline{U}_i}{\partial x_j} \tag{3.43}$$

$$P_\epsilon^2 = -\epsilon_{ij} \frac{\partial \overline{U}_i}{\partial x_j} \tag{3.44}$$

$$P_\epsilon^3 = -2\nu \overline{u_k \frac{\partial u_i}{\partial x_j}} \frac{\partial^2 \overline{U}_i}{\partial x_k \partial x_j} \tag{3.45}$$

$$P_\epsilon^4 = -2\nu \overline{\frac{\partial u_i}{\partial x_k} \frac{\partial u_i}{\partial x_j} \frac{\partial u_k}{\partial x_j}} \tag{3.46}$$

$$\Pi_\epsilon = -\frac{2\nu}{\rho} \frac{\partial}{\partial x_i} \left(\overline{\frac{\partial p}{\partial x_j} \frac{\partial u_i}{\partial x_j}} \right) \tag{3.47}$$

$$T_\epsilon = -\nu \frac{\partial}{\partial x_k} \left(\overline{u_k \frac{\partial u_i}{\partial x_j} \frac{\partial u_i}{\partial x_j}} \right) \tag{3.48}$$

$$D_\epsilon = \nu \nabla^2 \epsilon \tag{3.49}$$

$$\Upsilon_\epsilon = 2\nu^2 \overline{\left(\frac{\partial^2 u_i}{\partial x_j \partial x_k} \right)^2}. \tag{3.50}$$

In Eqs (3.43) and (3.44),

$$\epsilon_{ij}^c = 2\nu \overline{\frac{\partial u_k}{\partial x_i} \frac{\partial u_k}{\partial x_j}} \tag{3.51}$$

$$\epsilon_{ij} = 2\nu \overline{\frac{\partial u_i}{\partial x_k} \frac{\partial u_j}{\partial x_k}} \tag{3.52}$$

are referred to as the complimentary dissipation rate tensor and the dissipation rate tensor, respectively. For both of these quantities the trace satisfies $\epsilon_{ii} = 2\epsilon$.

In most cases the physical meaning of the terms in Eq. (3.42) is not known. The best that can be said in this regard is to classify the terms into general categories that are consistent with their mathematical form. Thus, it may be noticed that Π_ϵ, T_ϵ, and D_ϵ are in gradient form so that they are generally responsible for redistributing ϵ inside the flow domain. However, both Π_ϵ and D_ϵ are not zero at boundaries so they may also function as sources or sinks of dissipation. The term Υ_ϵ in Eq. (3.42) is strictly negative and thus contributes exclusively to the viscous loss of the dissipation rate. Remaining are the first four terms $P_\epsilon^1, P_\epsilon^2, P_\epsilon^3$, and P_ϵ^4 that for want of a more precise description of the underlying physics are referred to as production terms. Of course this is technically true only so long as they are positive, which is not always the case.

Modeling of the terms in Eq. (3.42) follows their general classifications. Some significant help in this regard comes from considering the ϵ equation in simpler circumstances, such as decaying isotropic turbulence (considered in Chapter 5) and homogeneous shear flow (considered in Chapter 6). In these instances, many of the terms in the ϵ equation are identically zero so that with the help of empirical data some progress can be made in developing models for the remaining terms. Additional insights into the terms in the

ϵ equation can be made by considering its budget as obtained using DNS data, a step that will be taken in Section 7.1.6.

3.4 Reynolds Stress Equation

As an alternative to developing constitutive laws for σ_t based on ideas about the nature of turbulent transport, one may consider modeling the exact differential equation describing the evolution of σ_t, or equivalently R. The use of an equation for R to determine the Reynolds stress turns attention toward modeling the physical processes in the flow that collectively combine to yield the distribution of R, rather than provide a direct estimate of what it should be.

Adopting index notation in this context, an equation for R_{ij} is derived by averaging and adding together Eq. (2.2) written for U_i multiplied by u_j and the same relation with the roles of i and j reversed. The result is

$$\frac{\partial R_{ij}}{\partial t} + \overline{U}_k \frac{\partial R_{ij}}{\partial x_k} = -R_{ik}\frac{\partial \overline{U}_j}{\partial x_k} - R_{jk}\frac{\partial \overline{U}_i}{\partial x_k} - \epsilon_{ij} - \frac{\partial \beta_{ijk}}{\partial x_k} + \Pi_{ij} + \nu\nabla^2 R_{ij} \tag{3.53}$$

where

$$\Pi_{ij} \equiv \overline{\frac{1}{\rho}p\left(\frac{\partial u_i}{\partial x_j} + \frac{\partial u_j}{\partial x_i}\right)} \tag{3.54}$$

and

$$\beta_{ijk} = \frac{1}{\rho}\overline{pu_i}\delta_{jk} + \frac{1}{\rho}\overline{pu_j}\delta_{ik} + \overline{u_i u_j u_k}. \tag{3.55}$$

The first two terms on the right-hand side of Eq. (3.53), which have a form similar to the production term in Eq. (3.37) and reduce to it under a contraction of indices, may be interpreted as production terms in the present context as well. The third term, ϵ_{ij}, is the general tensorial dissipation term defined in Eq. (3.52), while the next term, containing β_{ijk}, is in flux form. According to Eq. (3.55), β_{ij} is made up of two pressure work terms and a final expression that can be interpreted as a turbulent flux of Reynolds stress. The last two terms in Eq. (3.53) consist of the pressure-strain term term, Π_{ij}, and the viscous diffusion term. Contraction of the indices in Eq. (3.53) and multiplying by 1/2 gives exactly Eq. (3.37), as expected.

A potentially attractive feature of Eq. (3.53) is that the left-hand side of the equation, as well as the production terms, require no modeling. Thus, if Eq. (3.53) is solved in conjunction with Eq. (3.8), only the three quantities ϵ_{ij}, Π_{ij}, and β_{ijk} need to be modeled. In reality, experience shows that accurate solutions to Eq. (3.53) cannot generally be had without reasonably good models for the unclosed correlations.

By being a tensor, the dissipation implied by ϵ_{ij} will vary among its components and thus has an additional degree of complexity beyond the problem of estimating ϵ itself. The transport term β_{ijk} is no less inscrutable than its similar term in the K equation and this is reflected in the fact that it is generally not possible to find an accurate model for it. Of particular interest in Eq. (3.53) is the pressure-strain term Π_{ij}, which has no equivalent in Eq. (3.37). In fact, incompressibility guarantees that

$$\Pi_{11} + \Pi_{22} + \Pi_{33} = 0 \tag{3.56}$$

so the term Π_{ii} does not appear in the K equation. It will be seen later that Eq. (3.56) reflects the fact that Π_{ij} acts to redistribute energy between components, $\overline{u^2}$, $\overline{v^2}$, and $\overline{w^2}$ without a change in the total energy. If the individual terms of Π_{ii} are non-zero, then at least one must be positive and one negative with the resultant action causing $\overline{u^2}$, $\overline{v^2}$, and $\overline{w^2}$ to move closer toward an isotropic state where they are all equal. In fact, in isotropic turbulence each of the individual terms of Π_{ij} are identically zero as will be proven below, so that once an isotropic state is achieved the pressure-strain effect vanishes. Modeling Π_{ij} in general anisotropic conditions is a great challenge, but one of particular importance if the individual Reynolds stress components are to be accurately determined.

3.5 Vorticity Equation

The steps taken in deriving the mean velocity equation (3.8) were based on the goal of including the momentum flux term in the guise of the Reynolds stress tensor. An alternative formulation of the mean momentum equation, which was pursued to some extent by Taylor [2, 3], replaces the appearance of momentum transport in favor of vorticity transport. In this, the identity

$$\nabla \cdot \overline{\mathbf{u} \otimes \mathbf{u}} = \nabla K - \overline{\mathbf{u} \times \boldsymbol{\omega}} \tag{3.57}$$

is applied to the $\overline{\mathbf{U}}$ equation, resulting in

$$\frac{\partial \overline{\mathbf{U}}}{\partial t} + (\nabla \overline{\mathbf{U}})\overline{\mathbf{U}} = -\nabla(\overline{P}/\rho + K) + \nu \nabla^2 \overline{\mathbf{U}} + \overline{\mathbf{u} \times \boldsymbol{\omega}}, \tag{3.58}$$

where the expression $\overline{\mathbf{u} \times \boldsymbol{\omega}}$ appears. In view of the fact that

$$(\overline{\mathbf{u} \times \boldsymbol{\omega}})_i = \epsilon_{ijk} \overline{u_i \omega_j}, \tag{3.59}$$

it is seen that the last term in Eq. (3.58) is made up of the vorticity flux correlations $\overline{u_i \omega_j}$ and the Reynolds stresses do not make an appearance. It also may be noticed that while K appears in Eq. (3.58) it only does so in combination with the pressure. Numerical techniques for solving Eq. (3.58) together with Eq. (3.10) generally will yield a solution for $\overline{\mathbf{U}}$ and the combined quantity $\overline{P}/\rho + K$. Thus, if the pressure is only required on solid boundaries where $K = 0$, then the absence of knowledge about K does not impose a hardship.

The approach represented by Eq. (3.58) is one in which vorticity transport physics is embedded within a momentum balance. A more complete centering of the physics on vorticity can be had by developing an averaged form of the vorticity equation itself. In this case it is first helpful to use the identity Eq. (3.57) to replace $\nabla \mathbf{U}(\mathbf{U})$ in Eq. (3.1) as well as use the incompressible flow identity

$$\nabla^2 \mathbf{U} = -(\nabla \times \boldsymbol{\Omega}). \tag{3.60}$$

The resulting equation is

$$\frac{\partial \mathbf{U}}{\partial t} + \nabla K - \mathbf{U} \times \mathbf{\Omega} = -\frac{1}{\rho}\nabla P - \nu \nabla \times \mathbf{\Omega} \tag{3.61}$$

and taking a curl of this relation yields the vorticity equation

$$\frac{\partial \mathbf{\Omega}}{\partial t} - \nabla \times (\mathbf{U} \times \mathbf{\Omega}) = -\nu \nabla \times (\nabla \times \mathbf{\Omega}). \tag{3.62}$$

It is helpful to modify Eq. (3.62) by replacing the second term on the left-hand side by the identity

$$-\nabla \times (\mathbf{U} \times \mathbf{\Omega}) = -(\nabla \mathbf{U})\mathbf{\Omega} + (\nabla \mathbf{\Omega})\mathbf{U} \tag{3.63}$$

and the last term by the identity

$$-\nabla \times (\nabla \times \mathbf{\Omega}) = \nabla^2 \mathbf{\Omega}, \tag{3.64}$$

yielding the vorticity equation in incompressible flow as

$$\frac{\partial \mathbf{\Omega}}{\partial t} + (\nabla \mathbf{\Omega})\mathbf{U} = (\nabla \mathbf{U})\mathbf{\Omega} + \nu \nabla^2 \mathbf{\Omega}. \tag{3.65}$$

The left-hand side of this equation is the material derivative of vorticity, $D\mathbf{\Omega}/Dt$, which accounts for the changes in vorticity along fluid particle paths. The second term on the right-hand side of Eq. (3.65) is the viscous diffusion term that is responsible for the creation of vorticity at solid boundaries by frictional effects. For constant density flow this is the only source of vorticity so that, in particular, whatever vorticity is encountered away from boundaries must have had its origin at an upstream boundary point.

Having no parallel in the momentum equation is the first term on the right-hand side of Eq. (3.65) that accounts for vortex stretching and reorientation. This is a fundamental aspect of the vorticity field in three dimensions that plays a major role in transferring energy between scales and will be considered in some detail in the next section. Vortex stretching and reorientation can only modify vorticity that is pre-existing in the flow. Similarly, the convection term on the left-hand side of Eq. (3.65) can shift the position of vorticity that is already in the flow but cannot be a cause of its appearance in the flow.

The equation for mean vorticity is determined by taking an average of Eq. (3.65), and after some manipulation this can be put into the form

$$\frac{\partial \overline{\mathbf{\Omega}}}{\partial t} + (\nabla \overline{\mathbf{\Omega}})\overline{\mathbf{U}} = (\nabla \overline{\mathbf{U}})\overline{\mathbf{\Omega}} + \overline{(\nabla \mathbf{u})\boldsymbol{\omega}} + \nabla \cdot (\nu \nabla \overline{\mathbf{\Omega}} - \overline{\boldsymbol{\omega} \otimes \mathbf{u}}). \tag{3.66}$$

In the last term on the right-hand side viscous vorticity diffusion is augmented by turbulent vorticity diffusion given by $\overline{\boldsymbol{\omega} \otimes \mathbf{u}}$, similar to the way that viscous momentum diffusion is augmented by the Reynolds stress in Eq. (3.8). The first two terms on the right-hand side of Eq. (3.66) account for vortex stretching and reorientation by the mean and fluctuating fields, respectively.

The mean vorticity equation expressed in the form of Eq. (3.66) has the virtue that the vorticity transport term $\overline{\boldsymbol{\omega} \otimes \mathbf{u}}$ and the stretching correlation $\overline{(\nabla \mathbf{u})\omega}$ are included in a way that reflects their physical role in establishing the balance of mean vorticity. A formal way of modeling these correlations that is largely consistent with their physical meaning will be considered in Section 6.6. Note, however, that since the vorticity is solenoidal (i.e., divergence free), the second term on the right-hand side of Eq. (3.66) can be replaced via

$$\overline{(\nabla \mathbf{u})\omega} = \nabla \cdot \overline{\mathbf{u} \otimes \boldsymbol{\omega}} \tag{3.67}$$

so that it is possible to sidestep the need to model the fluctuating part of the stretching term. In fact, this is the same result as would happen by taking a curl of (3.58). In this formulation, however, the physical meaning of the stretching term is largely obscured.

3.5.1 Vortex Stretching and Reorientation

Vortex stretching and reorientation as represented by the term $(\nabla \mathbf{U})\boldsymbol{\Omega}$ appearing in Eq. (3.63) to a large extent is the physical process that lies at the heart of turbulent flow. It is the mechanism that explains the transfer of energy between scales and accounts for the acute sensitivity of turbulent motion to small perturbations so that it has the appearance of being random. For these reasons it is appropriate to take a closer look at how it is that the term $(\nabla \mathbf{U})\boldsymbol{\Omega}$ is related to the stretching and reorientation of vorticity.

Toward this end consider two material points $\mathbf{X}(t)$ and $\mathbf{Y}(t)$ convecting in the velocity field $\mathbf{U}(\mathbf{x}, t)$ that are separated by a small distance. Defining

$$\mathbf{r}(t) = \mathbf{Y}(t) - \mathbf{X}(t) \tag{3.68}$$

as a vector connecting the material points, then a unit vector in this direction is

$$\hat{\mathbf{r}}(t) = \mathbf{r}(t)/r(t) \tag{3.69}$$

where $r(t) = |\mathbf{r}(t)|$ is the distance between the points. $\mathbf{X}(t)$ and $\mathbf{Y}(t)$ satisfy

$$\frac{d\mathbf{X}}{dt} = \mathbf{U}(\mathbf{X}(t), t) \tag{3.70}$$

and

$$\frac{d\mathbf{Y}}{dt} = \mathbf{U}(\mathbf{Y}(t), t). \tag{3.71}$$

For small Δt and initial time $t = 0$, $\mathbf{X}(\Delta t)$ is well approximated by

$$\mathbf{X}(\Delta t) = \mathbf{X}(0) + \mathbf{U}(\mathbf{X}(0), 0)\Delta t \tag{3.72}$$

and $\mathbf{Y}(\Delta t)$ by

$$\mathbf{Y}(\Delta t) = \mathbf{Y}(0) + \mathbf{U}(\mathbf{Y}(0), 0)\Delta t. \tag{3.73}$$

Subtracting Eqs. (3.72) from (3.73) and using (3.68) gives

$$\mathbf{r}(\Delta t) = \mathbf{r}(0) + (\mathbf{U}(\mathbf{Y}(0), 0) - \mathbf{U}(\mathbf{X}(0), 0))\Delta t. \tag{3.74}$$

Since $\mathbf{Y}(0) = \mathbf{X}(0) + \mathbf{r}(0)$ and by assumption $r(0)$ is small, a Taylor series expansion gives

$$\mathbf{u}(\mathbf{Y}(0), 0) = \mathbf{U}(\mathbf{X}(0), 0) + (\nabla \mathbf{U})\mathbf{r}(0) + \ldots \tag{3.75}$$

and consequently Eq. (3.74) becomes

$$\mathbf{r}(\Delta t) = \mathbf{r}(0) + (\nabla \mathbf{U})\mathbf{r}(0)\ \Delta t. \tag{3.76}$$

Substituting for $\mathbf{r}$ in Eq. (3.76) using (3.69), dividing by $r(\Delta t)\Delta t$, adding $-\hat{\mathbf{r}}(0)/\Delta t$ to both sides of the equation, and rearranging terms gives

$$\frac{\hat{\mathbf{r}}(\Delta t) - \hat{\mathbf{r}}(0)}{\Delta t} = -\frac{\hat{\mathbf{r}}(0)}{r(\Delta t)} \frac{(r(\Delta t) - r(0))}{\Delta t} + \frac{r(0)}{r(\Delta t)}(\nabla \mathbf{U})\hat{\mathbf{r}}(0). \tag{3.77}$$

In the limit as $\Delta t \to 0$ it is found that

$$\frac{d\hat{\mathbf{r}}}{dt} + \frac{1}{r}\frac{dr}{dt}\hat{\mathbf{r}} = (\nabla \mathbf{U})\hat{\mathbf{r}} \tag{3.78}$$

where $(dr/dt)/r$ is the fractional rate of change of the length of an infinitesimal line segment oriented in the $\hat{\mathbf{r}}$ direction. Choosing $\hat{\mathbf{r}} = \boldsymbol{\Omega}/|\boldsymbol{\Omega}|$, which is aligned in the direction of the local vorticity, Eq. (3.78) becomes

$$(\nabla \mathbf{U})\boldsymbol{\Omega} = \frac{1}{r}\frac{dr}{dt}\boldsymbol{\Omega} + |\boldsymbol{\Omega}|\frac{d(\boldsymbol{\Omega}/|\boldsymbol{\Omega}|)}{dt} \tag{3.79}$$

after multiplying through by $|\boldsymbol{\Omega}|$. The left-hand side is the stretching term in Eq. (3.65) and is seen on the right-hand side to be decomposed into the sum of two physical processes involving the evolution of a local material line element. The first term is in the direction of the local vorticity vector and is proportional to the fractional rate at which a local fluid line element is stretching ($dr/dt > 0$) or contracting ($dr/dt < 0$). The effect on the local vorticity in this case is to make it either stronger or weaker depending on the sign of dr/dt.

The meaning of the second term on the right-hand side of Eq. (3.79) is made evident by noting that $\boldsymbol{\Omega} \perp d(\boldsymbol{\Omega}/|\boldsymbol{\Omega}|)/dt$, which follows from differentiating the identity

$$\frac{\boldsymbol{\Omega}}{|\boldsymbol{\Omega}|} \cdot \frac{\boldsymbol{\Omega}}{|\boldsymbol{\Omega}|} = 1. \tag{3.80}$$

This means that the role of this term in Eq. (3.79) is to represent the shearing of vorticity out of the $\boldsymbol{\Omega}$ direction into orthogonal directions. In this way the vorticity vector changes its orientation as the flow evolves. The total effect of $(\nabla \mathbf{U})\boldsymbol{\Omega}$ is thus to simultaneously stretch and reorient the vorticity field.

Numerical simulations of the vortex stretching process can be had via vortex filament methods in which short vortex segments are tracked through the flow field by moving their end points [4]. Stretching of the segments dominates over contraction and the number of elements rises if segments are subdivided as they grow in length. Such calculations show a robust tendency for vortex filaments to fold and bend. It is this mechanism that ensures a flux of energy to small scales while maintaining a fixed energy since vortex folding eliminates far-field velocity due to cancellation.

3.6 Enstrophy Equation

The enstrophy, $\zeta \equiv \overline{\boldsymbol{\omega} \cdot \boldsymbol{\omega}} = \overline{\omega_i^2}$, has a similar relationship to $\overline{\boldsymbol{\Omega}}$ as K does to $\overline{\mathbf{U}}$. Moreover, the identity

$$\zeta = \frac{\epsilon}{\nu} - \overline{\nabla \mathbf{u} : (\nabla \mathbf{u})^t}, \tag{3.81}$$

which is readily proven using index notation, suggests that ϵ and ζ have much in common. In fact, the last term in (3.81) is identically zero in homogeneous turbulence, as was noted previously in Eq. (3.33), implying that in this special circumstance ζ exactly equals ϵ/ν. Even in the presence of sizable mean shear, as in turbulent channel flow, evidence from DNS suggests that the approximation $\zeta \approx \epsilon/\nu$ is very good, as will be seen in Section 7.1.8. Consequently, it is generally acceptable to use $\nu\zeta$ interchangeably with ϵ when developing turbulence models. The advantage of using enstrophy in place of ϵ as an unknown comes from the opportunity it provides to consider the closure to the ζ equation instead of the ϵ equation. The former, though similar to the latter in many ways, has some subtle differences which make its modeling potentially easier. One obvious difference is that the ζ equation is lacking pressure terms.

The exact ζ equation is derived from (3.65) by multiplying it as written for Ω_i by ω_i and averaging. The result, after some significant manipulation, is

$$\frac{D\zeta}{Dt} = P^1_\zeta + P^2_\zeta + P^3_\zeta + P^4_\zeta + T_\zeta + D_\zeta - \Upsilon_\zeta, \tag{3.82}$$

where

$$P^1_\zeta = 2\overline{\omega_i\omega_k}\frac{\partial \overline{U}_i}{\partial x_k} \tag{3.83}$$

$$P^2_\zeta = 2\overline{\omega_i\frac{\partial u_i}{\partial x_k}}\overline{\Omega}_k \tag{3.84}$$

$$P^3_\zeta = -2\overline{u_k\omega_i}\frac{\partial \overline{\Omega}_i}{\partial x_k} \tag{3.85}$$

$$P^4_\zeta = 2\overline{\omega_i\omega_k\frac{\partial u_i}{\partial x_k}} \tag{3.86}$$

$$T_\zeta = -\frac{\partial}{\partial x_k}\overline{(u_k\omega_i\omega_i)} \tag{3.87}$$

$$D_\zeta = \nu\nabla^2\zeta \tag{3.88}$$

$$\Upsilon_\zeta = 2\nu\overline{\frac{\partial \omega_i}{\partial x_k}\frac{\partial \omega_i}{\partial x_k}}. \tag{3.89}$$

As in the case of the ϵ equation, the terms in (3.82) can be categorized as production as represented by Eqs. (3.83)–(3.86), turbulent transport given by Eq. (3.87), diffusion by Eq. (3.88), and dissipation of enstrophy by Eq. (3.89). A nice property of P^2_ζ and P^3_ζ is that they incorporate the transport and stretching correlations which would have been previously modeled for use in (3.66). The difficulty of analyzing the remaining terms is comparable to their counterparts in the ϵ equation.

References

1 Bernard, P.S. (2015) *Fluid Dynamics*, Cambridge University Press, Cambridge.

2 Taylor, G.I. (1915) Eddy motion in the atmosphere. *Phil. Trans. Roy. Soc. London*, 215, 1–26.

3 Taylor, G.I. (1932) The transport of vorticity and heat through fluids in turbulent motion. *Proc. Roy. Soc. London A*, 135, 685–705.

4 Bernard, P.S. (2006) Turbulent flow properties of large scale vortex systems. *Proc. Nat. Acad. Sci.*, 103, 10 174–10 179.

Problems

3.1 Prove the identity Eq. (3.15).

3.2 Derive the identity Eq. (3.33).

3.3 Show that the only source of vorticity in incompressible flow is due to frictional effects at the boundary.

3.4 Illustrate the influence of vortex stretching and reorientation on a local vorticity field $\mathbf{\Omega} = (\Omega, 0, 0)$ under the action of a velocity field $\mathbf{U}(\mathbf{x}) = (U(x,y), V(x,y), 0)$.

3.5 Write a code (e.g., using MATLAB) that simulates 2D turbulence by calculating the motion of smoothed point vortices in the unit square $0 \leq x, y \leq 1$ with periodic boundary conditions. Assume the smoothed vortices consist of Rankine vortices for which the u and v velocities at a point x, y due to a vortex located at xv, yv are given by

$$u(x,y) = -\frac{\Gamma}{2\pi}\frac{y - yv}{r^2}$$
$$v(x,t) = \frac{\Gamma}{2\pi}\frac{x - xv}{r^2} \tag{3.90}$$

where Γ is the circulation chosen randomly, though with equal numbers of + and − vortices in the simulation so the net circulation is zero. Here,

$$r^2 = \max((x - xv)^2 + (y - yv)^2, \sigma^2) \tag{3.91}$$

where σ is the radius for which the vortices have constant vorticity. Add a finite 2D array of periodic images of the vortices shifted by integer values in the x and y directions to the velocity calculation so as to approximately enforce periodicity. Move the vortices using a simple Euler scheme. Vortices traveling through one boundary should be brought back into the flow through the opposite boundary. After a startup period during which memory of the initial configuration of vortices is lost, run the simulation long enough to get "converged" velocity statistics. Then do the following:

(a) Make a quiver plot of the velocity field at one time step.
(b) Compute $\overline{U}$ and K and make a contour plot of each to assess the degree of homogeneity in the solution.
(c) Make contour plots of the two-point velocity correlations

$$\mathcal{R}_{11}(r,s) = \overline{u(0.5, 0.5)u(0.5 + r, 0.5 + s)}$$
$$\mathcal{R}_{12}(r,s) = \overline{u(0.5, 0.5)v(0.5 + r, 0.5 + s)} \tag{3.92}$$

where (r, s) is a vector that sweeps out the flow domain. From these results plot $f(r)$ and $g(r)$ as well as $g(r)$ computed from Eq. (4.33). Evaluate λ_f and Λ_f, and verify that $f(r)$ is zero well before the boundary. If not then adjust the numerical parameters by adding vortices and reducing σ. Comment on how well the isotropy relations are satisfied by this 2D model.

3.6 Use the code developed in 3.5 to study the dispersion of a steady point source of contaminant entering a 2D turbulent flow at rate 1/sec. Specifically, in conjunction with the time integration that marches the vortices forward in time, place N tracers into the flow at $x = 0.5, y = 0.5$ at each time step Δt, with each carrying $\Delta t/N$ of the contaminant. Move the tracers according to an Euler scheme in which they also take a random hop at every time step from a Gaussian distribution with variance $2D\Delta t$ so as to simulate molecular diffusion with diffusion coefficient D (e.g., see [1] for a discussion of such Monte-Carlo schemes). Do not enforce periodicity for the tracer motions. Once equilibrium concentrations of contaminants are reached in the vicinity of the source, compute concentration statistics as a function of time by collecting particles into the cells of a mesh and estimating the concentration as $N_{ij}\Delta t/(AN)$, where N_{ij} is the number of particles in the i, jth box and A is the area of a box. Study the relative effects of the turbulent dispersion versus that of molecular diffusion by comparing concentration statistics over a fixed time period for a series of different diffusivities.

3.7 Write a code using, for example, MATLAB that simulates the interaction between two vortex rings as they travel toward each other. To do this, create vortex rings a fixed distance apart that are each made up of NV straight vortex tubes linked end to end. Let $\mathbf{x}_i^l, \mathbf{x}_i^r, i = 1, \ldots, 2NV$ represent the left and right end points, respectively, of the vortex tubes. Define $\mathbf{s}_i = \mathbf{x}_i^r - \mathbf{x}_i^l$ to be a vector pointing along the length of each tube and $\mathbf{x}_i^c = (\mathbf{x}_i^l + \mathbf{x}_i^r)/2$ to be the center of each vortex tube. The velocity at any point $\mathbf{x}$ generated by the set of individual tubes is given by the Biot–Savart law [1]

$$\mathbf{u}(\mathbf{x}) = \sum_{i=1}^{2NV} \frac{\Gamma}{4\pi} \frac{\mathbf{s}_i \times (\mathbf{x} - \mathbf{x}_i^c)}{|\mathbf{x} - \mathbf{x}_i^c|^3}, \tag{3.93}$$

where Γ is the constant circulation of the vortex rings. Move the filaments in time, using either a simple Euler scheme, or for more accuracy, a fourth order Runge–Kutta scheme. Whenever individual vortices grow beyond a fixed length, subdivide them. Arrange the two vortex rings to travel toward each other from their original positions by reversing the order of tube ends in one of the rings. Animate the behavior of the rings and follow them until they break down into turbulent motion.

4

Turbulence at Small Scales

We begin our consideration of the physics of turbulent flow by characterizing the motions at small scale. This is the realm of motions where viscous forces act to smooth away gradients in the local velocity field and by so doing dissipate energy. Beginning with the universal equilibrium hypothesis of Kolmogorov [1, 2], it has long been speculated that the process by which dissipation occurs at high Reynolds numbers may be common to all turbulent flows, even if circumstances at large length scales that create the turbulent motion vary significantly from one flow to another. One consequence of this view of dissipation is that it is justified to consider the dissipative process as being largely isotropic since it is more or less insulated from the biases associated with large-scale shearing in the flow field. Positing a universal character to dissipative motions also motivates the development of simplified scaling arguments in which to characterize dissipation, in addition to taking advantage of the considerable mathematical simplifications that derive from the isotropy assumption. The circumstances for a given shearing under which the local isotropy [3] or universality [4] hypotheses can be justified have been extensively studied, and some instances where their applicability is called into question have been identified. In practical terms, however, despite the lingering uncertainties regarding the physics of dissipation, some of which will be pointed out in the following development, the modeling of dissipation in turbulence prediction schemes is based on the universal equilibrium idea.

Under the assumptions of homogeneity and isotropy the present analysis will take advantage of a number of mathematical consequences of isotropy, such as the theory of isotropic tensors [5], that go well beyond the basic idea of isotropy (e.g., equality of the normal Reynolds stresses) that were considered in Section 2.2. For example, isotropy means that the two-point correlation tensors $\mathcal{R}(\mathbf{r})$ and $\mathcal{S}(\mathbf{r})$ introduced in Eqs. (2.34) and (2.35) have special simplifying forms. Such results then allow the physics of dissipation to be considered from a more detailed and fruitful perspective than can be had from consideration of one-point statistics alone. In effect, one-point statistics summarize the local effect of the scales acting in the flow, while multi-point correlations can be manipulated to focus in on the properties of particular ranges of scales, and the latter analysis is greatly simplified if the turbulence is isotropic.

Our focus in this chapter is on the characterization of dissipation itself. In Chapter 5, these results are used to consider the dynamics of an isotropic turbulent flow whose energy changes in time from an initial state. Chapter 6 extends the analysis in this chapter and Chapter 5 further by including the imposition of a homogeneous shear that acts as a source of turbulence.

Turbulent Fluid Flow, First Edition. Peter S. Bernard.

Companion website: www.wiley.com/go/Bernard/Turbulent_Fluid_Flow

4.1 Spectral Representation of ϵ

The isotropic dissipation rate ϵ, defined in Eq. (3.32), that is an essential aspect of the kinetic energy balance, is determined locally in turbulent flow from events that transpire at small scales. Consequently, consideration of small-scale turbulence is largely with a view toward gaining understanding of how the local small-scale flow properties establish the dissipation rate. A useful perspective with which to consider ϵ is the same as that which was previously taken in regard to the energy in Eq. (2.75) wherein the relative contributions from motions at different scales were analyzed. Thus, our first goal is to derive a spectral decomposition of the dissipation rate.

For homogeneous turbulence, the two-point velocity correlation tensor $\mathcal{R}(\mathbf{x},\mathbf{y})$ defined in Eq. (2.29) depends only on the relative positions of $\mathbf{x}$ and $\mathbf{y}$ and not their absolute positions.[1] Thus

$$\mathcal{R}(\mathbf{x},\mathbf{y}) = \mathcal{R}(\mathbf{y}-\mathbf{x}), \tag{4.1}$$

where, introducing $\mathbf{r} = \mathbf{y} - \mathbf{x}$ and using the definition in Eq. (2.34), the right-hand side of this equation can be expressed as

$$\mathcal{R}(\mathbf{r})|_{\mathbf{r}=\mathbf{y}-\mathbf{x}}. \tag{4.2}$$

Consequently, Eq. (4.1) should be interpreted as meaning that

$$\mathcal{R}(\mathbf{x},\mathbf{y}) \equiv \mathcal{R}(\mathbf{r})|_{\mathbf{r}=\mathbf{y}-\mathbf{x}}. \tag{4.3}$$

Switching to index notation and introducing the definition of $\mathcal{R}_{ij}$, Eq. (4.1) becomes

$$\overline{u_i(\mathbf{x})u_j(\mathbf{y})} = \mathcal{R}_{ij}(\mathbf{y}-\mathbf{x}). \tag{4.4}$$

Differentiating Eq. (4.4) with respect to x_k, keeping Eq. (4.3) in mind, and applying the chain rule gives

$$\frac{\partial}{\partial x_k}\overline{u_i(\mathbf{x})u_j(\mathbf{y})} = \frac{\partial \mathcal{R}_{ij}}{\partial r_l}(\mathbf{y}-\mathbf{x})\frac{\partial r_l}{\partial x_k} \tag{4.5}$$

so that

$$\overline{\frac{\partial u_i}{\partial x_k}(\mathbf{x})u_j(\mathbf{y})} = -\frac{\partial \mathcal{R}_{ij}}{\partial r_k}(\mathbf{y}-\mathbf{x}). \tag{4.6}$$

Now computing $\partial/\partial y_l$ of Eq. (4.6) gives

$$\overline{\frac{\partial u_i}{\partial x_k}(\mathbf{x})\frac{\partial u_j}{\partial y_l}(\mathbf{y})} = -\frac{\partial^2 \mathcal{R}_{ij}}{\partial r_k \partial r_l}(\mathbf{y}-\mathbf{x}). \tag{4.7}$$

In the limit as $\mathbf{y} \to \mathbf{x}$, Eq. (4.7) becomes

$$\overline{\frac{\partial u_i}{\partial x_k}\frac{\partial u_j}{\partial x_l}}(\mathbf{x}) = -\frac{\partial^2 \mathcal{R}_{ij}}{\partial r_k \partial r_l}(0), \tag{4.8}$$

1 Note that in this chapter for notational simplicity the dependence of $\mathcal{R}$ and other quantities on t will be implied and not indicated explicitly.

where the left-hand side is clearly a constant independent of $\mathbf{x}$ in homogeneous turbulence. Setting $j = i$ and $l = k$ in Eq. (4.8) and using the definition of ϵ then gives the important relation

$$\frac{\epsilon}{\nu} = -\frac{\partial^2 \mathcal{R}_{ii}}{\partial r_l^2}(0). \tag{4.9}$$

One way to evaluate the right-hand side of Eq. (4.9) is to take advantage of Eq. (2.70). In particular, noting that

$$\frac{\partial \mathcal{R}_{ii}}{\partial r_l} = \int_{\Re^3} -\imath k_l \mathcal{E}_{ii}(\mathbf{k}, t) d\mathbf{k} \tag{4.10}$$

and then taking another derivative yields

$$\frac{\partial^2 \mathcal{R}_{ii}}{\partial r_l^2}(0) = -\int_{\Re^3} k^2 tr\mathcal{E}(\mathbf{k}, t) d\mathbf{k}, \tag{4.11}$$

where $k = |\mathbf{k}|$ in this relation.[2] Switching the integration to k shells, as was done in Eq. (2.73), and using Eq. (2.74) gives

$$\epsilon = 2\nu \int_0^\infty k^2 E(k) dk, \tag{4.12}$$

a relation which should be contrasted with Eq. (2.75).

An implication of Eqs. (2.75) and (4.12), which is illustrated in Figure 4.1, is that the motions making the greatest contributions to K and ϵ come from different wave number ranges. Each value of k corresponds to Fourier modes $e^{\imath \mathbf{k}\cdot\mathbf{r}}$ with $|\mathbf{k}| = k$, which may be viewed as essentially sine or cosine variations in the fluctuating velocity field over a distance $\sim 1/k$. Thus, small wave numbers are associated with variations in the

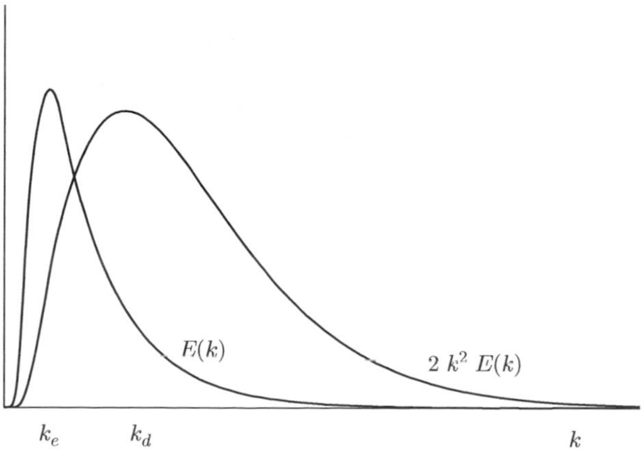

Figure 4.1 Spectral ranges of $E(k)$ and $2k^2E(k)$, with k_e and k_d marking their respective peaks.

2 Note that, depending on the context, the letter k is being used in a number of different ways in this discussion. For example, k_l means the lth component of the vector $\mathbf{k}$ while $k = |\mathbf{k}|$ is the magnitude of the vector $\mathbf{k}$ and r_k is the kth component of $\mathbf{r}$. Other meanings of k will be used below to fit in with common usage.

velocity over large distances, while large k corresponds to velocity variations over small distances. Evidently, Figure 4.1 may be interpreted as meaning that larger-scale motions tend to be most energetic, while small-scale motions are where dissipative processes take place.

The wave number at the peak in $E(k)$, say k_e, is generally reflective of the stirring mechanism from which turbulence is produced in a particular setting. In physical space it is helpful to think of eddies or vortical objects of the size $l_e = 1/k_e$ that contain a significant part of the turbulent energy. A larger wave number, say k_d, can be defined as corresponding to where $2k^2E(k)$ is at a peak. Here, the dissipative motions can be associated with the dynamics of fine-scale eddies or vortices. The fact that there is a separation between k_e and k_d suggests there must be a mechanism by which energy is brought to the small scales to be dissipated. The physics of this transfer process is an essential part of the turbulent flow phenomenon.

For large Reynolds numbers it is expected that the distance between k_e and k_d in wave number space will increase since k_d will be smaller for k_e fixed. For sufficiently high Reynolds numbers it may be imagined that a separation of scales develops in which the main physical process occurring between the peaks in energy and dissipation is a net transit of energy to small scales without significant dissipation. From these somewhat intuitive ideas about the interaction between energy, its dissipation, and the scale of turbulence, a number of predictions about the energy spectrum can be derived. An important step in this process is developing expressions for ϵ that tie its value to the scales of motion of the turbulent flow. This is done in the next section via an analysis of the mathematical consequences of isotropy as it pertains to two-point velocity correlation tensors.

4.2 Consequences of Isotropy

Two-point correlation tensors such as in Eqs. (2.34) and (2.35) take on special forms in the case of isotropic turbulence that open up the door to a much simplified analysis of the turbulent physics. As illustrated in Figure 4.2 for planar coordinates, in the absence of a preferred direction, correlations such as $\mathcal{R}_{11}(r\mathbf{e}_1)$, $\mathcal{R}_{22}(r\mathbf{e}_2)$, and $\mathcal{R}_{11}(-r\mathbf{e}_1)$ must be equal since they can be obtained from each other by rotating the coordinate axes or by reflection through the axes. For example, after rotating by 180° the streamwise velocity is now given by $-u$ and the left-hand side of the identity

$$\overline{(-u(0,0))(-u(-r,0))} = \overline{u(0,0)u(-r,0)} \tag{4.13}$$

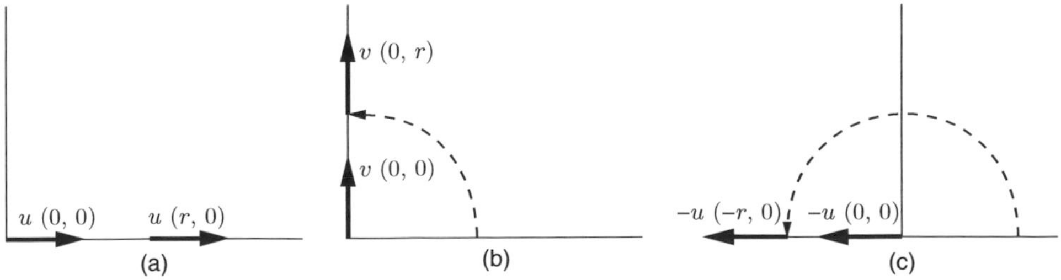

Figure 4.2 Rotational invariance in isotropic turbulence. The two-point correlations based on the velocities in (a), (b), and (c) all yield $f(r)$.

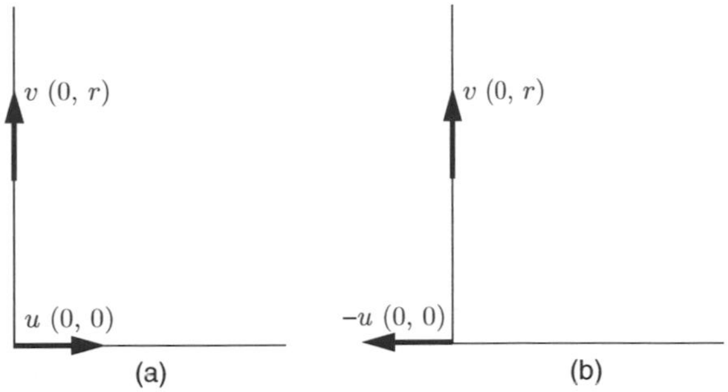

Figure 4.3 Antisymmetry of $\mathcal{R}_{12}(r\mathbf{e}_2)$ under reflection. The two-point correlations based on the velocity components in (a) and (b) in isotropic turbulence are equal.

is $\mathcal{R}_{11}(r\mathbf{e}_1)$ and the right-hand side of this relation is just $\mathcal{R}_{11}(-r\mathbf{e}_1)$. Here, just the r_1, r_2 dependence of u is indicated, consistent with Figure 4.2. It then follows from Eq. (2.38) that

$$f(r) = f(-r), \tag{4.14}$$

that is, f is an even function of r in homogeneous, isotropic turbulence. The same is true for $g(r)$ defined in Eq. (2.39).

In a similar vein, the constancy of $\mathcal{R}_{12}(r\mathbf{e}_2)$ upon reflection through the y axis implies that $\mathcal{R}_{12}(r\mathbf{e}_2) = -\mathcal{R}_{12}(r\mathbf{e}_2)$ so that $\mathcal{R}_{12}(r\mathbf{e}_2) = 0$ (see Figure 4.3). Systematic application of the isotropy condition leads to general formulas for the multi-point correlation tensors. A more direct route to the same end can be had through application of the formal mathematical theory of isotropic tensors [5]. From either perspective it is found that to be isotropic, the two-point double and triple velocity correlation tensors must take the forms

$$\mathcal{R}_{ij}(\mathbf{r}) = \overline{u^2}[R_1(r)r_ir_j + R_2(r)\delta_{ij}] \tag{4.15}$$

and

$$\mathcal{S}_{ij,l}(\mathbf{r}) = S_1(r)r_ir_jr_l + S_2(r)r_l\delta_{ij} + S_3(r)(r_j\delta_{il} + r_i\delta_{jl}), \tag{4.16}$$

respectively, where $R_1(r), R_2(r), S_1(r), S_2(r)$, and $S_3(r)$ are scalar functions of $r = |\mathbf{r}|$ and δ_{ij} is the Kronecker delta function.

Applying the definitions of $f(r)$ and $g(r)$ in Eq. (2.38) and (2.39) to Eq. (4.15) implies that $f = R_1r^2 + R_2$ and $g = R_2$ so that, after replacing R_1 and R_2, Eq. (4.15) becomes

$$\mathcal{R}_{ij}(\mathbf{r}) = \overline{u^2}\left[(f-g)\frac{r_ir_j}{r^2} + g\delta_{ij}\right], \tag{4.17}$$

or in direct notation

$$\mathcal{R}(\mathbf{r}) = \overline{u^2}\left[(f-g)\frac{\mathbf{r}\otimes\mathbf{r}}{r^2} + gI\right], \tag{4.18}$$

where I is the identity tensor. Note that in the interest of notational simplicity the dependence of f and g on r is not indicated explicitly. Moreover, even though it is not indicated, $\mathcal{R}, f, g$, and $\overline{u^2}$ may very well be time dependent.

In the case of $\mathcal{S}_{ij,l}$ scalar correlation functions $k(r), h(r)$, and $q(r)$ may be defined via

$$\mathcal{S}_{11,1}(r\mathbf{e}_1) = u_{rms}^3 k(r) \tag{4.19}$$

$$\mathcal{S}_{22,1}(r\mathbf{e}_1) = u_{rms}^3 h(r) \tag{4.20}$$

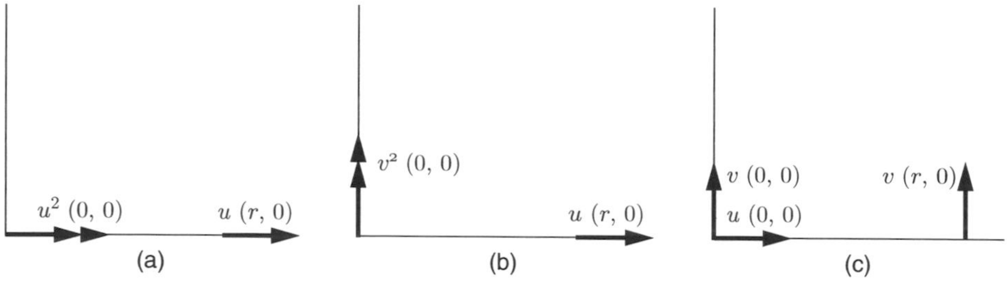

Figure 4.4 Definition of two-point triple velocity correlations: (a) $k(r)$, (b) $h(r)$, and (c) $q(r)$.

$$\mathcal{S}_{21,2}(r\mathbf{e}_1) = u_{rms}^3 q(r), \tag{4.21}$$

whose meaning is illustrated in Figure 4.4, and $u_{rms} = \sqrt{\overline{u^2}}$. Using Eq. (4.16) with Eqs. (4.19)–(4.21) gives

$$u_{rms}^3 k = S_1 r^3 + S_2 r + 2rS_3 \tag{4.22}$$

$$u_{rms}^3 h = S_2 r \tag{4.23}$$

$$u_{rms}^3 q = S_3 r. \tag{4.24}$$

Solving Eqs. (4.22)–(4.24) for S_1, S_2, and S_3, and substituting into Eq. (4.16) it is obtained that

$$\mathcal{S}_{ij,l}(\mathbf{r}) = u_{rms}^3 \left[(k - h - 2q)\frac{r_i r_j r_l}{r^3} + \delta_{ij} h \frac{r_l}{r} + q\left(\delta_{il}\frac{r_j}{r} + \delta_{jl}\frac{r_i}{r}\right)\right]. \tag{4.25}$$

Note that the appearance of the factor u_{rms}^3 in Eqs. (4.19)–(4.21) is by choice and not necessity. The intention is to make k, h, and q dimensionless in the most straightforward way. In fact, other scaling choices are possible. The need to select a scaling factor in this case has arisen from the fact that $\mathcal{S}_{ij,l}(0) = 0$. In contrast, $\mathcal{R}_{ij}(0) = \overline{u^2}\delta_{ij}$, so the coefficient $\overline{u^2}$ in Eq. (4.15) is predetermined.

The two-point, longitudinal, triple correlation coefficient $k(r)$ is antisymmetric, so that

$$k(r) = -k(-r), \tag{4.26}$$

as is illustrated in Figure 4.5, where it is seen that

$$\overline{u(0,0)^2 u(r,0)} = \overline{(-u(0,0))^2(-u(-r,0))} = -\overline{u(0,0)^2 u(-r,0)}. \tag{4.27}$$

This means that $k(0) = 0$ as are all its even derivatives at $r = 0$. In addition, since

$$u_{rms}^3 k(r) = \overline{u^2(x)u(x+r)}, \tag{4.28}$$

taking an r derivative and setting $r = 0$ yields

$$u_{rms}^3 \frac{dk}{dr}(0) = \overline{u^2\frac{\partial u}{\partial x}} = 1/3\ \frac{\partial \overline{u^3}}{\partial x} = 0 \tag{4.29}$$

in homogeneous turbulence. Thus, $dk/dr(0) = 0$ and if $k(r)$ is expanded in a Taylor series about $r = 0$, its leading order term is $\sim r^3 d^3k/dr^3(0)$ for small r.

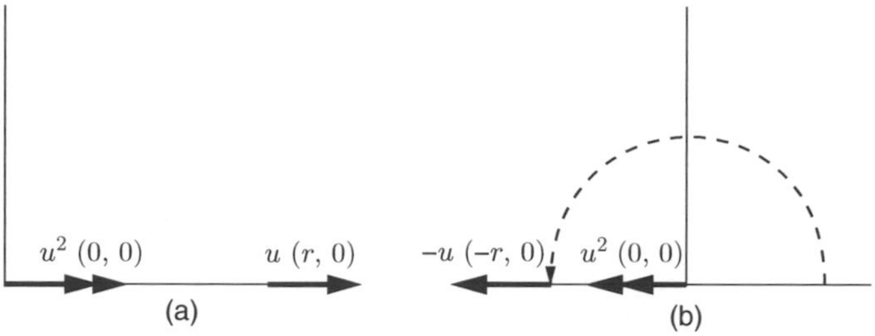

Figure 4.5 Antisymmetry of the two-point longitudinal triple velocity correlation. The correlations in (a) $k(r)$ and (b) $-k(-r)$ are equal.

Further simplification of Eqs. (4.17) and (4.25) follows from requiring that they be consistent with the incompressibility condition. For example, consider the identity

$$0 = \overline{u_i(\mathbf{x})\frac{\partial u_j}{\partial y_j}(\mathbf{y})} = \frac{\partial \mathcal{R}_{ij}}{\partial r_j}(\mathbf{r}), \tag{4.30}$$

that follows by a similar calculation as led to Eq. (4.5). Substituting Eq. (4.17) into (4.30) for $\mathcal{R}_{ij}$, and making use of identities such as

$$\frac{\partial}{\partial r_j}(r) = \frac{r_j}{r} \tag{4.31}$$

and

$$\frac{\partial}{\partial r_j}(r_i) = \delta_{ij}, \tag{4.32}$$

it may be shown that

$$g = f + \frac{r}{2}\frac{df}{dr}. \tag{4.33}$$

Numerical verification of this result from a simulation of isotropic turbulence is shown in Figure 4.6. Equation (4.33) can be used to eliminate $g(r)$ from Eq. (4.17), yielding

$$\mathcal{R}_{ij}(\mathbf{r}) = \overline{u^2}\left[\left(f + \frac{r}{2}\frac{df}{dr}\right)\delta_{ij} - \frac{r_i r_j}{r^2}\frac{r}{2}\frac{df}{dr}\right], \tag{4.34}$$

showing that $\mathcal{R}_{ij}$ depends on a single, as yet undetermined, scalar function, $f(r)$.

Similarly, the continuity equation can be applied to $\mathcal{S}_{ij,l}$ and after an equivalent calculation as led to Eq. (4.30), incompressibility implies that

$$\frac{\partial \mathcal{S}_{ij,l}}{\partial r_l}(\mathbf{r}) = 0. \tag{4.35}$$

Substituting for $\mathcal{S}_{ij,l}$ from Eq. (4.25) and performing a lengthy calculation gives

$$q = \frac{1}{4r}\frac{d(kr^2)}{dr} \tag{4.36}$$

and

$$h = -\frac{k}{2}. \tag{4.37}$$

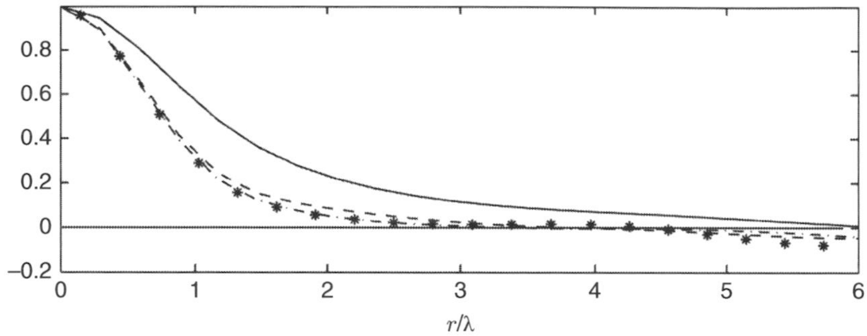

Figure 4.6 Confirmation of the isotropic identity Eq. (4.33) from a numerical simulation of isotropic turbulence using a vortex filament scheme [6]. $f(r)$, —; $g(r)$ based on v velocity, – –; $g(r)$ based on w velocity, – · –; ∗, evaluation of Eq. (4.33). Used with permission. Copyright (2006) National Academy of Sciences, USA.

Thus, it may be concluded that $\mathcal{S}_{ij,l}$ depends on the lone scalar function $k(r)$ through the relation

$$\mathcal{S}_{ij,l}(\mathbf{r}) = u_{rms}^3 \left[\left(k - r\frac{dk}{dr} \right) \frac{r_i r_j r_l}{2r^3} - \frac{k}{2}\delta_{ij}\frac{r_l}{r} + \frac{1}{4r}\frac{d(kr^2)}{dr}\left(\delta_{il}\frac{r_j}{r} + \delta_{jl}\frac{r_i}{r} \right) \right]. \tag{4.38}$$

The analysis of this section that culminates in Eqs. (4.34) and (4.38) shows that $\mathcal{R}_{ij}$ and $\mathcal{S}_{ij,l}$ each depend on a single scalar correlation function despite the fact that they are tensors with many components. A number of simplifications to the study of isotropic turbulence derive from these results and will be taken advantage of in the ensuing development. The formula contained in Eq. (4.34) will be seen in the next section to supply a relation for ϵ that is particularly useful in identifying the smallest scales of motion.

4.3 The Smallest Scales

In any particular turbulent flow it is generally possible to provide a Reynolds number that is descriptive of the length and velocity scales of the turbulence-producing mechanism. For example, if the fluid is being stirred by a propeller on a ship then the width of the blade or its radius and the rate of rotation of the blade establish a Reynolds number for the flow. Such a Reynolds number is tuned to the energy-containing scales of motion. At the same time, the length, time, and velocity scales of the smallest eddies in the flow, where viscous dissipation is expected to be dominant, is not a priori obvious. How small are the dissipation lengths for a given turbulent flow field? The answer to this question is of more than academic interest, since it must be used, for example, in developing a numerical mesh in direct numerical simulations of turbulence in order to guarantee that the flow is fully resolved. Similarly, the relationship between the sensing volume of a probe used in experimental work and the smallest scale at which the fluid velocity varies determines an acceptable Reynolds number range for which the probe can be used accurately.

As mentioned previously, the idea of a common structure to small-scale turbulent motion independent of how the turbulent flow comes about provides a helpful viewpoint from which to answer questions about the range of scales showing up in a particular flow field. The insight of Kolmogorov [1, 2] was to identify the parameters upon which a universal small-scale equilibrium should depend and then use these to derive basic scaling properties of the flow field.

Following the development of Kolmogorov, assume that in the high wave number spectral range beyond that associated with the energy -containing scales and the stirring mechanism, the turbulent motion can be succinctly characterized by a small number of relevant parameters. Within the dissipation range of scales assume that the Reynolds numbers is high enough for the universality idea to be relevant. In this case it may be assumed that the two main parameters governing the physics of the dissipative motion are the viscosity ν and the dissipation rate ϵ itself. Such a scenario defines the Kolmogorov dissipation length scale

$$\eta \equiv \left(\frac{\nu^3}{\epsilon}\right)^{1/4} \tag{4.39}$$

and the time scale

$$t_d = \left(\frac{\nu}{\epsilon}\right)^{1/2}. \tag{4.40}$$

Additionally, a velocity scale $v_d = (\nu\epsilon)^{1/4}$ can be defined by the ratio of Eqs. (4.39) and (4.40). It is evident from our previous discussion of k_d, in reference to Figure 4.1, that one expects that

$$\eta \sim \frac{1}{k_d}. \tag{4.41}$$

In fact, Eq. (4.41) does characterize the dissipation length scale, though it is more in the nature of a lower limit since it has been found that k_d and the bulk of the dissipation take place at scales that are some multiples of η. For example, experiments [7, 8] have shown that $k_d \approx \alpha/\eta$ for $\alpha = 0.1 \to 0.15$, with most of the dissipation taking place for $k < 0.5/\eta$. Furthermore, acceptable DNS can be performed using grids [9] with spacing in some directions equal to several times η.

Micro and macroscales, λ_f and Λ_f, were previously defined in Eqs. (2.45) and (2.47), respectively. It is of interest to see how these are related to η. Similarly, one can inquire as to how the velocity scale, v_d, is related to the root-mean square (*rms*) velocity components as typified by $u_{rms} = \sqrt{\overline{u^2}}$. Relations between these quantities can be derived by considering ϵ from two different points of view and then bringing these results together.

The first perspective on ϵ ties it to the microscale via its relationship to $\mathcal{R}_{ii}(\mathbf{r})$. Thus, contracting indices in Eq. (4.34) gives

$$\mathcal{R}_{ii} = \overline{u^2}[3f + rf'] \tag{4.42}$$

where r derivatives of f are henceforth denoted via a prime as in $f' = df/dr$. Differentiating Eq. (4.42) with respect to r_j yields

$$\frac{\partial \mathcal{R}_{ii}}{\partial r_j} = \overline{u^2}\left[4\frac{r_j}{r}f' + r_j f''\right] \tag{4.43}$$

and taking another derivative, setting $r = 0$, and substituting into Eq. (4.9) gives

$$\frac{\epsilon}{\nu} = -15\overline{u^2}f''(0). \tag{4.44}$$

In view of Eq. (2.45) it follows that

$$\epsilon = \frac{30\nu\overline{u^2}}{\lambda_f^2} = \frac{15\nu\overline{u^2}}{\lambda_g^2}, \tag{4.45}$$

where the second equality is a consequence of Eq. (4.33) (see Problems 4.5 and 4.6). Equation (4.45) is an important relation that connects the dissipation rate to well-defined scales of turbulent motion. In particular, it firmly establishes the relationship between the Taylor microscale and the dissipation rate.

Our interest now is in considering the dissipation rate from an alternative point of view. In particular, ϵ can be regarded as representing the rate at which energy is extracted from the energy-containing scales. The time necessary for a measurable amount of energy to be lost from such eddies, say t_e, should scale on their energy as represented by u_{rms}^2 and the physical extent of such eddies denoted by l_e. Note that the scales l_e and Λ both characterize large-scale motions and can be expected to be close to one another in magnitude. The rate of energy loss is thus u_{rms}^2 divided by the time scale $t_e \equiv l_e/u_{rms}$, which is often referred to as the *eddy turnover time*. t_e represents the life span of eddies so that there is a turnover in their population occurring at this rate. Alternatively, the energy loss in eddies can be viewed as being significant after they move over the distance l_e.

Since ϵ has units of energy/second, in terms of the eddy turnover time one is then led to the scaling:

$$\epsilon \sim \frac{u_{rms}^2}{l_e/u_{rms}} = \frac{u_{rms}^3}{l_e}. \tag{4.46}$$

Due to the fundamental importance of this relation to the analysis of turbulent flow, its validity and rationale have been widely considered [10]. Having expressed ϵ from two different perspectives in Eqs. (4.45) and (4.46), these may be brought together giving

$$\nu\frac{u_{rms}^2}{\lambda^2} \sim \frac{u_{rms}^3}{l_e} \tag{4.47}$$

and consequently

$$\frac{l_e}{\lambda} \sim R_\lambda, \tag{4.48}$$

where

$$R_\lambda = \frac{u_{rms}\lambda}{\nu} \tag{4.49}$$

is a turbulence Reynolds number and λ can represent either λ_f or λ_g. For many laboratory flows $R_\lambda < 100$, showing that in these cases there is up to two orders of magnitude difference in size between l_e and λ. Multiplying the numerator and denominator on the left-hand side of Eq. (4.48) by u_{rms}/ν gives the relation

$$R_e \sim R_\lambda^2 \tag{4.50}$$

or

$$\sqrt{R_e} \sim R_\lambda, \tag{4.51}$$

where

$$R_e = \frac{u_{rms} l_e}{\nu} \tag{4.52}$$

is a turbulence Reynolds number based on the physical size of the flow domain.

The ratio between λ and η can be found using Eqs. (4.39) and (4.45) and is

$$\frac{\eta}{\lambda} \sim \frac{1}{\sqrt{R_\lambda}} \sim \frac{1}{R_e^{1/4}}, \tag{4.53}$$

showing that η is generally smaller than λ, but not too far removed. From Eqs. (4.48) and (4.53) it follows that

$$\frac{l_e}{\eta} \sim R_\lambda^{3/2} \sim R_e^{3/4}, \tag{4.54}$$

which is the ratio of the largest to smallest scales in the flow. A similar calculation gives

$$\frac{u_{rms}}{v_d} \sim R_\lambda^{1/2} \sim R_e^{1/4} \tag{4.55}$$

as the ratio of velocities between the largest and smallest eddies.

Equation (4.54) can be made the basis for an important estimate of the numerical cost of computing turbulent flow. Since η is the smallest scale in the flow, a numerical simulation on a mesh would generally require a mesh spacing $\sim \eta$ to resolve the flow details. On the other hand, the spatial extent of the flow domain is $\sim l_e$, so in any one direction approximately l_e/η mesh points are required. A 3D mesh would then have to be $\sim (l_e/\eta)^3$ in size. In view of Eq. (4.54) the number of mesh points in a fully resolved turbulent flow simulation has a $R_e^{9/4}$ dependence on Reynolds number.

In a numerical calculation the shortest time period in need of resolution can be taken to be $\sim \eta/u_{rms}$, which is consistent with the role of η as the smallest length scale. At the same time, l_e/u_{rms} reflects time variations of the large-scale motions. The ratio of these two scales, $\sim l_e/\eta$, reflects roughly how many time steps must be computed in a turbulent flow simulation in order to obtain a reasonable amount of the motion, for example enough time to compute average quantities. According to Eq. (4.54) this means that $O(R_e^{3/4})$ time steps are required in the simulation and since there are $O(R_e^{9/4})$ grid points, the total computational effort involves $O(R_e^3)$ operations. This shows that the cost of doing a turbulence simulation increases quite rapidly with Reynolds number. Thus, even though simulations incorporating $O(10^{11})$ and even $O(10^{12})$ mesh points are currently feasible [11, 12] and have successfully represented flows with $R_e \sim 10^5$, to get simulations at $R_e \sim 10^6$ appropriate to many engineering problems would require three orders of magnitude gain in computational power beyond current teraflop machines (e.g., visit the website top500.org). For still larger Reynolds numbers and in the presence of complex geometries the outlook is even more bleak. This explains why there is little expectation that DNS will become a practical tool for simulating high Reynolds number flows anytime soon.

4.4 Inertial Subrange

A second far-reaching idea of Kolmogorov was that an "inertial subrange" consisting of a section of wave number space between k_e and k_d exists where energy is in transit through the spectrum toward small scales with neither significant energy production or dissipation taking place. Such a range of wave numbers should appear at Reynolds numbers so high that the direct effect on dissipation of ν occurs at scales that are distant from those where energy transfer toward the dissipation range is the dominant flow physics. A consequence of this model is that the energy spectrum function $E(k)$, which has units of $(\text{length})^3/(\text{sec})^2$, will have to scale according to

$$E(k) \sim k^{-5/3}\epsilon^{2/3} \tag{4.56}$$

if it is to be dimensionality consistent. Defining a Kolmogorov constant C_K, this becomes

$$E(k) = C_K k^{-5/3}\epsilon^{2/3}. \tag{4.57}$$

The prediction of a $-5/3$ spectrum is amenable to verification through experimental and DNS calculations if they have a sufficiently high Reynolds number.

Though Eq. (4.57) is a statement concerning the 3D energy spectrum, much of the effort that has considered its validity has been performed in terms of the 1D spectrum that is measured from time series data at a fixed point. Before considering results concerning the $-5/3$ law in Eq. (4.57) it is thus necessary to discuss the connection between the 1D and 3D spectrums.

4.4.1 Relations Between 1D and 3D Spectra

The connection between the 1D energy spectrum $\widehat{E}_{11}(\omega)$ computed from a time series and $E(k)$ is made by first developing a relationship between $E(k)$ and the spatial analogue of $\widehat{E}_{11}(\omega)$, namely, the 1D spatial spectrum $E_{11}(k_1)$ that forms a Fourier transform pair with $f(r)$. Thus, analogously to Eqs. (2.76) and (2.77) define

$$E_{11}(k_1) = \frac{\overline{u^2}}{\pi}\int_{-\infty}^{\infty} e^{-\imath k_1 x} f(x)dx \tag{4.58}$$

and

$$\overline{u^2} f(x) = \frac{1}{2}\int_{-\infty}^{\infty} e^{\imath k_1 x} E_{11}(k_1)dk_1. \tag{4.59}$$

The connection between $E_{11}(k_1)$ and $E(k)$ requires first making a connection between $E(k)$ and the energy spectrum tensor $\mathcal{E}_{ij}(\mathbf{k}, t)$ followed by a relationship between $\mathcal{E}_{ij}(\mathbf{k}, t)$ and $f(r)$ and then taking advantage of Eq. (4.58).

For isotropic turbulence it is the case that

$$\mathcal{E}_{ij}(\mathbf{k}) = E_1(k)k_i k_j + E_2(k)\delta_{ij}, \tag{4.60}$$

where E_1 and E_2 are scalar functions of $k = |\mathbf{k}|$. From Eq. (2.70) and the incompressibility condition it follows that

$$0 = \frac{\partial \mathcal{R}_{ij}}{\partial r_j}(\mathbf{r}) = -\int \imath k_j \mathcal{E}_{ij}(\mathbf{k}) e^{-\imath \mathbf{k}\cdot\mathbf{r}} d\mathbf{k} \tag{4.61}$$

for all $\mathbf{r}$, which can only be true if

$$k_j\mathcal{E}_{ij}(\mathbf{k}) = 0. \tag{4.62}$$

Applying this condition to Eq. (4.60) yields

$$E_2(k) = -k^2E_1(k) \tag{4.63}$$

so that

$$\mathcal{E}_{ij}(\mathbf{k}) = E_1(k)(k_ik_j - k^2\delta_{ij}). \tag{4.64}$$

Contracting indices gives

$$\mathcal{E}_{ii}(\mathbf{k}) = -2E_1(k)k^2, \tag{4.65}$$

which shows that $\mathcal{E}_{ii}(\mathbf{k})$ depends only on k.

A consequence of Eq. (4.65) is that the integration over the solid angle in Eq. (2.74) can be carried out, yielding a factor 4π equal to the total solid angle. Consequently,

$$E(k) = 2\pi k^2\mathcal{E}_{ii}(\mathbf{k}) \tag{4.66}$$

and, in view of Eq. (4.65),

$$E(k) = -4\pi k^4E_1(k). \tag{4.67}$$

Thus, Eqs. (4.64) and (4.67) give

$$\mathcal{E}_{ij}(\mathbf{k}) = \frac{E(k)}{4\pi k^2}\left(\delta_{ij} - \frac{k_ik_j}{k^2}\right) \tag{4.68}$$

as the energy spectrum tensor in isotropic turbulence, which is the desired relationship between $\mathcal{E}$ and E. If $E_1(k)$ is a well-behaved function near $k = 0$ (e.g., it approaches a finite constant there), then Eq. (4.67) implies that

$$E(k) \sim k^4 \tag{4.69}$$

as $k \to 0$, and when used in Eq. (4.68) the analyticity of $\mathcal{E}_{ij}$ at $k = 0$ is assured. If $E_1(k)$ were to be singular at the origin, and in particular that $E_1 \sim k^{-2}$, then $E(k) \sim k^2$ as $k \to 0$, which fits in with an alternative theoretical analysis of isotropic turbulence [13] that has received credible support in experimental work, as will be described in the next chapter.

To connect $\mathcal{E}$ with $f(r)$ note that from Eq. (4.42) it follows that $\mathcal{R}_{ii}(\mathbf{r})$ is a function of r only and consequently for isotropic turbulence it can be written that

$$\mathcal{E}_{ii}(k) = \frac{1}{(2\pi)^3}\int \mathcal{R}_{ii}(r)e^{\imath\mathbf{k}\cdot\mathbf{r}}d\mathbf{r}. \tag{4.70}$$

Shifting to spherical coordinates, the integration in the θ and ϕ directions for shells of fixed r can be carried out explicitly (see Problem 4.7) with the result that

$$\mathcal{E}_{ii}(k) = \frac{2}{(2\pi)^2}\int_0^\infty \mathcal{R}_{ii}(r)r^2\frac{\sin kr}{kr}dr. \tag{4.71}$$

Substituting for $\mathcal{E}_{ii}$ using Eq. (4.66) gives

$$E(k) = \frac{1}{\pi}\int_0^\infty \mathcal{R}_{ii}(r)kr\sin kr dr \tag{4.72}$$

and now substituting Eq. (4.42) results in

$$E(k) = \frac{\overline{u^2}}{\pi} \int_0^\infty (3f(r) + rf'(r))\ kr \sin krdr. \tag{4.73}$$

Integrating the term containing f' in this equation by parts and collecting terms gives

$$E(k) = \frac{\overline{u^2}}{\pi} \int_0^\infty f(r)(kr \sin kr - k^2r^2 \cos kr)dr \tag{4.74}$$

under the condition that

$$\lim_{r\to\infty} r^2 f(r) = 0. \tag{4.75}$$

In fact, whether or not Eq. (4.75) holds is tied to the possibility that $E(k)$ is proportional to k^4 or k^2 at small k, as will now be shown.

Substituting a Taylor series for sin kr into Eq. (4.73) gives

$$E(k,t) = \left[\frac{\overline{u^2}}{\pi} \int_0^\infty (3f + rf')r^2 dr\right] k^2 - \left[\frac{\overline{u^2}}{6\pi} \int_0^\infty (3f + rf')r^4 dr\right] k^4 + ... \tag{4.76}$$

In view of the identity

$$3r^2 f + r^3 f' = \frac{d(r^3 f)}{dr}, \tag{4.77}$$

the first term in Eq. (4.76) satisfies

$$\frac{\overline{u^2}}{\pi} \int_0^\infty (3f + rf')r^2 dr = 0 \tag{4.78}$$

if

$$\lim_{r\to\infty} r^3 f(r) = 0. \tag{4.79}$$

Clearly, Eq. 4.79 implies that Eq. (4.75) holds and so the model that predicts that Eqs. (4.69) and (4.74) would be valid in this case. In the theory of Saffman [13, 14] it is assumed that Eq. (4.78) does not hold so that the limit in Eq. (4.79) is not zero, but rather an invariant in the decay of isotropic turbulence that has a constant finite value.

Under the assumption that Eq. (4.69) is valid, it is of interest to derive the inverse equation to Eq. (4.74). This requires imitating the steps leading to Eq. (4.71) by carrying out the integration on a shell of fixed k in Eq. (2.70) after taking advantage of Eq. (4.66). The result is

$$\mathcal{R}_{ii}(r) = 2 \int_0^\infty \frac{\sin kr}{kr} E(k)dk \tag{4.80}$$

and using Eq. (4.42), (4.80) gives

$$\overline{u^2}\frac{d(r^3 f)}{dr} = 2 \int_0^\infty \frac{r\ \sin kr}{k} E(k)dk. \tag{4.81}$$

Integrating Eq. (4.81) from $0 \to r$ yields

$$\overline{u^2} f(r) = 2 \int_0^\infty E(k) \left(\frac{\sin kr}{k^3 r^3} - \frac{\cos kr}{k^2 r^2}\right) dk. \tag{4.82}$$

Thus Eqs. (4.74) and (4.82) represent transforms between f and E.

To link up $E(k)$ with $E_{11}(k_1)$, note that since Eq. (4.33) holds, it may be shown using Eq. (4.58) that

$$\frac{\overline{u^2}}{\pi}\int_{-\infty}^{\infty} e^{-\iota k_1 x} g(x)dx = \frac{1}{2}\left(E_{11} - k_1\frac{\partial E_{11}}{\partial k_1}\right). \tag{4.83}$$

Noting from Eq. (4.17) that $\mathcal{R}_{ii} = \overline{u^2}(f + 2g)$ and adding Eq. (4.58) to twice Eq. (4.83) gives

$$2E_{11}(k_1) - k_1\frac{\partial E_{11}}{\partial k_1} = \frac{1}{\pi}\int_{-\infty}^{\infty} e^{-\iota x k_1}\mathcal{R}_{ii}(x,0,0)dx. \tag{4.84}$$

Differentiating both sides of Eq. (4.84) with respect to k_1 and then applying the symmetry relations for $f(r)$ and $g(r)$ to simplify the integral on the right-hand side, it is found that

$$\frac{\partial E_{11}}{\partial k_1}(k_1) - k_1\frac{\partial^2 E_{11}}{\partial k_1^2}(k_1) = -\frac{2}{\pi}\int_0^{\infty} x \sin xk_1 \mathcal{R}_{ii}(x)dx \tag{4.85}$$

where the fact that $\mathcal{R}_{ii}$ is a function of r only has been used once again. Finally, using (4.72) it follows that

$$E(k_1) = \frac{1}{2}k_1^2\frac{\partial^2 E_{11}}{\partial k_1^2} - \frac{1}{2}k_1\frac{\partial E_{11}}{\partial k_1}, \tag{4.86}$$

which relates the 1D and 3D spatial spectra in isotropic turbulence. By solving the differential equation in Eq. (4.86) for E_{11}, its inverse relation is derived in the form

$$E_{11}(k_1) = \int_{k_1}^{\infty} dk\frac{E(k)}{k}\left(1 - \frac{k_1^2}{k^2}\right) \tag{4.87}$$

in which the condition that $\lim_{k_1\to\infty} E_{11}(k_1, t) = 0$ is used to determine the constants of integration [15].

Equation (4.86) implies that the 1D spatial energy spectrum in the inertial range will also obey a $-5/3$ law as in Eq. (4.57), but with a Kolmogorov constant C_K that is 18/55 smaller than the 3D value (see Problem 4.9). This suggests that verification of the $-5/3$ law is justified by examining 1D spectra which can be obtained from a single velocity component in a single direction such as $u(x)$. Since obtaining such data experimentally is problematical, however, there is an incentive to develop a connection between $E_{11}(k_1)$ and $\widehat{E}_{11}(\omega)$ since the latter is readily available from time series data at a fixed point in turbulent flow. This is now considered.

4.4.2 1D Spatial and Time Series Spectra

A standard means of relating time sequences of streamwise velocity to data distributed along a straight line in the flow direction is to imagine that the velocity field at a given instant in time convects downstream at the local mean velocity as if it were frozen. The idea is referred to as Taylor's hypothesis [16] and has been subjected to tests using physical experiments and DNS data [17]. On the whole such studies call into question the validity of the hypothesis, yet nonetheless its use remains a part of the analysis of spectra, as will be considered here.

For Taylor's hypothesis, the fluctuating velocity field measured at a point x_0 at time $t_0 + \tau$ is assumed to be the same as the fluctuating velocity u at the upstream point $x_0 - \overline{U}\tau$ at time t_0, where $\overline{U}$ is a local mean velocity that stays relatively fixed over the time interval of length T where $\tau \le T$. Thus, $u(x_0, t)$ over the time interval $t_0 \le t \le t_0 + T$ corresponds to $u(x, t_0)$ for $x_0 \ge x \ge x_0 - \overline{U}T$. Thus, under Taylor's hypothesis the formal connection is made (for fixed y, z) that

$$u(x, t+\tau) = u(x - \overline{U}\tau, t) \tag{4.88}$$

so that

$$\overline{u(x,t)u(x,t+\tau)} = \overline{u(x,t)u(x-\overline{U}\tau,t)} \tag{4.89}$$

in which case, according to the definitions of $f(x)$ in Eq. (2.38) and $\mathcal{R}_E(\tau)$ in Eq. (2.48)

$$\mathcal{R}_E(\tau) = f(-\overline{U}\tau) = f(\overline{U}\tau), \tag{4.90}$$

where the second equality takes advantage of the symmetry of $f(r)$. Setting $x = \overline{U}\tau$ Eq. (4.90) becomes

$$f(x) = \mathcal{R}_E(x/\overline{U}). \tag{4.91}$$

Substituting Eq. (4.91) into (4.58) and changing the x integration to τ integration so as to copy Eq. (2.76), it follows from examination of the exponential term that

$$k_1\overline{U} = 2\pi\omega. \tag{4.92}$$

Introducing the definition Eq. (2.78) it is thus seen that $\widehat{E}_{11}(\omega)$ is related to $E_{11}(k_1)$ via

$$E_{11}(k_1) \equiv \frac{\overline{U}}{2\pi}\widehat{E}_{11}\left(\frac{k_1\overline{U}}{2\pi}\right) \tag{4.93}$$

or, conversely,

$$\widehat{E}_{11}(\omega) \equiv \frac{2\pi}{\overline{U}}E_{11}\left(\frac{2\pi\omega}{\overline{U}}\right). \tag{4.94}$$

To summarize, it has been shown that the mappings in Eqs. (4.91) and (4.93) substituted into Eqs. (4.58) and (4.59) and using Eq. (4.92) are identical to Eqs. (2.76) and (2.77).

The utility of Eqs. (4.93) and (4.94) is that they imply that the measured time spectra as given in $\widehat{E}_{11}(\omega)$ can be considered to also represent a determination of $E_{11}(k_1)$ which has been seen previously to share a $-5/3$ law with $E(k)$ if such a law holds. Of course, this connection relies on the questionable validity of Taylor's hypothesis.

Measured spectra, such as that displayed in Figure 4.7, which includes a compendium of data from different studies and covering a wide range of Reynolds numbers, support the occurrence of an inertial range law with slope equal to or close to $-5/3$. Results similar to Figure 4.7 occur for many other experiments that range from laboratory flows to meteorological flows of very large scale [18]. For the data at high Reynolds numbers plotted in Figure 4.7, the inertial range is seen to extend over several decades of wave numbers. That spectra from different flows collapse to a common spectral form at higher wavenumbers provides support for Kolmogorov's hypothesis regarding the universality of the small scales of turbulence. The extent of the inertial subrange is seen to increase with R_λ. It is generally not possible to arrive at a definitive estimate of the slope of the

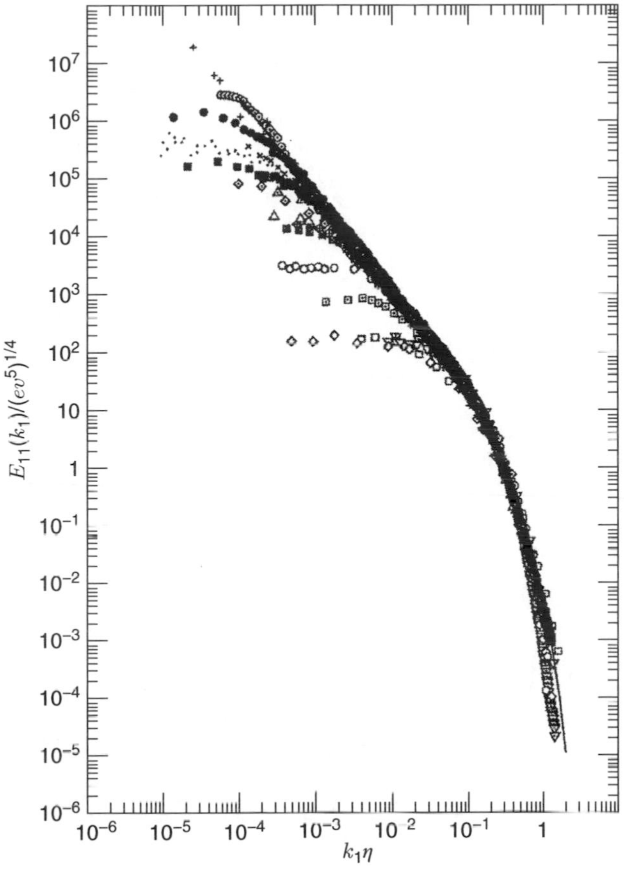

Figure 4.7 Experimental tests of the −5/3 law [3]. Reprinted with the permission of Cambridge University Press.

spectrum from experimental data, such as that in Figure 4.7, since there is generally too much scatter in the data. Nonetheless, analysis of the data yields a best estimate of the Kolmogorov constant for the 1D spectrum that is approximately 1/2. This means that for the 3D spectrum the constant is expected to be approximately $C_K = 1.4$ [3, 18]. More recent estimates of C_K using simulations of isotropic turbulence at Reynolds numbers as high as $R_\lambda = 1000$ [19] yield the estimate $C_K = 1.58$.

Numerical simulations of isotropic turbulence offer a means of studying the appropriateness of a −5/3 law with less scatter in the spectra than in physical experiments. One significant result in such a study is shown in Figure 4.8, where the compensated spectrum $E(k)/(\epsilon^{2/3}k^{-5/3})$ is shown for a calculation of isotropic turbulence in a periodic unit cube at several Reynolds numbers, with the highest value being $R_\lambda = 1201$ computed on a 4096^3 mesh. To the extent that a −5/3 law is satisfied the compensated spectrum should be constant. The figure shows that the spectra merge to form an increasingly long region that is not precisely constant but instead has a slightly negative slope of −0.1. This suggests that a more appropriate spectrum for the inertial range might be $\sim k^{-5/3-0.1}$. A number of other notable features of the spectrum include the pronounced peak that

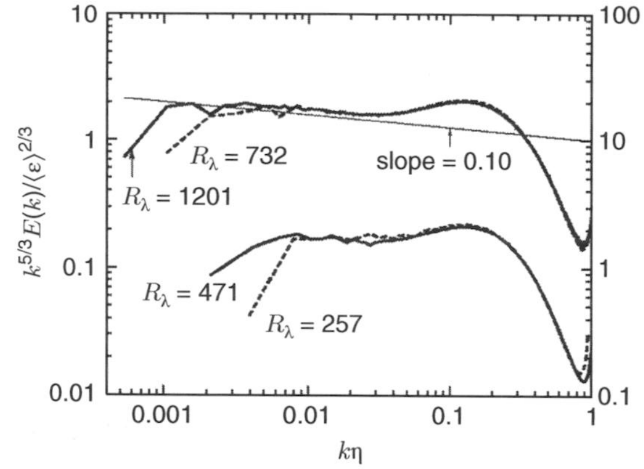

Figure 4.8 Compensated energy spectrum as given in [21]. With increasing R_λ the simulations used 512^3, 1024^3, 2048^3, and 4096^3 meshes. Scales on the left and right are for the upper and lower curves, respectively. Reproduced from *Physics of Fluids*, Vol. 15, pp. L21–L24, 2003, with the permission of AIP Publishing.

forms for wave numbers just larger than those in the inertial range, a phenomenon that is referred to as the *bottleneck effect*. A number of explanations have been provided to explain this phenomenon. For example, it may be the result of energy transferring to scales just above the dissipation range where there is insufficient small-scale vortices to efficiently dissipate energy so that it accumulates to form a bump in the spectrum [20]. For very high Reynolds numbers there is evidence that the bottleneck effect vanishes [19].

Among the explanations for the departure of small-scale turbulence from a $-5/3$ law, perhaps the most fundamental is the recognition that the distribution of the turbulence dissipation rate is highly intermittent in space and time. For example, as shown in Figure 4.9, a plot of the dissipation field ϵ from a simulation of isotropic turbulence using a vortex filament scheme shows that it is concentrated into a random subset of the

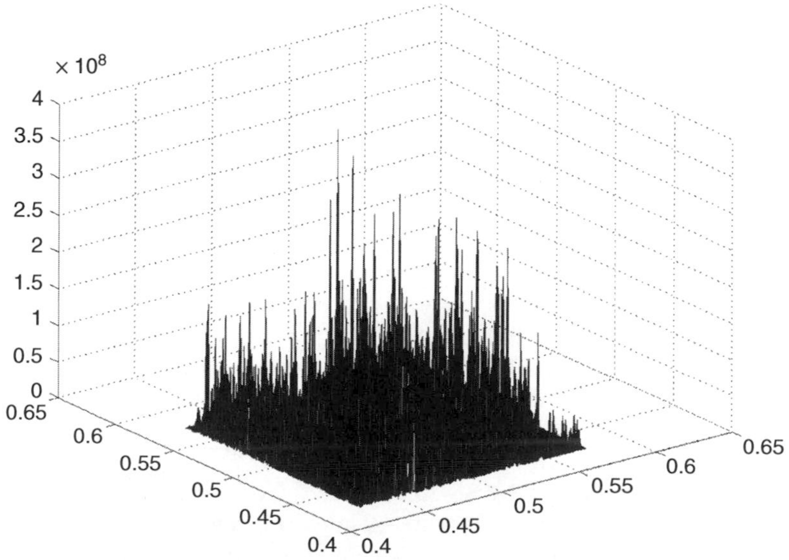

Figure 4.9 Dissipation rate on a plane showing intermittency within a region of isotropic turbulence computed in a vortex filament simulation of flow in a periodic box.

field. In the locations where dissipation is high, it might be 1000 times higher than the surrounding dissipation rate in this example. Such behavior is antithetical to the simple scaling argument behind the $-5/3$ law, one which, in effect, assumes that dissipation is spread somewhat uniformly over all space.

As mentioned in Chapter 3, it is also the case that the energy transfer process between large and small scales is not in the character of a one-way cascade of energy to small scales. Backscatter is also present where it is imagined that larger vortices form from smaller ones thus transferring energy to large scales. In effect, the current view is that the processes affecting energy transfer to both large and small scales coexist having the character of an equilibrium process rather than the disequilibrium implied by a one-way energy cascade to small scales. Explaining why and how intermittency should impact the inertial range spectrum as well as the nature of the equilibrium transfer process are important problems in turbulence theory that remain to be solved in full [22].

4.5 Structure Functions

Much of the discussion in this chapter has concerned the spectral analysis of isotropic turbulence both at the smallest dissipative scales and in the inertial range. It is also possible to approach the analysis of these fundamental properties of turbulent flow in a physical space analysis that was first developed by Kolmogorov [2, 23]. In this, structure functions of the nth order are considered that are defined as

$$S_n(\mathbf{x}, \mathbf{r}, t) \equiv \overline{|\mathbf{u}(\mathbf{x}+\mathbf{r}, t) - \mathbf{u}(\mathbf{x}, t)|^n}. \tag{4.95}$$

As an alternative to Eq. (4.95), structure functions can be defined for individual velocity components as in the longitudinal structure function

$$S_n^L(r) \equiv \overline{|u(x+r) - u(x)|^n} \tag{4.96}$$

and the transverse structure function

$$S_n^T(r) \equiv \overline{|v(x+r) - v(x)|^n}. \tag{4.97}$$

For small separations in r it is expected that the structure functions are affected by the dynamics of the dissipation range scales while for somewhat larger values of r they give an indication of the inertial range motions.

For the particular case of $n = 2$ in isotropic turbulence, with the help of the relation in Eq. (4.80) it may be shown that

$$\overline{|\mathbf{u}(\mathbf{x}+\mathbf{r}, t) - \mathbf{u}(\mathbf{x}, t)|^2} = 4\int_0^\infty E(k,t)\left(1 - \frac{\sin kr}{kr}\right) dk. \tag{4.98}$$

At high Reynolds numbers it is reasonable to evaluate this integral using Eq. (4.57) for the energy spectrum since the integrand is small for both small and large k. For the former, the factor $\sin kr/kr \approx 1$ so the term in parenthesis in the integrand of Eq. (4.98) should be small while for large k the energy spectrum is in the dissipation range and is also small. The integration then yields the useful approximate relation

$$\overline{|\mathbf{u}(\mathbf{x}+\mathbf{r}, t) - \mathbf{u}(\mathbf{x}, t)|^2} = 4.82 C_K (\epsilon r)^{2/3}. \tag{4.99}$$

Some insight into the structure functions for general n can be had by a scaling analysis [4, 23]. For the case of $S_n^L(r)$ it may be assumed on dimensional grounds that the velocity differences can be expressed as

$$|\mathbf{u}(x+r) - \mathbf{u}(x)| \sim \xi(\epsilon_r r)^{1/3} \tag{4.100}$$

where ξ is a random variable that does not depend on ϵ and r, and ϵ_r is the average of ϵ over a sphere of radius r. Substituting Eq. (4.100) into Eq. (4.96) for the case of the longitudinal structure function motivates the approximation that

$$S_n^L(r) = C_n r^{n/3} \overline{\epsilon_r^{n/3}}. \tag{4.101}$$

A number of analyses have proposed that $\overline{\epsilon_r^{n/3}}$ obeys a power law r^α, where $\alpha = \alpha(n/3)$ is a function of n. If so, then $S_n^L(r)$ obeys a power law with exponent $n/3 + \alpha(n/3)$ so that the fluctuating dissipation acts to modify the nominal $n/3$ power law. This suggests that for Eq. (4.99) to be true, it would have to follow that $\alpha(2/3) = 0$. Similarly, for $n = 3$ Kolmogorov specifically obtained

$$S_3^L(r) = -\frac{4}{5} r \overline{\epsilon_r} \tag{4.102}$$

which implies that $\alpha(1) = 0$. Shown in Figure 4.10 are plots of the compensated structure functions

$$S_2^L / (\overline{\epsilon_r} r)^{2/3} \tag{4.103}$$

and

$$-S_3^L / (\overline{\epsilon_r} r), \tag{4.104}$$

respectively, that are designed to be constant where the power law is given by the $n/3$ exponent. The horizontal coordinate is r/η with η defined in Eq. (4.39) being the Kolmogorov dissipation length scale.

In both cases it is seen that the compensated structure functions achieve nearly constant values in the region beyond $100/\eta$ which is appropriate to the inertial subrange. The constant values achieved by the curves match the coefficients in Eqs. (4.101) and

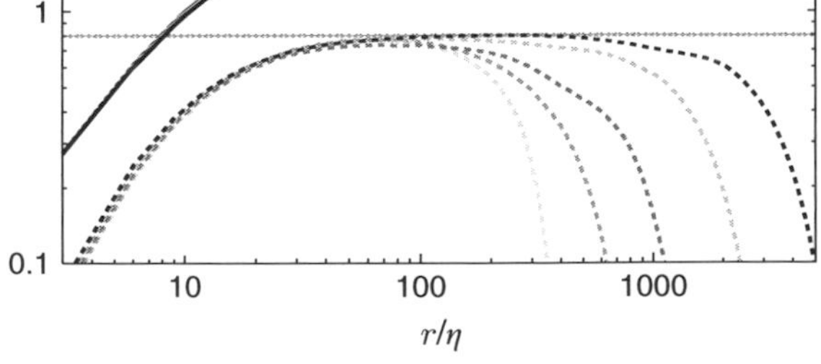

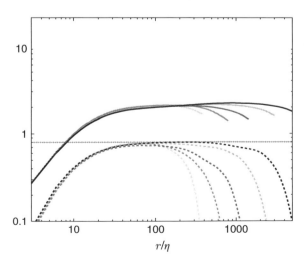

Figure 4.10 Compensated longitudinal structure functions computed in isotropic turbulence [24]. $S_2^L/(\epsilon_r r)^{2/3}$, top curves; $-S_3^L/(\epsilon_r r)$, bottom curves. Constant, dotted line is 0.8. The sequence of curves in each group covers an increasing r/η domain correspond to $R_\lambda = 167, 257, 471, 732, 1131$.

(4.102) [24]. However, for both relations there is a slight tilt to the nearly constant regions so that the observed power laws are close to an $r^{n/3}$ behavior, but do not exactly have this trend. This suggests that $\alpha(2/3)$ and $\alpha(1)$ do not vanish. In fact, measurements of the structure functions for increasing n show that the departure from $r^{n/3}$ behavior becomes more significant. The trends in $S_n^L(r)$ are the physical space analogue to the discrepancies in the $-5/3$ wave number spectrum that are in Figure 4.8. As in that case it is believed that these are a consequence of intermittency and ongoing research attempts to investigate the statistical properties of such quantities as ϵ_r in order to predict the measured behavior of $\alpha(n/3)$.

4.6 Chapter Summary

The aspect of turbulent flow considered in this chapter is the chaotic motion at small scales that plausibly has characteristics that can be found in all turbulent flows regardless of how they are created or maintained. To some extent this is the equivalent for turbulent flow of the motions at the molecular level that exists within all fluids and is responsible for such macroscopic properties as viscosity and specific heats. For turbulence, however, unless the Reynolds number is very high, it is not clear that there will be a sufficiently large separation between small- and large-scale motions to justify the idea that the small scales share a universal behavior. Moreover, even if the small scales do have a universal character in a wide range of circumstances, there still remains much to find out about such properties of flows as the spectrum and dissipation rate.

Many of the topics introduced in this chapter continue to be the object of great interest within the field of turbulence since progress in such areas may one day have direct benefit in improving turbulence modeling strategies. It can be expected that DNS studies of ever increasing size and Reynolds number will continue to pursue an understanding of the physics of isotropic turbulence at small scales in isolation from the influences of the large scales of motion. Already clear evidence is mounting from simulations with $R_\lambda = 1000$ and higher for the presence of an inertial range spectrum with exponent slightly smaller than $-5/3$. These and many other aspects of small-scale turbulent flow, such as the behavior of structure functions, are now and will continue to be clarified by numerical simulations that have the capability of giving definitive information free of modeling or measurement error. The next chapter extends the present discussion to introduce essential aspects of the dynamics of homogeneous isotropic turbulence during its decay from an initial state. Out of such considerations some of the most basic ideas in turbulence modeling arise.

References

1 Kolmogorov, A.N. (1941) The local structure of turbulence in an incompressible viscous fluid for very large Reynolds numbers. *C. R. Acad. Sci. URSS*, 30, 301–305.

2 Kolmogorov, A.N. (1941) Dissipation of energy in locally isotropic turbulence. *C. R. Acad. Sci. URSS*, 32, 19–21.

3 Saddoughi, S.G. and Veeravalli, S.V. (1994) Local isotropy in turbulent boundary layers at high Reynolds number. *J. Fluid Mech.*, 268, 333–372.

4 Sreenivasan, K.R. and Antonia, R.A. (1997) The phenomenology of small-scale turbulence. *Ann. Rev. Fluid Mech.*, 29, 435–472.
5 Robertson, H.P. (1940) The invariant theory of isotropic turbulence. *Proc. Camb. Phil. Soc.*, 36, 209–223.
6 Bernard, P.S. (2006) Turbulent flow properties of large scale vortex systems. *Proc. Nat. Acad. Sci.*, 103, 10 174–10 179.
7 Stewart, R.W. and Townsend, A.A. (1951) Similarity and self-preservation in isotropic turbulence. *Phil. Trans. Roy. Soc. London*, 243A, 359–386.
8 Grant, H.L., Stewart, R.W., and Moilliet, A. (1962) Turbulence spectra from a tidal channel. *J. Fluid Mech.*, 12, 241–263.
9 Moin, P. and Mahesh, K. (1998) Direct numerical simulations: a tool in turbulence. *Ann. Rev. Fluid Mech.*, 30, 539–578.
10 Vassilicos, J.C. (2015) Dissipation in turbulent flows. *Ann. Rev. Fluid Mech.*, 47, 95–114.
11 Lee, M. and Moser, R.D. (2015) Direct numerical simulation of channel flow up to $Re_\tau \approx 5200$. *J. Fluid Mech.*, 774, 395–415.
12 Ishihara, T., Morishita, K., Yokokawa, M., Uno, A., and Kaneda, Y. (2016) Energy spectrum in high-resolution direct numerical simulations of turbulence. *Phys. Rev. Fluids*, 1, 082 403.
13 Saffman, P. (1967) The large-scale structure of homogeneous turbulence. *J. Fluid Mech.*, 27, 581–593.
14 Saffman, P. (1967) Note on decay of homogeneous turbulence. *Phys. Fluids*, 10, 1349–1349.
15 Hinze, J.O. (1975) *Turbulence*, McGraw-Hill, N.Y.
16 Taylor, G.I. (1938) The spectrum of turbulence. *Proc. Roy. Soc. London A*, 164, 476–490.
17 He, G., Jin, G., and Yang, Y. (2017) Space-time correlations and dynamic coupling in turbulent flows. *Ann. Rev. Fluid Mech.*, 49, 51–70.
18 Sreenivasan, K.R. (1995) On the universality of the Kolmogorov constant. *Phys. Fluids*, 7, 2778–2784.
19 Donzis, D.A. and Sreenivasan, K.R. (2010) The bottleneck effect and the Kolmogorov constant in isotropic turbulence. *J. Fluid Mech.*, 657, 171–188.
20 Katul, G.G., Manes, C., Porporato, A., Bou-Zeid, E., and Chamecki, M. (2015) Bottlenecks in turbulent kinetic energy spectra predicted from structure function inflections using the Von Kármán-Howarth equation. *Phys. Rev. E*, 92, 033 009.
21 Kaneda, Y., Ishihara, T., Yokokawa, M., Itakura, K., and Uno, A. (2003) Energy dissipation rate and energy spectrum in high resolution direct numerical simulations of turbulence in a periodic box. *Phys. Fluids*, 15, L21–L24.
22 Lesieur, M. (2008) *Turbulence in Fluids*, Springer, Dordrecht, 4th edn.
23 Kolmogorov, A.N. (1962) A refinement of previous hypotheses concerning the local structure of turbulence in a viscous incompressible fluid at high Reynolds number. *J. Fluid Mech.*, 13, 82–85.
24 Ishihara, T., Gotoh, T., and Kaneda, Y. (2009) Study of high-Reynolds number isotropic turbulence by direct numerical simulation. *Ann. Rev. Fluid Mech.*, 41, 165–180.

Problems

4.1 For

$$E(k) = \frac{1}{\sqrt{2\pi}}\overline{u^2}\lambda_f(k\lambda_f)^4 e^{-(k\lambda_f)^2/2}, \tag{4.105}$$

which occurs in the final period of isotropic decay (see Chapter 5), find k_e and k_d. What can be said about the distance between these locations?

4.2 According to Eq. (4.17), in the r_1, r_2 plane, $\mathcal{R}_{12}(0, 0, r_3) = 0$, yet $\mathcal{R}_{12}(r_1, r_2, r_3) \neq 0$ if r_1 and r_2 are both not zero. Develop a physical argument to explain why this is so.

4.3 Derive Eq. (4.33).

4.4 Derive Eq. (4.38) from (4.25).

4.5 Use Eq. (4.8) to prove that

$$\overline{\left(\frac{\partial u}{\partial y}\right)^2} = 2\overline{\left(\frac{\partial u}{\partial x}\right)^2}, \tag{4.106}$$

so that the mean-square shearing in isotropic turbulence is twice the mean-square stretching.

4.6 Use Eq. (4.106) to prove that

$$\lambda_f = \sqrt{2}\lambda_g. \tag{4.107}$$

4.7 Carry out the θ and ϕ integration in Eq. (4.70) to then derive Eq. (4.71).

4.8 Solve the differential equation (4.86) to obtain Eq. (4.87).

4.9 Show using Eq. (4.86) that the Kolmogorov constant C_K in the 3D inertial range spectrum for $E(k)$ is 55/18 times that of C_K in the 1D inertial range spectrum corresponding to $E_{11}(k_1)$.

4.10 Derive Eq. (4.98) from Eq. (4.80).

5

Energy Decay in Isotropic Turbulence

With the preparation provided in the previous chapter we now consider some aspects of the decay process by which turbulent kinetic energy is dissipated by the action of viscosity. This discussion considers the idealized case of the decay of homogeneous, isotropic turbulence, which contains all the essential physical arguments relating to dissipation without additional complicating factors. Moreover, the isotropy condition yields equations in the simplest possible form. The results of this analysis of the decay process form a basic part of virtually all turbulence models applied to general flow situations where the turbulence is neither homogeneous nor isotropic.

5.1 Energy Decay

The problem of interest here consists of a physical flow domain occupying either a large region far away from boundaries or turbulent flow in a box assuming periodic boundary conditions. In the latter case, the box is assumed to be of sufficient size so that periodicity has no influence on the evolution of the flow. The minimal requirement for this to occur is that correlations such as $f(r)$ and $k(r)$ have decayed to zero with r well within the domain. It will be seen below, however, that in any decay scenario, after the energy has fallen sufficiently, it is inevitable that the turbulent length scales will increase, leading to a conflict with the finite domain within which periodicity is applied. This is a complicating factor that needs to be taken into account in studies that explore isotropic decay via numerical simulations [1].

For the flow of interest here it is assumed that at an initial time the fluid in the domain is uniformly stirred, creating a homogeneous turbulent field whose mean statistics obey all isotropic conditions. The problem then is to determine how the flow proceeds through time back to a quiescent state, which it must do in the absence of a production mechanism (e.g., the device used to stir the flow field in the first place). The possibility of creating either a physical experiment or computational study whose initial state matches that of the abstract decay problem is an open question. In practice it is known that decay rates are sensitive to the initial forcing mechanism, and this has complicated efforts at predicting the nature of universal laws governing the decay process, if they exist.

Complete statistical specification of the initial turbulent velocity field requires having knowledge of all its multi-point correlations. For example, in the present case, besides specifying K, ϵ, and other one-point statistics at the initial time, it should theoretically be necessary to specify the initial state of $f(r)$, $k(r)$, and an infinite number of higher order

Turbulent Fluid Flow, First Edition. Peter S. Bernard.

Companion website: www.wiley.com/go/Bernard/Turbulent_Fluid_Flow

moments. Specification of $f(r)$ implies that λ_f is known, which illustrates the fact that all scales associated with the motion are implicitly available through the multi-point correlations. Note that throughout this chapter it is to be assumed that variables such as K and ϵ, and functions such as f and k can be time dependent even if not indicated explicitly.

A good place to see what information is needed in order to perform a calculation of the decay is to consider the terms in the K and ϵ equations specialized to the present case. These equations consist of

$$\frac{dK}{dt} = -\epsilon \tag{5.1}$$

determined from Eq. (3.38) and the ϵ equation (3.42) in the form

$$\frac{d\epsilon}{dt} = P_\epsilon^4 - \Upsilon_\epsilon = -2\nu\overline{\frac{\partial u_i}{\partial x_l}\frac{\partial u_i}{\partial x_j}\frac{\partial u_l}{\partial x_j}} - 2\nu^2\overline{\left(\frac{\partial^2 u_i}{\partial x_j \partial x_l}\right)^2}. \tag{5.2}$$

Since $\epsilon = \nu\zeta$ in isotropic turbulence, as noted in the discussion surrounding Eq. (3.81), an alternative to Eq. (5.2) is to adopt Eq. (3.82) to isotropic turbulence, giving

$$\frac{d\epsilon}{dt} = \nu P_\zeta^4 - \nu\Upsilon_\zeta = 2\nu\overline{\omega_i\omega_k\frac{\partial u_i}{\partial x_k}} - 2\nu^2\overline{\frac{\partial \omega_i}{\partial x_k}\frac{\partial \omega_i}{\partial x_k}}. \tag{5.3}$$

The physics of isotropic decay as revealed by these equations consists of the decay of K at the rate specified by ϵ in Eq. (5.1) and the evolution of ϵ determined by the balance of the two terms on the right-hand side of either Eq. (5.2) or Eq. (5.3). The latter terms represent the effect of vortex stretching – in either P_ϵ^4 or νP_ζ^4 – and dissipation of ϵ – in either $-\Upsilon_\epsilon$ or $-\nu\Upsilon_\zeta$. Despite the formidable appearance of the terms in the ϵ equation, they are amenable to great simplification after the isotropy condition is applied to them. This will also aid in determining, in fact, that the P_ϵ, or equivalently νP_ζ, terms are strictly positive and thus represent production of dissipation.

The first step in simplifying Eq. (5.2) is to express the velocity derivative correlations in terms of $\mathcal{R}_{ij}$ and $S_{il,i}$ given in Eqs. (2.30) and (2.32), by extending the identity in Eq. (4.6) to include additional derivatives. In the case of the last correlation in Eq. (5.2) a straightforward calculation yields

$$\overline{\frac{\partial^2 u_i}{\partial x_j \partial x_l}(\mathbf{x})\frac{\partial^2 u_i}{\partial y_j \partial y_l}(\mathbf{y})} = \frac{\partial^4 \mathcal{R}_{ii}}{\partial r_j^2 \partial r_l^2}(\mathbf{y}-\mathbf{x}), \tag{5.4}$$

where $\mathbf{r} = \mathbf{y} - \mathbf{x}$. Taking the limit as $\mathbf{y} \to \mathbf{x}$ then gives

$$\overline{\left(\frac{\partial^2 u_i}{\partial x_j \partial x_l}\right)^2} = \frac{\partial^4 \mathcal{R}_{ii}}{\partial r_j^2 \partial r_l^2}(0). \tag{5.5}$$

To find a similar simplification of the triple velocity derivative correlation in Eq. (5.2), differentiate $S_{il,i}$ in Eq. (2.32) according to $\partial^3/\partial x_j \partial x_l \partial y_j$, yielding

$$\overline{u_l(\mathbf{x})\frac{\partial^2 u_i}{\partial x_j \partial x_l}(\mathbf{x})\frac{\partial u_i}{\partial y_j}(\mathbf{y})} + \overline{\frac{\partial u_i}{\partial x_l}(\mathbf{x})\frac{\partial u_l}{\partial x_j}(\mathbf{x})\frac{\partial u_i}{\partial y_j}(\mathbf{y})} = -\frac{\partial^3 S_{il,i}}{\partial r_l \partial r_j^2}(\mathbf{y}-\mathbf{x}), \tag{5.6}$$

where the continuity equation (2.1) has been used. Letting $\mathbf{y} \to \mathbf{x}$ gives

$$\overline{\frac{\partial u_i}{\partial x_l}\frac{\partial u_l}{\partial x_j}\frac{\partial u_i}{\partial x_j}} = \frac{\partial^3 S_{il,i}}{\partial r_l \partial r_j^2}(0) \tag{5.7}$$

since after using the incompressibility condition again

$$\overline{u_l \frac{\partial^2 u_i}{\partial x_j \partial x_l} \frac{\partial u_i}{\partial x_j}} = \frac{1}{2} \frac{\partial}{\partial x_l} \left[\overline{u_l \left(\frac{\partial u_i}{\partial x_j} \right)^2} \right] = 0 \tag{5.8}$$

in homogeneous turbulence.

To summarize the results thus far, according to Eqs. (5.5) and (5.7), Eq. (5.2) may be written as

$$\frac{d\epsilon}{dt} = -2\nu \frac{\partial^3 S_{il,i}}{\partial r_l \partial r_j^2}(0) - 2\nu^2 \frac{\partial^4 \mathcal{R}_{ii}}{\partial r_j^2 \partial r_l^2}(0). \tag{5.9}$$

To simplify this relation further, substitute Eq. (4.34) for $\mathcal{R}_{ij}$ in the last term, yielding

$$\frac{\partial^4 \mathcal{R}_{ii}}{\partial r_j^2 \partial r_l^2}(\mathbf{r}) = \overline{u^2} \left[\frac{24}{r} f'''(r) + 11 f^{iv}(r) + r f^{v}(r) \right]. \tag{5.10}$$

To evaluate this at $r = 0$, as required by Eq. (5.9), note that a Taylor series expansion of $f'''(r)$ gives

$$f'''(r) = r f^{iv}(0) + \frac{r^3}{3!} f^{vi}(0) + \ldots \tag{5.11}$$

since f is an even function of r. Consequently,

$$\lim_{r \to 0} \frac{f'''(r)}{r} = f^{iv}(0), \tag{5.12}$$

and finally, from Eqs. (5.10) and (5.12), it follows that

$$\frac{\partial^4 \mathcal{R}_{ii}}{\partial r_j^2 \partial r_l^2}(0) = 35 \overline{u^2} f^{iv}(0). \tag{5.13}$$

A similar calculation may be done using Eq. (4.38) to simplify the first term on the right-hand side of Eq. (5.9). The result is

$$\frac{\partial^3 S_{il,i}}{\partial r_l \partial r_j^2}(0) = \frac{35}{2} u_{rms}^3 k'''(0). \tag{5.14}$$

Later it will be seen that $k'''(0)$ tends to be negative, justifying the view that P_ϵ^4 represents production of ϵ via vortex stretching. Applying this and the previous result to Eq. (5.9) gives

$$\frac{d\epsilon}{dt} = -35 \nu u_{rms}^3 k'''(0) - 70 \nu^2 \overline{u^2} f^{iv}(0). \tag{5.15}$$

Thus, despite the complexity of Eq. (5.2), it is seen that the ϵ equation in isotropic turbulence depends on only the two time-dependent scalars $k'''(0)$ and $f^{iv}(0)$ besides K and ϵ.

Additional use of the formalism in Eq. (4.6) reveals that

$$\overline{\left(\frac{\partial u}{\partial x} \right)^3} = u_{rms}^3 k'''(0) \tag{5.16}$$

and

$$\overline{\left(\frac{\partial^2 u}{\partial x^2} \right)^2} = u_{rms}^2 f^{iv}(0), \tag{5.17}$$

relations which help expose the meaning of $k'''(0)$ and $f^{iv}(0)$. The skewness of the velocity derivative field $\partial u/\partial x$ is defined by

$$S_K \equiv -\frac{\overline{\left(\frac{\partial u}{\partial x}\right)^3}}{\overline{\left(\frac{\partial u}{\partial x}\right)^2}^{\frac{3}{2}}} \tag{5.18}$$

and this provides an opportunity to further refine the meaning of $k'''(0)$. Thus, according to Eqs. (4.8), (4.106), and (4.107),

$$\overline{\left(\frac{\partial u}{\partial x}\right)^2} = \frac{u_{rms}^2}{\lambda_g^2}. \tag{5.19}$$

From Eqs. (5.16) and (5.19) it follows that Eq. (5.18) becomes

$$k'''(0) = -S_K/\lambda_g^3 = -S_K\left(\frac{\epsilon}{15u_{rms}^2\nu}\right)^{3/2} \tag{5.20}$$

where the second equality comes from Eq. (4.45).

In a similar vein the *palenstrophy* coefficient is defined via

$$G = \frac{\overline{u^2}\,\overline{\left(\frac{\partial^2 u}{\partial x^2}\right)^2}}{\overline{\left(\frac{\partial u}{\partial x}\right)^2}^2}. \tag{5.21}$$

In this case, Eqs. (5.17) and (5.19) imply that

$$f^{iv}(0) = G/\lambda_g^4 = G\left(\frac{\epsilon}{15u_{rms}^2\nu}\right)^2. \tag{5.22}$$

Substituting Eqs. (5.20) and (5.22) into (5.15), and using the fact that $K = \frac{3}{2}u_{rms}^2$ gives the ϵ equation for homogeneous isotropic turbulence in the standard form

$$\frac{d\epsilon}{dt} = S_K^* R_T^{\frac{1}{2}}\frac{\epsilon^2}{K} - G^*\frac{\epsilon^2}{K}, \tag{5.23}$$

where the convenient definitions

$$S_K^* = \frac{7}{3\sqrt{15}}S_K, \tag{5.24}$$

$$G^* = \frac{7}{15}G, \tag{5.25}$$

and

$$R_T = \frac{K^2}{\nu\epsilon}, \tag{5.26}$$

have been made. Here, R_T is a dimensionless parameter which may be interpreted as a Reynolds number, as will be seen in the next section.

The coupled system of equations (5.1) and (5.23) represent two equations in the four unknowns, K, ϵ, S_K^*, and G^*, and thus is not closed. The initial state of the turbulence is specified by assigning values to each of these variables at $t = 0$, say $K_0, \epsilon_0, S_{K_0}^*$, and

G_0^*. Alternatively, in view of Eqs. (5.20) and (5.22), initial forms for $f(r)$ and $k(r)$ can be specified from which S_{K_0} and G_0 can be obtained.

5.1.1 Turbulent Reynolds Number

The parameter R_T defined in Eq. (5.26) can be regarded as a turbulent Reynolds number formed from the velocity scale $\sqrt{K}$ and length scale $K^{3/2}/\epsilon$. In view of Eq. (4.46) it is clear that this length scale is equivalent to that associated with the eddy turnover process. In the present context the eddy turnover time is $T_t = K/\epsilon$, which in view of the fact that

$$\frac{1}{T_t} = \frac{\epsilon}{K} = -\frac{1}{K}\frac{dK}{dt} \tag{5.27}$$

can be interpreted as the fractional rate at which energy is being dissipated. Consequently, T_t is the time scale over which a significant fraction of the turbulent kinetic energy dissipation might occur. For example, if K were decaying exponentially according to $K(t) = K(0)e^{-t/\tau}$, then a calculation with Eq. (5.27) gives $T_t = \tau$, so that in this case T_t is the time over which K falls by the factor $1/e$.

Since R_T can be written as the quotient $(K/\epsilon)/(\nu/K)$ it has a second interpretation as the ratio of turbulent and viscous time scales T_t/T_μ, where $T_\mu = \nu/K$ is a characteristic time scale over which viscous dissipation can be expected to act. Large R_T means that the turbulence is very energetic and far from being dissipated, since the time scale over which significant energy is being lost, T_t, is much larger than that over which the smallest dissipation scales can act. In this case, many small dissipative time units must come and go before there is a significant change in the energy. When R_T is small, the energy must be mainly in the dissipation range since the rate at which the energy of the flow drops matches the rate at which it is being dissipated. This is a sign that the turbulence may be considered to be weak. This also explains why R_T decreases to zero during the decay of isotropic turbulence. The appearance of R_T in the stretching term in Eq. (5.23) suggests that the latter will tend to be large when the flow is energetic and turbulence has not yet filled out the dissipative range of scales.

Another useful turbulence Reynolds number is that defined using the microscale, λ_g. In this case

$$R_\lambda = \frac{\lambda_g u_{rms}}{\nu}, \tag{5.28}$$

which can be related to R_T via Eq. (4.45), yielding

$$R_T = \frac{3}{20} R_\lambda^2. \tag{5.29}$$

R_T and R_λ may thus be used interchangeably to characterize the degree to which a homogeneous field is turbulent. $R_\lambda > 100$ is a useful measure of the range of turbulence that is no longer weak and $R_\lambda = 1000$ is strong turbulence while $R_\lambda < 1$ signifies very weak turbulence. In fact, flow with $R_\lambda < 1$ is referred to as the "final period" and represents the last stage of turbulence before the flow relaminarizes. It has been studied in some detail by Batchelor and Townsend [2] and others. Our interest in the decay process centers on following the history of a turbulent field as it changes from an initial state with large R_T to the final period where $R_T < 1$.

Before considering the decay process further, it is useful to note that Eqs. (5.1) and (5.23) can be combined into a single equation for R_T. This is derived by first computing from Eq. (5.26),

$$\frac{dR_T}{dt} = \frac{2K}{\nu\epsilon}\frac{dK}{dt} - \frac{K^2}{\nu\epsilon^2}\frac{d\epsilon}{dt}, \tag{5.30}$$

and then substituting Eqs. (5.1) and (5.23) to obtain

$$\frac{dR_T}{dt} = -\frac{2K}{\nu} - S_K^* \sqrt{R_T}\frac{K}{\nu} + G^*\frac{K}{\nu}. \tag{5.31}$$

Since K and ϵ are always positive, a dimensionless time can be unambiguously defined via

$$\tau(t) = \int_0^t \frac{\epsilon(t')}{K(t')}dt', \tag{5.32}$$

where for convenience this is constructed having $\tau(0) = 0$. In view of Eq. (5.1), (5.32) integrates exactly to yield

$$\tau(t) = \ln(K(0)/K(t)). \tag{5.33}$$

Since Eq. (5.1) implies that K decays monotonically to zero, $\tau(t)$ is well defined and, in particular, $\tau \to \infty$ as $t \to \infty$.

Equation (5.32) is a mapping from t to τ. Let $t(\tau)$ be the inverse mapping of τ to t, so that $\tau(t(\tau)) = \tau$. Defining

$$R_T^*(\tau) = R_T(t(\tau)), \tag{5.34}$$

or equivalently

$$R_T^*(\tau(t)) = R_T(t), \tag{5.35}$$

a calculation gives

$$\frac{dR_T}{dt} = \frac{dR_T^*}{d\tau}\frac{d\tau}{dt} = \frac{\epsilon}{K}\frac{dR_T^*}{d\tau} \tag{5.36}$$

using Eq. (5.32). Substituting Eq. (5.36) into (5.31) gives an equation for R_T^* in the form

$$\frac{dR_T^*}{d\tau} = R_T^*\left(G^* - 2 - S_K^*\sqrt{R_T^*}\right). \tag{5.37}$$

Thus, as an alternative to solving the decay problem via the coupled system of equations (5.1) and (5.23) there is the option of solving the single differential equation (5.37). Of course, G^* and S_K^* are time dependent and must be found by additional considerations. Thus, no matter which way the decay problem is approached, further progress in solving for K and ϵ requires that additional assumptions be made so that a closed system of equations can be deduced.

5.2 Modes of Isotropic Decay

The previous section has gone as far as possible in rigorously deriving a mathematical framework for predicting the isotropic decay problem. Any further progress depends on applying additional assumptions concerning the physics of the decay process to achieve closed relations that predict either all or part of the regimes of the decay. The limits of isotropic decay both at low Reynolds numbers in the final period and at high Reynolds numbers have been the subject of considerable analysis. Several competing theories attempt to characterize decay in these different Reynolds number realms and a variety of experiments and numerical simulations have been performed in order to discover what laws the decay rate follows. A common element that unites this work is the belief that the decay satisfies a power law in time of the form $K \sim (t - t_0)^{-n}$ with t_0 a virtual time origin. Different theories claim different values for the exponent n as well as other statistical quantities of the decay process.

One approach to the decay problem is to assume full or partial similarity [3, 4] as the flow evolves. In the high Reynolds number realm a $K \sim t^{-1}$ decay law is predicted. In contrast, a formalism developed by Saffman [5] leads to a $K \sim t^{-6/5}$ decay law (see also Problem 5.5) as a result of examining the consequences of initiating homogeneous turbulence via the stirring of the fluid via a random field of impulsive forces. The verification of these or any other theory depends on data for the energy decay rate and the values of various flow statistics including, in particular, moments of the velocity correlation function $f(r)$. The capabilities of numerical and physical studies in deciding between the different decay theories is affected by a number of intrinsic limitations to such approaches that place qualifications on their results. For example, as mentioned previously, numerical simulations in a finite periodic domain are constrained by the inevitable interactions between the physical length scale of the turbulence and the box size. Physical experiments depend on such devices as equating decay time to distance downstream of a grid and they are affected by probe errors and flow anisotropy. All methodologies for investigating isotropic turbulence must contend with the difficulty of establishing unbiased homogeneous isotropic initial conditions, particularly since studies show that properties of the initial state have a tendency to linger within the subsequent decay.

For the purposes of this exposition, further analysis of the decay problem given by Eqs. (5.1) and (5.23) will be pursued by invoking self-similarity since this provides a convenient viewpoint with which to consider a number of the physical aspects of the decay process. Where appropriate, some consequences deriving from alternative viewpoints will be considered.

5.3 Self-Similarity

The isotropic decay problem assumes homogeneous conditions over all space so that there is no externally imposed geometric length scale upon which the decay process should depend. If this decay is to be self-similar it requires that while the length scale of the turbulence changes with time, its structural features as reflected in the multi-point velocity correlation functions retain the same form independent of time. While there may not be an imposed length scale on the isotropic decay, nonetheless the turbulent field has two intrinsic length scales associated with the isotropic motion, namely, scales in the high wave number dissipation range depending on viscosity and an integral length scale of turbulence associated with the larger energy-containing eddies. The fact of two different scales raises some questions as to the likelihood that a similarity solution can exist. Nonetheless it is instructive to investigate the hypothesis of self-similarity to see where it leads.

Proceeding formally, the question at hand is whether or not there may exist functions $\tilde{f}(s)$ and $\tilde{k}(s)$ and a length scale $L(t)$ so that the two-point double and triple velocity correlation functions can be written as

$$f(r,t) = \tilde{f}(r/L(t)) \tag{5.38}$$

and

$$k(r,t) = \tilde{k}(r/L(t)) \tag{5.39}$$

during the decay. The turbulent decay has complete similarity or is "self-similar" or "self-preserving" if Eqs. (5.38) and (5.39) hold for all r and has "partial" or "incomplete" similarity if they are satisfied for a limited range of r. It may also be the case that similarity only holds for certain Reynolds number ranges during the decay. Both complete and partial similarity have been treated in the literature and a complete understanding of the role of similarity remains to be fully understood. Here, the implications of complete self-similarity are considered and it will be seen that with this assumption the isotropic decay problem becomes solvable, though some caution must be exercised in interpreting the physical meaning of the solutions, particularly as they apply to the entire decay process starting from large R_T.

Substituting Eq. (5.38) into (2.45) yields

$$\frac{\lambda_f^2}{L^2} = \frac{-2}{\tilde{f}''(0)} \tag{5.40}$$

where $\tilde{f}''(0)$ is a constant in time under the self-similarity assumption in Eq. (5.38). Equation (5.40) implies that $L \sim \lambda_f$, so the choice of L is not arbitrary. In view of Eq. (4.107) without loss of generality it is convenient to take

$$L = \lambda_g. \tag{5.41}$$

In this case it also follows from Eq. (5.39) that

$$k'''(0) = \tilde{k}'''(o)/\lambda_g^3. \tag{5.42}$$

Comparing Eq. (5.42) with Eq. (5.20) shows that

$$-S_K = \tilde{k}'''(0) \tag{5.43}$$

and, similarly, Eqs. (5.38) and (5.22) imply that

$$G = \tilde{f}^{iv}(0). \tag{5.44}$$

In other words, both S_K and G are constants during self-similar decay. In this case, the self-similar decay of isotropic turbulence is governed by the *closed* system of equations

$$\frac{dK}{dt} = -\epsilon \tag{5.45}$$

$$\frac{d\epsilon}{dt} = S_{K_0}^* R_T^{\frac{1}{2}} \frac{\epsilon^2}{K} - G_0^* \frac{\epsilon^2}{K} \tag{5.46}$$

where $S^*_{K_0}$ and G^*_0 are given by Eqs. (5.43) and (5.44) as constant values during the decay process. $S^*_{K_0}$ and G^*_0 may either be assigned values or deduced from given forms of $\tilde{f}$ and $\tilde{k}$.

In the case of complete similarity, Eq. (5.37) becomes the single solvable equation for $R^*_T(\tau)$:

$$\frac{dR^*_T}{d\tau} = R^*_T\left(G^*_0 - 2 - S^*_{K_0}\sqrt{R^*_T}\right). \tag{5.47}$$

Our focus now is on investigating the properties of solutions to Eq. (5.47) or, equivalently, (5.45) and (5.46) from an initial state with $R^*_T(0) >> 1$ to a time when $R^*_T << 1$.

5.3.1 Fixed Point Analysis

A useful means [4] for discovering the general properties of solutions to Eq. (5.47) is to consider its "fixed points", that is, those values of R^*_T where the right-hand side is zero and hence R^*_T does not change in time. To the extent that such points represent "attracting" solutions in the sense that solutions with different initial values of R_T travel towards them as $t \to \infty$, they can be used to gain insights into the general behavior of solutions to isotropic decay. An important aspect of this analysis is that even if the decay were not exactly self-similar so that G^*_0 and $S^*_{K_0}$ were to change somewhat in time, it can be shown [6] that the fixed points of the equation are stable nodes of the dynamical system so that the character of the solution to the K and ϵ equations is unaffected by small changes in G^* and S^*_K during the decay. Consequently, the results of this section are likely to provide a meaningful description of isotropic decay even when the self-similar hypothesis is not applied to Eqs. (5.1) and (5.23).

Denoting the fixed points of Eq. (5.47) as $R^*_{T_\infty}$, a calculation gives

$$R^*_{T_\infty}\left(G^*_0 - 2 - S^*_{K_0}\sqrt{R^*_{T_\infty}}\right) = 0, \tag{5.48}$$

which is satisfied by either

$$R^*_{T_\infty} = 0 \tag{5.49}$$

or

$$R^*_{T_\infty} = \left(\frac{G^*_0 - 2}{S^*_{K_0}}\right)^2. \tag{5.50}$$

Both of these solutions are attracting since solutions starting from arbitrary values of R^*_T will move toward one or the other of the two fixed points depending on the value of G^*_0. For example, Eq. (5.49) is reached for all initial states if $G^*_0 \leq 2$, while Eq. (5.50) is reached for all initial conditions if $G^*_0 > 2$. These behaviors are evident from examination of Eq. (5.47). Thus, when $G^*_0 < 2$, the term in parenthesis on the right-hand side is always negative and R^*_T must decay to zero. For $G^*_0 > 2$, the right-hand side of Eq. (5.47) is positive if $R_{T_0} < R^*_{T_\infty}$ and negative if $R_{T_0} > R^*_{T_\infty}$. In either case the tendency of the right-hand side of the equation is to raise or lower R^*_T until it converges to $R^*_{T_\infty}$ given by Eq. (5.50).

The two fixed-point solutions represent, in essence, different equilibrium states of turbulence during isotropic decay. Assuming only that self-similarity is maintained during the decay, one or other of these states is achieved depending on the initial value of G^*. Before considering the properties of the equilibrium solutions and what they suggest about the nature of the decay process, it should be noted that the solution for which Eq. (5.50) holds is one in which the asymptotic state of the flow field at large t does not have $R_T^* = 0$. This suggests that if the decay is completely self-similar and is constrained by the fact that it must end up in the final period solution, where $R_T^* \rightarrow 0$, then it cannot be that $G_0^* > 2$. On the other hand, it will be seen below that solutions for $G_0^* < 2$ cannot reasonably account for decay from an initial large Reynolds number to the final period. Thus, the conclusion is reached that self-similarity throughout the decay process is not feasible. This result is also anticipated by the fact that the similarity length scale had to be $\sim \lambda_g$, which describes dissipative motions in the final period but cannot be expected to characterize a vortex stretching dominated process at high Reynolds numbers. These observations do not necessarily mean that the separate equilibria do not have physical significance, rather it means they must apply separately to different regimes of the overall decay process. If so, at some point in time in the decay process there must be a shift between them in which G_0^* changes value.

5.3.2 Final Period of Isotropic Decay

The equilibrium solution with $R_{T_\infty}^* = 0$ represents the end point of the decay process in which there is no motion. Prior to reaching this state $R_T << 1$ but is not exactly zero and it is of interest to examine the properties of turbulence in this very weakened condition. Rewriting Eq. (5.46) in the form

$$\frac{d\epsilon}{dt} = \left(\frac{S_{K_0}^* R_T^{\frac{1}{2}}}{G_0^*} - 1 \right) G_0^* \frac{\epsilon^2}{K}, \tag{5.51}$$

it is clear that the term containing $R_T^{\frac{1}{2}}$, representing vortex stretching, can be neglected when

$$\frac{S_{K_0}^* R_T^{\frac{1}{2}}}{G_0^*} << 1. \tag{5.52}$$

Low Reynolds number experiments have measured typical values of $S_K \approx 0.5$ and $G \approx 3$ so that $S_K^* \approx 0.3$ and $G_0^* \approx 1.4$, and in particular, $G_0^* < 2$. Equation (5.52) is reasonably well satisfied when $R_T < 0.1$. This implies that in the vicinity of $R_T = 0$, the coupled equations for K and ϵ reduce to

$$\frac{dK}{dt} = -\epsilon \tag{5.53}$$

and

$$\frac{d\epsilon}{dt} = -G_0^* \frac{\epsilon^2}{K}. \tag{5.54}$$

These have the exact solution

$$\frac{K}{K_0} = \left(1 + \frac{t}{\alpha T_{t_0}} \right)^{-\alpha} \tag{5.55}$$

$$\frac{\epsilon}{\epsilon_0} = \left(1 + \frac{t}{\alpha T_{t_0}}\right)^{-1-\alpha} \tag{5.56}$$

where

$$\alpha = \frac{1}{G_0^* - 1}, \tag{5.57}$$

and K_0, ϵ_0, and $T_{t_0} = K_0/\epsilon_0$ are the initial values of the flow properties. After t advances several multiples of T_{t_0}, so that $t/(\alpha T_{t_0}) >> 1$, then Eqs. (5.55) and (5.56) become simple power laws

$$K \sim t^{-\alpha} \tag{5.58}$$

and

$$\epsilon \sim t^{-1-\alpha}, \tag{5.59}$$

respectively. Thus, for $G_0^* < 2$, so that $\alpha > 1$, and R_T small, the self-similar solution for K and ϵ consists of power laws with exponent depending on G_0^*.

In the final sections of this chapter some aspects of the two-point correlation functions $f(r)$ and $k(r)$ during isotropic decay will be considered. It suffices for the present to mention that physical experiments [2], such as that illustrated in Figure 5.1, as well as the theoretical discussion considered below, suggest that in the final period of decay f is given by the Gaussian function

$$f(r,t) = e^{-\frac{r^2}{2\lambda_g^2}}. \tag{5.60}$$

Substituting this into Eq. (5.44) yields $G = 3$, so that $G_0^* = 7/5$ according to Eq. (5.25), and, by (5.57), $\alpha = 5/2$. According to Eq. (5.58) this means that K obeys a $-5/2$ decay law in the final period if the decay is self-preserving. The experimental results given

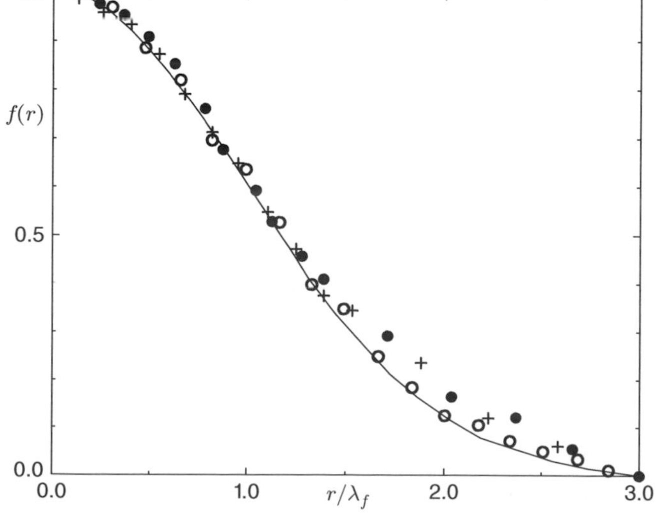

Figure 5.1 Measured and predicted $f(r/\lambda_f)$ in the final period [2]. With permission of the Royal Society.

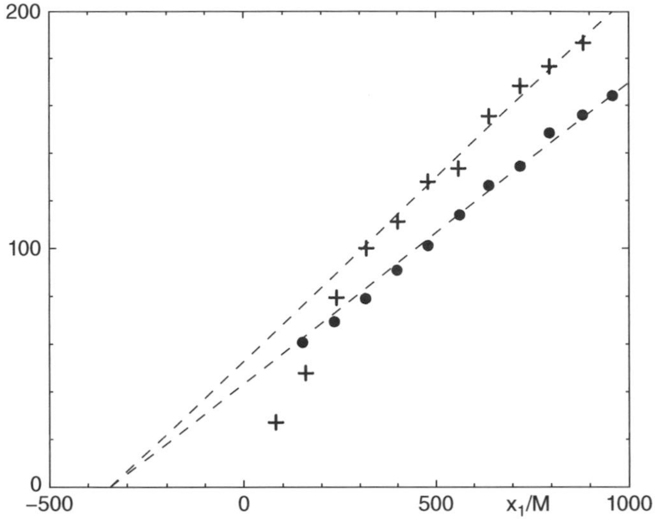

Figure 5.2 Confirmation of the −5/2 decay law in the final period [2]. •, $(U_m^2/\overline{u^2})^{2/5}$; +, λ_f^2. With permission of the Royal Society.

in Figure 5.2, point towards showing that this is indeed the case. In this experiment, turbulence is generated downstream of a mesh of bars with spacing M. The decay time is associated with distance along the wind tunnel x_1 via the mapping $t = x_1/U_m$, where U_m is the mean velocity of the uniform flow. The linear behavior of $(U_m^2/\overline{u^2})^{2/5}$ in the figure implies that $\overline{u^2}$ satisfies a −5/2 decay law and so too will K in isotropic turbulence. It can be concluded from this that the equilibrium solution associated with Eq. (5.49) when $G_0^* = 7/5$ is consistent with data for the final period solution.

Rearranging Eq. (4.45) and using the definition of K gives

$$\lambda_g^2 = \frac{10\nu K}{\epsilon} \tag{5.61}$$

and substituting Eq. (5.58) and (5.59) into Eq. (5.61) shows that in the final period

$$\lambda^2 \sim t, \tag{5.62}$$

where λ can be either λ_f or λ_g, so that λ is growing proportional to $\sqrt{t}$. This can be explained by the expectation that the small-scale motions are vanishing faster than the large-scale motions, causing the scale of the surviving turbulence to increase. As the fluid reverts to a laminar state, $f(r)$ becomes 1 and λ moves to ∞. The behavior of λ in the final period is supported experimentally, as seen in Figure 5.2, where λ^2 (in this case λ_f) varies linearly with x_1. Another interesting facet of the final period is that according to Eqs. (4.33) and (5.60)

$$g(r,t) = \left(1 - \frac{r^2}{2\lambda_g^2}\right) e^{-\frac{r^2}{2\lambda_g^2}}, \tag{5.63}$$

a function which contains a negative correlation when $r > \sqrt{2}\lambda_g$.

While it is clear from this discussion that $G_0^* < 2$ in the final period when R_T is small, it cannot be the case that this value also holds for large R_T. To see this, it is a matter of demonstrating that the solution to Eq. (5.47) is highly non-physical when $G_0^* < 2$ and R_{T_0} is large. Equation (5.47) can either be solved numerically (e.g., using *ode45* in MATLAB) or it can be analyzed via its exact solution

$$R_T^* = R_{T_0}^* \left[\frac{(G_0^* - 2)\exp((G_0^* - 2)\tau/2)}{G_0^* - 2 - S_{K_0}^* \sqrt{R_{T_0}^*}(1 - \exp((G_0^* - 2)\tau/2))} \right]^2 \tag{5.64}$$

when $G_0^* \neq 2$ and

$$R_T^* = R_{T_0}^* \left[\frac{1}{1 + S_{K_0}^* \sqrt{R_{T_0}^*}\tau/2} \right]^2 \tag{5.65}$$

when $G_0^* = 2$. When $G_0^* < 2$, $R_{T_0}^*$ is large, and τ is large enough, Eq. (5.64) gives

$$R_T^* \sim e^{(G_0^* - 2)\tau}, \tag{5.66}$$

which will tend to be small regardless of its initial value $R_{T_0}^*$. Thus, if a solution starts with large $R_{T_0}^*$, R_T^* itself will rapidly become small as τ increases, and in view of Eq. (5.33) this happens when $K << K_0$, that is, most of the energy has gone. An example of this solution is given in Figure 5.3, where Eqs. (5.45) and (5.46) scaled by K_0 and ϵ_0 have been solved numerically. It is seen that at the outset the decay consists of an unphysical precipitous drop in K. Time is scaled in this case by the initial eddy turnover time T_{t_0}, so that it is seen that K falls 5 orders of magnitude in one such time unit. In the same time period R_T swiftly falls to values representative of the final period. Taken together, the conclusion is reached that the fixed-point solution associated with Eq. (5.49) cannot apply to the initial, high Reynolds number stage of isotropic decay.

In contrast to the $-5/2$ decay law predicted by the self-similar theory, the approach in [5] leads to a $-3/2$ decay law. While the classical data in Figure 5.2 give support to the

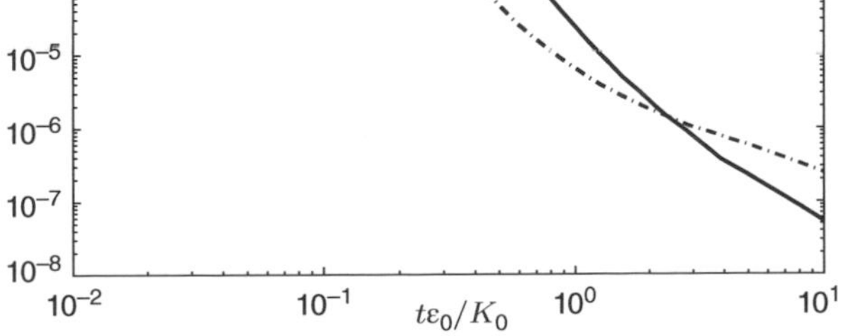

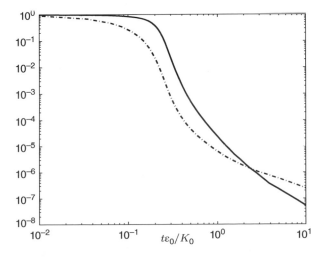

Figure 5.3 Self-similar decay corresponding to $G_0^* = 7/5$ and $R_{T_0}^* = 1000$: —, K/K_0; $- \cdot -$, R_T/R_{T_0}.

−5/2 law, the same data have been used to support other decay laws, including the −3/2 law [1], with this effort helped by the choice of virtual origin. As pointed out previously, the rise in length scale, particularly λ_f in this case, is a significant obstacle to acquiring a definitive result. For example, some studies [7, 8] suggest a decay rate of −2 is expected in computations on a periodic domain due to the interference of the finite box size when λ_f becomes large enough. This can color the interpretation of results in the final period determined from computational studies.

5.3.3 High Reynolds Number Equilibrium

Now consider the decay associated with the second equilibrium solution given by Eq. (5.50). When R_T reaches this value during self-similar decay with $G_0^* > 2$, Eq. (5.46) reduces to

$$\frac{d\epsilon}{dt} = (G_0^* - 2)\frac{\epsilon^2}{K} - G_0^* \frac{\epsilon^2}{K} \tag{5.67}$$

and thus

$$\frac{d\epsilon}{dt} = -2\frac{\epsilon^2}{K}. \tag{5.68}$$

According to Eq. (5.50), $R_{T_\infty} >> 1$ requires that G_0^* be relatively large as long as S_{K_0} does not vary significantly away from its typical measured values. As an example, $R_{T_\infty} = 10^4$ coincides with $G_0^* \approx 25$. In this case the right-hand side of Eq. (5.67) consists of a near balance between vortex stretching, with coefficient $G_0^* - 2$, and viscous dissipation, with coefficient $-G_0^*$. The slight edge toward the latter leaves a net dissipation with coefficient −2 so that Eq. (5.68) results. An important point is that the high Reynolds number equilibrium decay contains nearly equal contributions from vortex stretching and dissipation of dissipation. In this, the net dissipation rate assumes a universal value independent of initial conditions, including the value of G_0^*.

Similar to the analysis of the $R_{T_\infty} = 0$ case that leads to the solution in Eqs. (5.58) and (5.59), Eqs. (5.45) and (5.68) can be solved, this time yielding an asymptotic power law decay of the form [4]

$$K \sim t^{-1} \tag{5.69}$$

$$\epsilon \sim t^{-2}, \tag{5.70}$$

where, unlike the $R_{T_\infty} = 0$ case, the decay rate is independent of the initial conditions. In general, $R_{T_0} \neq R_{T_\infty}$, so a transient period occurs during which time the coefficient $S_{K_0}\sqrt{R_T}$ in Eq. (5.46) changes to $S_{K_0}\sqrt{R_{T_0}} = G_0^* - 2$. The larger the discrepancy between R_{T_0} and R_{T_∞}, the farther the initial solution is from equilibrium and the longer the time interval until equilibrium is reached. However, if $G_0^* - 2$ is large, then so too is $e^{(G_0^*-2)\tau/2}$ for relatively small τ. The exponential term in Eq. (5.64) then dominates the numerator and denominator and thus cancels, so that in lieu of Eq. (5.50) it follows from Eq. (5.64) that $R_T \approx R_{T_\infty}$ in a relatively short time, that is, just a few eddy turnover times.

To illustrate these results, Figure 5.4 shows the computed behavior of K, ϵ and R_T in a log–log plot for the relatively high value $G_0^* = 5$, where $R_{T_0} = 1000$ is purposefully taken

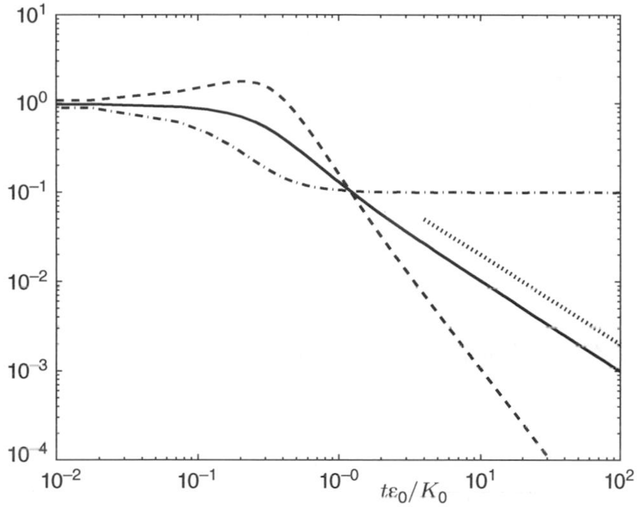

Figure 5.4 Self-similar decay corresponding to $G_0^* = 5$ and $R_{T_0}^* = 1000$: —, K/K_0; −−, ϵ/ϵ_0; − · −, R_T/R_{T_0}; · · ·, line of slope −1.

to be large and the value of $S_K^* = 0.3$, typically seen in experiments, is used. In contrast to the low G_0^* solution, the solution for K converges to the t^{-1} decay law and it does so relatively quickly within which time only a modest energy loss has occurred. R_T has an initially rapid drop, followed by convergence to the non-zero value given by Eq. (5.50). While R_T is falling, it may be noticed that K is relatively constant and ϵ is increasing. Consistent with these trends λ_g computed from Eq. (5.61) decreases during the initial decay, as shown in Figure 5.5. This is the natural outcome of having insufficient energy in the dissipation range scales at the outset of the simulations. The figure shows that the flow rapidly adjusts to the imbalance by sending energy to small scales via vortex stretching so that λ_g decreases and ϵ rises. After just a few eddy turnover times the

Figure 5.5 λ_g in isotropic decay corresponding to the conditions in Fig. 5.4.

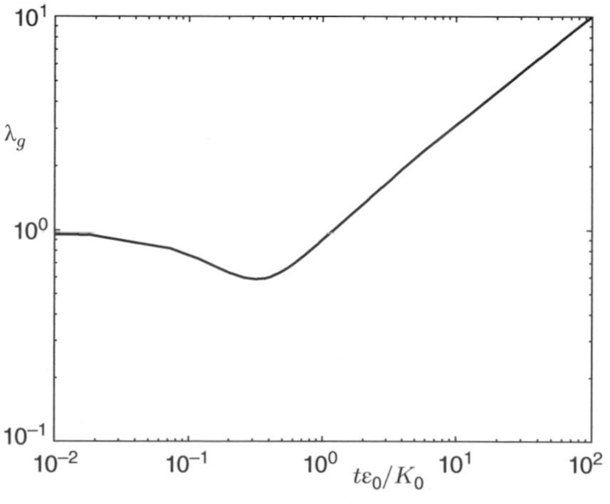

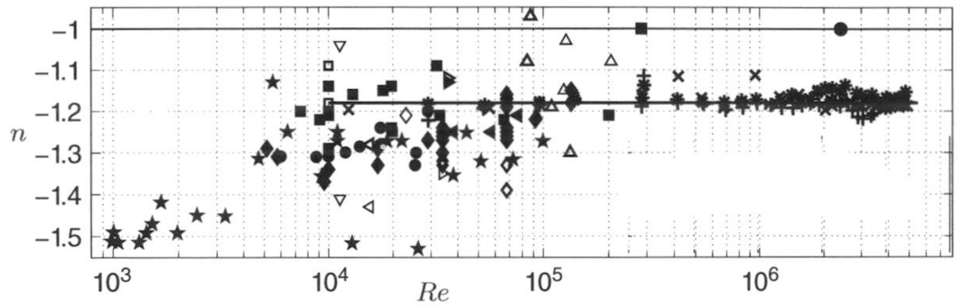

Figure 5.6 Measured power law exponents in decaying homogeneous turbulence from numerous experiments [9]. Filled symbols represent traditional turbulence behind a grid of bars. Open symbols are other turbulence sources. The symbols with ×,*, and + are from three different probes used in decaying turbulence behind a grid of bars for which the Reynolds number was changed only by altering the viscosity of the working fluid. Used by permission of AIP.

equilibrium between dissipation and stretching given in Eq. (5.50) is achieved and the energy begins to decay according to Eq. (5.68). By this time, as the turbulence weakens, λ_g is rising and ϵ is falling. Eventually, the cumulative energy decay is such that the high Reynolds number equilibrium can no longer be tenable and the decay must shift towards that of the final period. At this time self-similarity must be lost and G_0^* begins to decrease to 7/5.

While Figures 5.1 and 5.2 suggest that the $-5/2$ decay law in the final period is a true physical effect, the trends in the experimental record for decay at high Reynolds number have considerable scatter, as seen in Figure 5.6, where results from many experiments spanning a wide range of Reynolds numbers are gathered together [9]. Within this plot is the result of recent experiments for Reynolds numbers R_M varying from 1000 to 5×10^6, where $R_M = MU/\nu$ and M is the forcing scale. For these experiments M is the mesh spacing for flow past a grid of bars. The average decay rate from this study is -1.18, with experimental uncertainty of 0.02, which lends support for a $K \sim t^{-6/5}$ decay law. For the many individual trials of the experiment the decay rate showed little variation with Reynolds number and held largely constant down the length of the wind tunnel. These results may show the effect of flow inhomogeneities in the measurement region, which was relatively close to the generating grid [10].

Among the many other measurements in Figure 5.6 a small subset of the data supports the t^{-1} decay law at high Reynolds number. At somewhat lower Reynolds numbers many estimates of the decay rate are less than -1.2. Some sets of data from particular facilities may be trending toward the -1 decay law with increasing Reynolds number while others do not show definite trends. Since the -1 decay law can only be expected to appear after a sufficient interval for vortex stretching and dissipation to equilibrate, it remains unclear if the duration of many of the experiments is sufficiently long for a final determination of the decay exponent in the high Reynolds number limit. Complicating the significance of specific power law exponents measured during decay is the fact that they are known to generally depend on initial conditions [10] as well having a tendency to change values as the flow shifts from initial to final period decay laws, a phenomenon confirmed in physical experiments [11]. These behaviors are also readily reproduced in the transient flows observed in the numerical solutions to Eqs. (5.45) and (5.46) [4].

5.4 Implications for Turbulence Modeling

Though the analysis and modeling of the ϵ equation in the previous sections is primarily limited to isotropic decay, nonetheless the formalism established there is routinely taken to be the starting point for modeling the equivalent terms appearing in the ϵ equation under general circumstances. The two main results that were developed previously consist of the isotropic formulas

$$P_\epsilon^4 = S_K^* R_T^{\frac{1}{2}} \frac{\epsilon^2}{K} \tag{5.71}$$

and

$$\Upsilon_\epsilon = G^* \frac{\epsilon^2}{K}. \tag{5.72}$$

When these relations are expropriated for use in modeling the general ϵ equation the traditional route is to assume that the factors multiplying ϵ^2/K in Eqs. (5.71) and (5.72) satisfy

$$S_K^* R_T^{\frac{1}{2}} - G^* = -C_{\epsilon_2} \tag{5.73}$$

where C_{ϵ_2} is a constant. This leads to the model

$$P_\epsilon^4 - \Upsilon_\epsilon = (S_K^* R_T^{\frac{1}{2}} - G^*)\frac{\epsilon^2}{K} = -C_{\epsilon_2}\frac{\epsilon^2}{K}. \tag{5.74}$$

Note that this relation when used in Eq. (5.23) yields an equation identical to Eq. (5.68), except that the constant C_{ϵ_2} appearing on the right-hand side is not necessarily equal to 2. The rationale for assuming that Eq. (5.73) holds is based on a scaling argument, which is interesting to examine in the light of the previous results.

As may be inferred from Eqs. (5.2) and (5.3) $\Upsilon_\epsilon = \nu \Upsilon_\zeta$ in isotropic turbulence by virtue of the identity

$$\overline{\left(\frac{\partial^2 u_i}{\partial x_j \partial x_l}\right)^2} = \overline{\left(\frac{\partial \omega_l}{\partial x_j}\right)^2}, \tag{5.75}$$

which may be proven by substituting the definition of the vorticity on the right-hand side. Then from Eq. (5.72) and the fact that $\epsilon = \nu\zeta$, it follows that

$$G^* \sim \frac{\Upsilon_\zeta/\zeta}{\epsilon/K} \tag{5.76}$$

which says that G^* is the ratio of the fractional rate of change of enstrophy, namely Υ_ζ/ζ, to the fractional rate of change of energy given by ϵ/K. In effect, both numerator and denominator are inverse time scales.

Since enstrophy dissipation is influenced most strongly by small-scale phenomena it can be hypothesized that the numerator in Eq. (5.76) scales with the Kolmogorov dissipation time scale $(\nu/\epsilon)^{1/2}$ given in Eq. (4.40). Substituting this into Eq. (5.76) gives

$$G^* \sim \sqrt{R_T} \tag{5.77}$$

for large Reynolds numbers. This result is consistent with Eq. (5.73) though the latter goes well beyond Eq. (5.77) to assume a very specific linear relation between G^* and

$\sqrt{R_T}$, one that leads to exact cancellation of the vortex stretching term as seen by substituting for G^* using Eq. (5.73) in (5.23), giving

$$\frac{d\epsilon}{dt} = \frac{\epsilon^2}{K}\left(S_K^*\sqrt{R_T} - (S_K^*\sqrt{R_T} + C_{\epsilon_2})\right) = -C_{\epsilon_2}\frac{\epsilon^2}{K}. \tag{5.78}$$

In other words, the coefficient of $\sqrt{R_T}$ in Eq.(5.73) is chosen to be precisely what is needed to cancel the effect of vortex stretching from the ϵ equation.

In view of the discussion in the previous section it is clear that Eq. (5.73) is tantamount to imposing an equilibrium structure on the turbulent decay process, in fact one which forces an energy equilibrium decay law of the form

$$K \sim t^{-\frac{1}{C_{\epsilon_2}-1}}, \tag{5.79}$$

to occur. In other words, the choice of C_{ϵ_2} sets the decay rate. For $C_{\epsilon_2} = 11/6$ the Saffman $t^{-6/5}$ decay law is achieved. Any of the other experimental values can be reached similarly if desired. In all such cases, the flow decays at the given rate without a transitional phase involving the action of vortex stretching. In contrast, if vortex stretching is given an independent role, then the t^{-1} decay law will develop after a sufficient time interval.

Another questionable consequence of Eq. (5.78) is encountered in the case of vanishing viscosity. Here, Eq. (5.78) reduces to

$$\frac{d\epsilon}{dt} = 0 \tag{5.80}$$

as $\nu \to 0$, implying that ϵ is constant and thus so is ζ. This outcome is unphysical since it is expected that ζ must grow rapidly due to vortex stretching in the absence of dissipation. In fact, such behavior is observed in simulations of inviscid turbulent flow [12]. Moreover, in the inviscid limit it can be shown that Eq. (5.23) can be written as

$$\frac{d\zeta}{dt} = S_{K_\infty}^* \zeta^{3/2}, \tag{5.81}$$

where $S_{K_\infty}^*$ is the infinite Reynolds number limit of the skewness factor. The solution to Eq. (5.81) is

$$\zeta(t) = \zeta(0)\left(1 - \frac{S_{K_\infty}^*\sqrt{\zeta(0)}t}{2}\right)^{-2}, \tag{5.82}$$

which implies that in the absence of viscosity ζ will blow up at the finite time $t = 2/(S_{K_\infty}^*\sqrt{\zeta(0)})$. While this argument does not constitute proof about the behavior of ζ in turbulent flow, it does suggest that the dissipation model in Eq. (5.73) unnecessarily limits the range of physical phenomena that can be captured. More realism is attained by assuming a modification of Eq. (5.73) which allows for a residual effect of vortex stretching. This point is returned to in the discussion of homogeneous shear flow in Chapter 6.

5.5 Equation for Two-Point Correlations

The analysis of isotropic decay carried out in the previous section concentrates on tracing the history of K and ϵ as they change in time. Only minimal information about the flow structure was needed, in fact, just the skewness and palenstrophy coefficient that are related to the two-point correlation functions. To proceed to a more extensive analysis of the decay problem that includes analyzing the time dependence of G^* and S_K^* it is necessary to include dynamical information about multi-point correlations. This means introducing an equation for the time history of the two-point velocity correlation tensor $\mathcal{R}(\mathbf{r}, t)$ and then considering its form during isotropic decay. From such an analysis it is also possible to consider the spectral properties of the turbulence during the decay process.

An equation governing $\mathcal{R}_{ij}(\mathbf{x}, \mathbf{y}, t)$ for arbitrary incompressible flow is derived by taking the average of $u_i(\mathbf{x}, t)$ times the jth component of the Navier–Stokes equation in Eq. (2.2) at $\mathbf{y}$ and adding to this the same quantity with i and j and $\mathbf{x}$ and $\mathbf{y}$ reversed. The result is

$$\begin{aligned}
&\overline{\rho u_i(\mathbf{x}, t)\frac{\partial U_j}{\partial t}(\mathbf{y}, t)} + \overline{\rho u_j(\mathbf{y}, t)\frac{\partial U_i}{\partial t}(\mathbf{x}, t)} \\
&+ \overline{\rho u_i(\mathbf{x}, t)U_k(\mathbf{y}, t)\frac{\partial U_j}{\partial y_k}(\mathbf{y}, t)} + \overline{\rho u_j(\mathbf{y}, t)U_k(\mathbf{x}, t)\frac{\partial U_i}{\partial x_k}(\mathbf{x}, t)} = \\
&- \overline{u_i(\mathbf{x}, t)\frac{\partial p}{\partial y_j}(\mathbf{y}, t)} - \overline{u_j(\mathbf{y}, t)\frac{\partial p}{\partial x_i}(\mathbf{x}, t)} \\
&+ \overline{\mu u_i(\mathbf{x}, t)\nabla^2 U_j(\mathbf{y}, t)} + \overline{\mu u_j(\mathbf{y}, t)\nabla^2 U_i(\mathbf{x}, t)}.
\end{aligned} \tag{5.83}$$

Using the definition of $\mathcal{R}_{ij}$ given in Eq. (2.30) it follows that the first two terms on the left-hand side of Eq. (5.83) may be written as

$$\overline{\rho u_i(\mathbf{x}, t)\frac{\partial U_j}{\partial t}(\mathbf{y}, t)} + \overline{\rho u_j(\mathbf{y}, t)\frac{\partial U_i}{\partial t}(\mathbf{x}, t)} = \rho\frac{\partial \mathcal{R}_{ij}}{\partial t}(\mathbf{x}, \mathbf{y}, t) \tag{5.84}$$

since terms such as $\overline{u_i(\mathbf{x}, t)\partial \overline{U}_j(\mathbf{y}, t)/\partial t} \equiv 0$. The next two terms in Eq. (5.83), coming from the advection term, give

$$\begin{aligned}
&\overline{\rho u_i(\mathbf{x}, t)U_k(\mathbf{y}, t)\frac{\partial U_j}{\partial y_k}(\mathbf{y}, t)} + \overline{\rho u_j(\mathbf{y}, t)U_k(\mathbf{x}, t)\frac{\partial U_i}{\partial x_k}(\mathbf{x}, t)} = \\
&\overline{\rho u_i(\mathbf{x}, t)u_k(\mathbf{y}, t)}\frac{\partial \overline{U}_j}{\partial y_k}(\mathbf{y}, t) + \overline{\rho u_j(\mathbf{y}, t)u_k(\mathbf{x}, t)}\frac{\partial \overline{U}_i}{\partial x_k}(\mathbf{x}, t) + \\
&\rho\overline{U}_k(\mathbf{y}, t)\overline{u_i(\mathbf{x}, t)\frac{\partial u_j}{\partial y_k}(\mathbf{y}, t)} + \rho\overline{U}_k(\mathbf{x}, t)\overline{u_j(\mathbf{y}, t)\frac{\partial u_i}{\partial x_k}(\mathbf{x}, t)} + \\
&\overline{\rho u_i(\mathbf{x}, t)u_k(\mathbf{y}, t)\frac{\partial u_j}{\partial y_k}(\mathbf{y}, t)} + \overline{\rho u_j(\mathbf{y}, t)u_k(\mathbf{x}, t)\frac{\partial u_i}{\partial x_k}(\mathbf{x}, t)}.
\end{aligned} \tag{5.85}$$

The first two terms on the right-hand side of Eq. (5.85) are equal to

$$\rho\mathcal{R}_{ik}(\mathbf{x}, \mathbf{y}, t)\frac{\partial \overline{U}_j}{\partial y_k}(\mathbf{y}, t) + \rho\mathcal{R}_{jk}(\mathbf{y}, \mathbf{x}, t)\frac{\partial \overline{U}_i}{\partial x_k}(\mathbf{x}, t). \tag{5.86}$$

Furthermore, differentiation of Eq. (2.30) gives

$$\frac{\partial \mathcal{R}_{ij}}{\partial y_k}(\mathbf{x},\mathbf{y},t) = \overline{u_i(\mathbf{x},t)\frac{\partial u_j}{\partial y_k}(\mathbf{y},t)} \tag{5.87}$$

and similarly for x_k derivatives, so that the third and fourth terms on the right-hand side of Eq. (5.85) take the form of convection terms

$$\rho\overline{U}_k(\mathbf{y},t)\frac{\partial \mathcal{R}_{ij}}{\partial y_k}(\mathbf{x},\mathbf{y},t) + \rho\overline{U}_k(\mathbf{x},t)\frac{\partial \mathcal{R}_{ij}}{\partial x_k}(\mathbf{x},\mathbf{y},t). \tag{5.88}$$

As far as the last two terms on the right-hand side of Eq. (5.85) are concerned, they may be written using Eq. (2.31) as

$$\overline{\rho u_i(\mathbf{x},t)u_k(\mathbf{y},t)\frac{\partial u_j}{\partial y_k}(\mathbf{y},t)} = \rho\frac{\partial \mathcal{S}_{jk,j}}{\partial y_k}(\mathbf{y},\mathbf{x},t) \tag{5.89}$$

and

$$\overline{\rho u_j(\mathbf{y},t)u_k(\mathbf{x},t)\frac{\partial u_i}{\partial x_k}(\mathbf{x},t)} = \rho\frac{\partial \mathcal{S}_{ik,j}}{\partial x_k}(\mathbf{x},\mathbf{y},t) \tag{5.90}$$

where the fact that

$$\overline{u_i(\mathbf{x},t)\frac{\partial u_j}{\partial x_j}(\mathbf{x},t)u_k(\mathbf{y},t)} = 0 \tag{5.91}$$

has been used as implied by incompressibility.

To treat the contribution to Eq. (5.83) from the terms containing pressure, introduce the two-point pressure-velocity correlation vector

$$\mathcal{K}_i(\mathbf{x},\mathbf{y},t) = \overline{u_i(\mathbf{x},t)p(\mathbf{y},t)} \tag{5.92}$$

and see that

$$\overline{u_i(\mathbf{x},t)\frac{\partial p}{\partial y_j}(\mathbf{y},t)} + \overline{u_j(\mathbf{y},t)\frac{\partial p}{\partial x_i}(\mathbf{x},t)} = \frac{\partial \mathcal{K}_i}{\partial y_j}(\mathbf{x},\mathbf{y},t) + \frac{\partial \mathcal{K}_j}{\partial x_i}(\mathbf{y},\mathbf{x},t). \tag{5.93}$$

Finally, the viscous terms in Eq. (5.83) become

$$\overline{\mu u_i(\mathbf{x},t)\nabla^2 U_j(\mathbf{y},t)} + \overline{\mu u_j(\mathbf{y},t)\nabla^2 U_i(\mathbf{x},t)} = \mu\frac{\partial^2 \mathcal{R}_{ij}}{\partial y_k^2}(\mathbf{x},\mathbf{y},t) + \mu\frac{\partial^2 \mathcal{R}_{ij}}{\partial x_k^2}(\mathbf{x},\mathbf{y},t). \tag{5.94}$$

Putting the above results together it has been shown that Eq. (5.83) becomes

$$\begin{aligned}
&\frac{\partial \mathcal{R}_{ij}}{\partial t}(\mathbf{x},\mathbf{y},t) + \overline{U}_k(\mathbf{y},t)\frac{\partial \mathcal{R}_{ij}}{\partial y_k}(\mathbf{x},\mathbf{y},t) + \overline{U}_k(\mathbf{x},t)\frac{\partial \mathcal{R}_{ij}}{\partial x_k}(\mathbf{x},\mathbf{y},t)\\
&\quad = -\mathcal{R}_{ik}(\mathbf{x},\mathbf{y},t)\frac{\partial \overline{U}_j}{\partial y_k}(\mathbf{y},t) - \mathcal{R}_{jk}(\mathbf{y},\mathbf{x},t)\frac{\partial \overline{U}_i}{\partial x_k}(\mathbf{x},t)\\
&\qquad - \frac{\partial \mathcal{S}_{jk,i}}{\partial y_k}(\mathbf{y},\mathbf{x},t) - \frac{\partial \mathcal{S}_{ik,j}}{\partial x_k}(\mathbf{x},\mathbf{y},t) - \frac{1}{\rho}\frac{\partial \mathcal{K}_i}{\partial y_j}(\mathbf{x},\mathbf{y},t)\\
&\qquad - \frac{1}{\rho}\frac{\partial \mathcal{K}_j}{\partial x_i}(\mathbf{y},\mathbf{x},t) + \nu\frac{\partial^2 \mathcal{R}_{ij}}{\partial y_k^2}(\mathbf{x},\mathbf{y},t) + \nu\frac{\partial^2 \mathcal{R}_{ij}}{\partial x_k^2}(\mathbf{x},\mathbf{y},t).
\end{aligned} \tag{5.95}$$

When $\mathbf{x} = \mathbf{y}$, $\mathcal{R}_{ij}(\mathbf{x}, \mathbf{x}, t) = R_{ij}(\mathbf{x}, t)$, and it may be shown that Eq. (5.95) becomes identical to Eq. (3.53). This connection suggests that the first two terms on the right-hand side of Eq. (5.95) are "production" terms. The remaining terms acquire meaning by noting their similarity to the corresponding terms in Eq. (3.53).

The formidable complexity of Eq. (5.95) can be reduced somewhat by applying the relation to the specific case of homogeneous, isotropic turbulence. Since $\overline{U}_k(\mathbf{y}, t) = \overline{U}_k(\mathbf{x}, t)$ in homogeneous turbulence, and using results like Eq. (5.87), it follows that the two convection terms on the left-hand side of Eq. (5.95) sum to zero. Uniformity of $\overline{U}_i$ also implies that the two production terms on the right-hand side of Eq. (5.95) are zero.

The simplification for homogeneous turbulence used in Eq. (4.1) can be generalized to include the statements that

$$\mathcal{S}_{ij,k}(\mathbf{x}, \mathbf{y}, t) = \mathcal{S}_{ij,k}(\mathbf{y} - \mathbf{x}, t) \tag{5.96}$$

and

$$\mathcal{K}_i(\mathbf{x}, \mathbf{y}, t) = \mathcal{K}_i(\mathbf{y} - \mathbf{x}, t), \tag{5.97}$$

where for convenience the same symbols $\mathcal{R}_{ij}$, $\mathcal{S}_{ij,k}$, and $\mathcal{K}_i$ on the right-hand side are adopted; their applicability to homogeneous turbulence is implied by the appearance of one less argument than their more general counterparts. Using these relations, it follows that

$$\frac{\partial \mathcal{S}_{jk,i}}{\partial y_k}(\mathbf{y}, \mathbf{x}, t) = -\frac{\partial \mathcal{S}_{jk,i}}{\partial r_k}(\mathbf{x} - \mathbf{y}, t) \tag{5.98}$$

and

$$\frac{\partial \mathcal{S}_{ik,j}}{\partial x_k}(\mathbf{x}, \mathbf{y}, t) = -\frac{\partial \mathcal{S}_{ik,j}}{\partial r_k}(\mathbf{y} - \mathbf{x}, t), \tag{5.99}$$

and that

$$\frac{\partial \mathcal{K}_i}{\partial y_j}(\mathbf{x}, \mathbf{y}, t) = \frac{\partial \mathcal{K}_i}{\partial r_j}(\mathbf{y} - \mathbf{x}, t) \tag{5.100}$$

and

$$\frac{\partial \mathcal{K}_j}{\partial x_i}(\mathbf{y}, \mathbf{x}, t) = \frac{\partial \mathcal{K}_j}{\partial r_i}(\mathbf{x} - \mathbf{y}, t). \tag{5.101}$$

Putting together the various results, it is found that the two-point velocity correlation tensor in homogeneous turbulence is governed by the equation

$$\begin{aligned}\frac{\partial \mathcal{R}_{ij}}{\partial t}(\mathbf{r}, t) = {} & \frac{\partial \mathcal{S}_{jk,i}}{\partial r_k}(-\mathbf{r}, t) + \frac{\partial \mathcal{S}_{ik,j}}{\partial r_k}(\mathbf{r}, t) \\ & -\frac{1}{\rho}\frac{\partial \mathcal{K}_i}{\partial r_j}(\mathbf{r}, t) - \frac{1}{\rho}\frac{\partial \mathcal{K}_j}{\partial r_i}(-\mathbf{r}, t) + 2\nu\frac{\partial^2 \mathcal{R}_{ij}}{\partial r_k^2}(\mathbf{r}, t).\end{aligned} \tag{5.102}$$

Contracting the indices in Eqs. (5.100) and (5.101), noting the definition of $\mathcal{K}_i$ in Eq. (5.92) and using the incompressibility condition gives in both cases

$$\frac{\partial \mathcal{K}_i}{\partial r_i}(\mathbf{r}, t) = 0. \tag{5.103}$$

Now taking a trace of Eq. (5.102) and using (5.103) gives

$$\frac{\partial \mathcal{R}_{ii}}{\partial t}(\mathbf{r},t) = \frac{\partial S_{ik,i}}{\partial r_k}(-\mathbf{r},t) + \frac{\partial S_{ik,i}}{\partial r_k}(\mathbf{r},t) + 2\nu\frac{\partial^2 \mathcal{R}_{ii}}{\partial r_k^2}(\mathbf{r},t), \tag{5.104}$$

which shows that the time rate of change of the trace of the two-point velocity correlation tensor depends on a balance between viscous diffusion, given in the last term, and the two terms depending on the two-point triple velocity correlation tensor. The latter represent the process by which vortex stretching brings energy to small dissipative scales.

5.6 Self-Preservation and the Kármán–Howarth Equation

Under the stricture of isotropy Eq. (5.104) can be simplified with the use of Eqs. (4.34) and (4.38). After a lengthy but straightforward calculation, there results the Kármán–Howarth equation [13] given by

$$\frac{\partial(\overline{u^2}f)}{\partial t} = (\overline{u^2})^{\frac{3}{2}}\left(\frac{\partial k}{\partial r} + \frac{4}{r}k\right) + 2\nu\overline{u^2}\left(\frac{\partial^2 f}{\partial r^2} + \frac{4}{r}\frac{\partial f}{\partial r}\right), \tag{5.105}$$

which represents a relation between $f(r,t)$, $k(r,t)$, and $u_{rms}(t)$. Note that by substituting Taylor series for f and k into Eq. (5.105) and using Eq. (4.45), Eqs. (5.1) and (5.23) can be derived by gathering terms depending on like powers of r. Thus, all the isotropy information in the K and ϵ equations is contained within Eq. (5.105).

As it stands, Eq. (5.105) is intractable, but by adding the additional constraint that the decay be self-similar so that Eqs. (5.38) and (5.39) hold, the Kármán–Howarth equation becomes

$$\begin{aligned} &2\eta^{-4}\frac{d}{d\eta}\left(\eta^4\frac{d\tilde{f}}{d\eta}\right) + \eta\frac{d\tilde{f}}{d\eta}\left(\frac{7}{3}G_0 - 5\right) + 10\tilde{f} \\ &\quad = R_\lambda\left(\frac{7}{6}S_{K_0}\eta\frac{d\tilde{f}}{d\eta} - \eta^{-4}\frac{d(\eta^4\tilde{k})}{d\eta}\right) \end{aligned} \tag{5.106}$$

where $\eta = r/\lambda_g$ is a similarity variable. This equation constitutes a single ordinary differential equation for the two unknown functions $\tilde{f}(\eta)$ and $\tilde{k}(\eta)$ with R_λ acting as a time-dependent parameter.

If isotropic decay were completely self-similar at all times, then Eq. (5.106) would have to be satisfied throughout the decay, regardless of how R_λ varies. The only way this can occur is if each side of the equation vanished independently of the other, since otherwise the left-hand side of Eq. (5.106) would be multi-valued as R_λ changed. Thus, for complete self-similarity it follows that

$$2\eta^{-4}\frac{d}{d\eta}\left(\eta^4\frac{d\tilde{f}}{d\eta}\right) + \eta\frac{d\tilde{f}}{d\eta}\left(\frac{7}{3}G_0 - 5\right) + 10\tilde{f} = 0 \tag{5.107}$$

and

$$\frac{7}{6}S_{K_0}\eta\frac{d\tilde{f}}{d\eta} - \eta^{-4}\frac{d(\eta^4\tilde{k})}{d\eta} = 0, \tag{5.108}$$

assuming $R_\lambda \neq 0$.

Equation (5.107) is an example of the confluent hypergeometric equation [14–16] and has solution

$$\tilde{f}(\eta) = M\left(\frac{1}{G_0^* - 1}, \frac{5}{2}, -\frac{5(G_0^* - 1)}{4}\eta^2\right) \tag{5.109}$$

where M is the confluent hypergeometric function. Integration of Eq. (5.108) gives

$$\tilde{k}(\eta) = \frac{7}{6} S_{K_0} \frac{1}{\eta^4} \int_0^\eta s^5 \frac{d\tilde{f}}{ds} ds, \tag{5.110}$$

with $\tilde{f}$ given by Eq. (5.109) so that $\tilde{k}(\eta)$ can be determined by evaluating the integral. It is interesting to note from Eqs. (5.109) and (5.110) that $\tilde{f}$ and $\tilde{k}$ are entirely specified once G_0 and S_{K_0} are chosen. Thus, within the constraint of complete self-similarity, a unique pair of functions $\tilde{f}$ and $\tilde{k}$ is associated with constant values of G_0 and S_{K_0} and vice versa.

It has already been shown that complete similarity is unattainable in isotropic decay, so it is clear that $\tilde{f}$ and $\tilde{k}$ must change in time in a real flow. Nonetheless, it is still useful to consider the meaning of Eq. (5.106) in the vicinity of the high and low Reynolds number equilibrium solutions. First consider the case of small R_T in the neighborhood of $R_{T_\infty} = 0$. Here, the right-hand side of Eq. (5.106) is small and so (5.107) is well satisfied even when not invoking complete self-similarity. Since $G_0^* = 7/5$ for the final period, the solution to Eq. (5.107) given by (5.109) is [17]

$$\tilde{f}(\eta) = M\left(\frac{5}{2}, \frac{5}{2}, -\frac{\eta^2}{2}\right) = e^{-\frac{\eta^2}{2}}, \tag{5.111}$$

in agreement with Eq. (5.60).

Assuming that the decay is self-similar in the neighborhood of $R_T = 0$, then Eq. (5.110) holds as well and after integrating this relation to get $\tilde{k}(\eta)$ it is found that

$$\tilde{k}(\eta) = \frac{7}{6} S_{K_0} \frac{1}{\eta^4} \left[(\eta^5 + 5\eta^3 + 15\eta)e^{-\eta^2/2} - 15\sqrt{\frac{\pi}{2}} erf\left(\frac{\eta}{\sqrt{2}}\right)\right], \tag{5.112}$$

where

$$erf(\eta) = \frac{2}{\sqrt{\pi}} \int_0^\eta e^{-s^2} ds \tag{5.113}$$

is the error function [18]. A plot of Eq. (5.112) is given in Figure 5.7, where $\tilde{k}$ is seen to have a much slower decay for large η than the Gaussian form of $\tilde{f}$. In fact, Eq. (5.112) shows that $\tilde{k}(\eta) \sim \eta^{-4}$ as $\eta \to \infty$.

In the case of the high Reynolds number equilibrium, R_λ is a non-zero constant. Thus, in the equilibrium state, Eq. (5.106) does not have the separability requirement leading to Eqs. (5.107) and (5.108) so that it remains as a single equation in two unknown functions. This fits with the fact that self-similarity cannot account for the complete decay from large R_T to zero. It also may be concluded that unlike the system of self-similar K and ϵ equations which can plausibly be taken as a model of the exact equations, the self-similar Kármán–Howarth equation is not of significant use for analyzing high Reynolds number flow even if the flow is in a regime that is *close* to similarity, but not exactly self-similar.

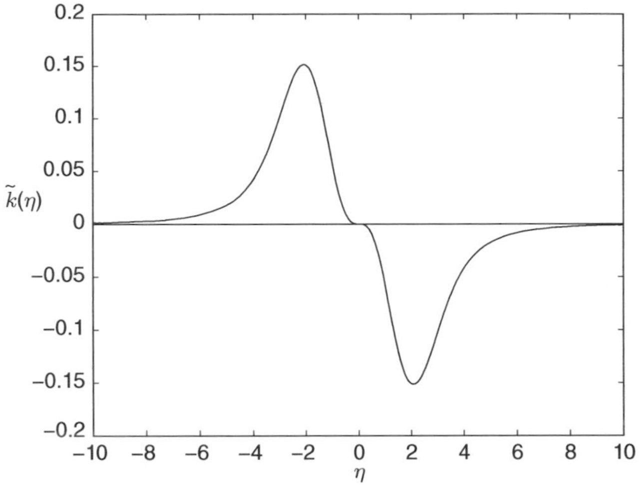

Figure 5.7 $\tilde{k}$ in the final period.

5.7 Energy Spectrum Equation

Some additional insights into isotropic decay can be gained by considering the equation governing the energy spectrum tensor $\mathcal{E}_{ij}$ defined in Eq. (6.9). This may be derived by applying a Fourier transform to the terms in Eq. (5.102), yielding

$$\frac{\partial \mathcal{E}_{ij}}{\partial t}(\mathbf{k}, t) = T_{ij}(\mathbf{k}, t) + P_{ij}(\mathbf{k}, t) - 2k^2 \nu \mathcal{E}_{ij}(\mathbf{k}, t) \tag{5.114}$$

where

$$T_{ij}(\mathbf{k}, t) = (2\pi)^{-3} \int_{\mathfrak{R}^3} e^{\imath \mathbf{r}\cdot\mathbf{k}} S_{ij}(\mathbf{r}, t) d\mathbf{r} \tag{5.115}$$

is the Fourier transform of the transfer term $S_{ij}(\mathbf{r}, t)$, defined as

$$S_{ij}(\mathbf{r}, t) \equiv \frac{\partial S_{jk,i}}{\partial r_k}(-\mathbf{r}, t) + \frac{\partial S_{ik,j}}{\partial r_k}(\mathbf{r}, t), \tag{5.116}$$

and

$$P_{ij}(\mathbf{k}, t) \equiv -(2\pi)^{-3} \int_{\mathfrak{R}^3} e^{\imath \mathbf{r}\cdot\mathbf{k}} \left[\frac{1}{\rho} \frac{\partial \mathcal{K}_i}{\partial r_j}(\mathbf{r}, t) + \frac{1}{\rho} \frac{\partial \mathcal{K}_j}{\partial r_i}(-\mathbf{r}, t) \right] d\mathbf{r} \tag{5.117}$$

is the transformed pressure velocity term. T_{ij} describes the rate at which a given wave number (or scale of motion) is gaining or losing energy by transfer of energy from or to the other scales. The physical process driving the energy exchange has to do with vortex stretching and reorientation, and T_{ij} is the form this process takes in Fourier space. P_{ij}, which will later be seen to be zero in isotropic turbulence, represents the influence of the pressure field in bringing initially anisotropic turbulence back toward isotropy. Finally, the viscous dissipation rate term in Eq. (5.114) is just $2\nu k^2$ times the energy spectrum tensor as follows from a straightforward calculation.

Our interest now lies in using Eq. (5.114) to derive a relation for the energy spectral density $E(k, t)$. Upon contraction of the indices in Eq. (5.114) the pressure terms drop out and after integrating over a spherical shell so as to convert $\mathcal{E}_{ii}$ to E according to the definition in Eq. (2.74), it is found that

$$\frac{\partial E}{\partial t}(k,t) = T(k,t) - 2\nu k^2 E(k,t) \tag{5.118}$$

where

$$T(k,t) = \frac{1}{2}\int_{|\mathbf{k}|=k} T_{ii}(\mathbf{k},t)d\Omega \tag{5.119}$$

is the transfer term and $T_{ii}(\mathbf{k},t)$ is defined in Eq. (5.115) as the Fourier transform of $S_{ii}(\mathbf{r},t)$.

Substituting the isotropic form $S_{ij,k}$ in Eq. (4.38) into Eq. (5.116) gives

$$S_{ii}(r,t) = u_{rms}^3 \frac{1}{r^2}\frac{d}{dr}\left(r^3 \frac{dk}{dr}(r,t) + 4r^2 k(r,t)\right) \tag{5.120}$$

so that S_{ii} is only a function of r in isotropic turbulence. Note that to distinguish the function $k(r,t)$ appearing in Eq. (5.120) from the wave number k in such relations as Eq. (5.118), in this section the former will always be written with its arguments (r,t). In view of Eq. (5.120), (5.115) can be simplified to

$$T_{ii}(k,t) = \frac{1}{2\pi^2}\int_0^\infty S_{ii}(r,t) r^2 \frac{\sin kr}{kr} dr, \tag{5.121}$$

where it is seen that T_{ii} is a function of k and not $\mathbf{k}$. Substituting Eq. (5.120) into (5.121), integrating by parts twice and then substituting the result into Eq. (5.119) leads to

$$T(k,t) = \frac{u_{rms}^3}{\pi}\int_0^\infty k[(3-k^2r^2)\sin kr - 3kr\cos kr]k(r,t)dr. \tag{5.122}$$

In this way it is seen that the two-point triple velocity correlation $k(r,t)$ determines the rate of transfer of energy between scales.

Integrating Eq. (5.118) from $k = 0 \to \infty$ and using Eqs. (2.75), (4.12), and (5.1) shows that

$$\int_0^\infty T(k,t)dk = 0. \tag{5.123}$$

In other words, the net amount of energy transfer between scales is zero, so that whatever loss of energy some scales encounter is counterbalanced by a gain in other scales.

Whenever $f(r,t)$ is known $E(k,t)$ can be obtained from Eq. (4.73). Then $T(k,t)$ can be obtained directly from Eq. (5.118), without the need to evaluate Eq. (5.122). As an important example, consider the final decay period where f is given by Eq. (5.60). Substituting this into Eq. (4.73) and carrying out the integration gives

$$E(k,t) = \frac{u_{rms}^2 \lambda_g}{\sqrt{2\pi}}(k\lambda_g)^4 e^{-\frac{1}{2}(k\lambda_g)^2}. \tag{5.124}$$

It is apparent that this formula is consistent with the previously mentioned condition Eq. (4.67). It is a simple matter to substitute Eq. (5.124) into (5.118) and calculate

$$T(k,t) = E(k,t)\frac{u_{rms}}{\lambda_g}((k\lambda_g)^2 - 5), \tag{5.125}$$

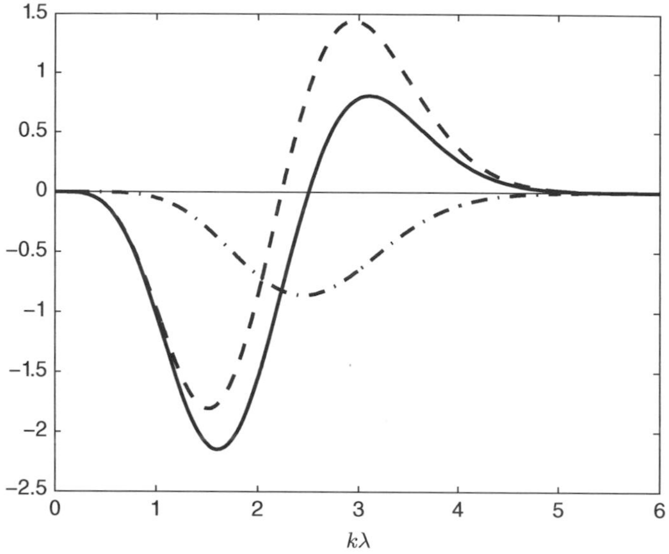

Figure 5.8 Energy spectrum budget in final period. —, $\partial E/\partial t$; ––, transfer term; – · –, dissipation. In this illustration $R_\lambda = 10$, with λ denoting λ_g.

so that in the final period of decay scales for which $k < \sqrt{5}/\lambda_g$ lose energy to those for which $k > \sqrt{5}/\lambda_g$. Figure 5.8 contains a budget of the energy spectrum equation (5.118) in the final period showing that energy given up at the large scales is passed to smaller scales. The peak in energy is at $k_e = 2/\lambda_g$ while that of dissipation is at $k_d = \sqrt{6}/\lambda_g$. In this solution, the energy and dissipation range scales are not well separated, such as would occur at high Reynolds numbers. In particular, there is no inertial range associated with Eq. (5.124). As the microscale rises in the final period consistent with Figure 5.5, the balance in Figure 5.8 shifts toward smaller wave numbers or, equivalently, larger scales.

5.8 Energy Spectrum Equation via Fourier Analysis of the Velocity Field

In this section we pursue an alternative analysis of the transfer physics based on a Fourier analysis of the Navier–Stokes equation. This is to be contrasted with the approach leading to Eq. (5.122) in which the expression for the transfer term is based on the two-point triple longitudinal correlation function $k(r,t)$. The Fourier analysis leads to an expression for $T(k,t)$ that shows how energy transfer arises by interactions between scales at specific combinations of wave numbers. In the approach pursued here an equation for $E(k,t)$ is derived by converting the Fourier transform of Eq. (2.2) for a finite cubic domain into an energy equation and then taking the limit as the domain increases to encompass all of space.

5.8.1 Fourier Analysis on a Cubic Region

Consider the velocity field expressed via the Fourier series in Eq. (2.50) so that

$$u_i(\mathbf{x}, t) = \sum_{\mathbf{k}} \hat{u}_i(\mathbf{k}, t) e^{\imath \mathbf{k}\cdot\mathbf{x}}, \tag{5.126}$$

where the wave number vector $\mathbf{k} = 2\pi\mathbf{n}/L$, and $\mathbf{n}$ denotes the set of integer triples (n_1, n_2, n_3) running from $-\infty \rightarrow +\infty$. Now proceed by applying the Fourier transform to each of the terms in Eq. (2.2). This is straightforward except for the advection term, which is now considered. A Fourier transform of the non-linear advection term in Eq. (2.2) gives

$$\widehat{u_j \frac{\partial u_i}{\partial x_j}}(\mathbf{k}, t) = \frac{1}{L^3} \int_{\mathcal{V}_L} u_j(\mathbf{x}, t) \frac{\partial u_i}{\partial x_j}(\mathbf{x}, t) e^{-\imath \mathbf{k}\cdot\mathbf{x}} d\mathbf{x}. \tag{5.127}$$

Noting from Eq. (5.126) the identity

$$\frac{\partial u_i}{\partial x_j}(\mathbf{x}, t) = \imath \sum_{\mathbf{k}} k_j \hat{u}_i(\mathbf{k}, t) e^{\imath \mathbf{k}\cdot\mathbf{x}} \tag{5.128}$$

and substituting Eqs. (5.126) and (5.128) into the right-hand side of Eq. (5.127) yields

$$\begin{aligned} \widehat{u_j \frac{\partial u_i}{\partial x_j}}(\mathbf{k}, t) &= \imath \sum_{\mathbf{l}} \sum_{\mathbf{m}} l_j \hat{u}_i(\mathbf{l}, t) \hat{u}_j(\mathbf{m}, t) \frac{1}{L^3} \int_{\mathcal{V}_L} d\mathbf{x}\ e^{-\imath(-\mathbf{l}+\mathbf{k}-\mathbf{m})\cdot\mathbf{x}} \\ &= \imath \sum_{\mathbf{l}} l_j \hat{u}_i(\mathbf{l}, t) \hat{u}_j(\mathbf{k} - \mathbf{l}, t), \end{aligned} \tag{5.129}$$

where the identity Eq. (2.54) has been used. Equation (5.129) has the interesting property that the contributions to the advection term at $\mathbf{k}$ only come from triads of wave numbers. Thus the influence of $\mathbf{l}$ on $\mathbf{k}$ also includes $\mathbf{k} - \mathbf{l}$ and no other wave numbers. These wave numbers form the vertices of a triangle, as shown in Figure 5.9.

Collecting the above results together, the transformed version of Eq. (2.2) is

$$\frac{\partial \hat{u}_i}{\partial t}(\mathbf{k}, t) = -\imath \sum_{\mathbf{l}} l_j \hat{u}_i(\mathbf{l}, t) \hat{u}_j(\mathbf{k} - \mathbf{l}, t) + \imath k_i \frac{\hat{p}(\mathbf{k}, t)}{\rho} - \nu k^2 \hat{u}_i(\mathbf{k}, t), \tag{5.130}$$

Figure 5.9 Triads of wave numbers.

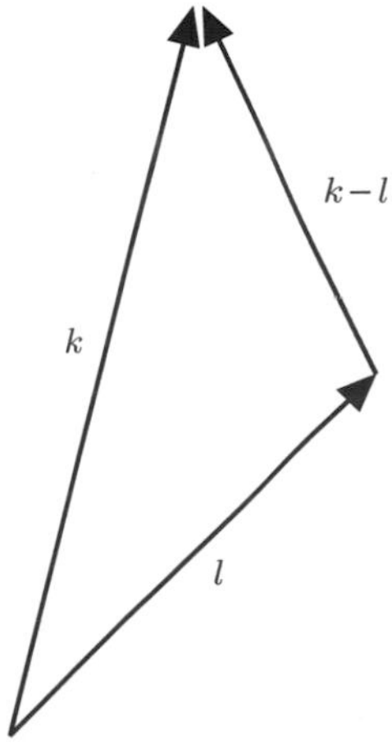

where pressure has the Fourier series

$$p(\mathbf{x}, t) = \sum_{\mathbf{k}} \hat{p}(\mathbf{k}, t) e^{\imath \mathbf{k}\cdot\mathbf{x}} \tag{5.131}$$

and inverse transform

$$\hat{p}(\mathbf{k}, t) = \frac{1}{L^3} \int_{\mathcal{V}_L} p(\mathbf{x}, t) e^{-\imath \mathbf{k}\cdot\mathbf{x}} d\mathbf{x}. \tag{5.132}$$

A Fourier transform of the continuity equation (2.1) yields

$$k_i \hat{u}_i(\mathbf{k}, t) = 0, \tag{5.133}$$

meaning that in incompressible flow $\hat{u}_i(\mathbf{k}, t)$ is perpendicular to $\mathbf{k}$. In other words, for any value of $\mathbf{k}$, $\hat{u}_i(\mathbf{k}, t)$ is oriented tangent to the surface of the sphere of radius $|\mathbf{k}|$ centered at the origin.

Taking a dot product of Eq. (5.130) with $\mathbf{k}$ and using Eq. (5.133) yields

$$\hat{p}(\mathbf{k}, t) = \rho \frac{k_i}{k^2} \sum_{\mathbf{l}} l_j \hat{u}_i(\mathbf{l}, t) \hat{u}_j(\mathbf{k} - \mathbf{l}, t). \tag{5.134}$$

In view of Eq. (5.133)

$$l_j \hat{u}_j(\mathbf{k} - \mathbf{l}, t) = k_j \hat{u}_j(\mathbf{k} - \mathbf{l}, t). \tag{5.135}$$

Using this in Eq. (5.134) and substituting for $\hat{p}$ in Eq. (5.130) yields the $\hat{u}_i$ equation in the convenient form

$$\frac{\partial \hat{u}_i}{\partial t}(\mathbf{k}, t) + \nu k^2 \hat{u}_i(\mathbf{k}, t) = -\imath P_{ij}(\mathbf{k}) k_m \sum_{\mathbf{l}} \hat{u}_j(\mathbf{l}, t) \hat{u}_m(\mathbf{k} - \mathbf{l}, t), \tag{5.136}$$

where

$$P_{ij}(\mathbf{k}) = \delta_{ij} - \frac{k_i k_j}{k^2} \tag{5.137}$$

is a projection operator. P_{ij} maps any vector into its component perpendicular to the $\mathbf{k}$ direction. For example, for an arbitrary vector $\mathbf{v}$, $P_{ij}(\mathbf{k}) v_j = v_i - (v_j k_j) k_i / k^2$ is perpendicular to k_i since $(P_{ij}(\mathbf{k}) v_j)(k_i) = 0$. The appearance of P_{ij} in Eq. (5.136) is consistent with the fact that the two other terms in the equation are perpendicular to $\mathbf{k}$ as may be deduced from Eq. (5.133). In other words, if the left-hand side of the equation is normal to $\mathbf{k}$, so too must be the right-hand side.

An equivalent form of (5.136), which is often used in theoretical analyses, is given by

$$\frac{\partial \hat{u}_i}{\partial t}(\mathbf{k}, t) + \nu k^2 \hat{u}_i(\mathbf{k}, t) = M_{ijm}(\mathbf{k}) \sum_{\mathbf{l}} \hat{u}_j(\mathbf{l}, t) \hat{u}_m(\mathbf{k} - \mathbf{l}, t), \tag{5.138}$$

where

$$M_{ijm} = -\frac{\imath}{2}(k_m P_{ij}(\mathbf{k}) + k_j P_{im}(\mathbf{k})). \tag{5.139}$$

This is derived by noting that the right-hand side of Eq. (5.136) is left unchanged if the dummy indices j and m are switched, and summation on $\mathbf{l}$ is replaced by the equivalent summation on $\mathbf{l}' \equiv \mathbf{k} - \mathbf{l}$. The right-hand side of Eq. (5.138) represents the average of these two equivalent expressions for the right-hand side of Eq. (5.136).

A dynamical equation for the discrete energy spectrum component $E_{\mathbf{k}} = |\hat{\mathbf{u}}(\mathbf{k},t)|^2/2$ that was introduced previously in Section 2.2.4 may be derived by taking the average of the sum of Eq. (5.136) times $\hat{u}_i^*(\mathbf{k},t)$ and the complex conjugate of Eq. (5.136) times $\hat{u}_i(\mathbf{k},t)$. The result is

$$\frac{\partial E_{\mathbf{k}}}{\partial t}(t) + 2\nu k^2 E_{\mathbf{k}}(t) = \frac{1}{2}M_{ijm}(\mathbf{k})\sum_{\mathbf{l}}(\overline{\hat{u}_i(-\mathbf{k})\hat{u}_j(\mathbf{l})\hat{u}_m(\mathbf{k}-\mathbf{l})} - \overline{\hat{u}_i(\mathbf{k})\hat{u}_j(\mathbf{l})\hat{u}_m(-\mathbf{k}-\mathbf{l})}), \tag{5.140}$$

where $\mathbf{l}$ has been replaced with $-\mathbf{l}$ in the second term for later convenience. The term on right-hand side of Eq. (5.140) accounts for the energy transfer between wave numbers. The triadic nature of such exchanges is evident in these expressions.

5.8.2 Limit of Infinite Space

To acquire an alternative form of the dynamical equation for $E(k,t)$ consider the limiting form of Eq. (5.140) as $L \to \infty$. With the goal of taking this limit define

$$E_{ij}^L(\mathbf{k},t) = \left(\frac{L}{2\pi}\right)^3 \overline{\hat{u}_i(\mathbf{k})\hat{u}_j(-\mathbf{k})}, \tag{5.141}$$

which becomes, after using Eqs. (2.57) and (2.62),

$$E_{ij}^L(\mathbf{k},t) = \left(\frac{1}{2\pi}\right)^3 \int_{\mathcal{V}_L} \mathcal{R}_{ij}(\mathbf{r},t)e^{-\imath\mathbf{k}\cdot\mathbf{r}}d\mathbf{r}. \tag{5.142}$$

In the limit as $L \to \infty$ the right-hand side of Eq. (5.142) becomes the Fourier transform of $\mathcal{R}_{ij}(\mathbf{r},t)$ so it has been established that

$$\lim_{L\to\infty} E_{ij}^L(\mathbf{k},t) = \mathcal{E}_{ij}(\mathbf{k},t). \tag{5.143}$$

In this limit the values of $\mathbf{k}$ become closer and closer together, and at the end of the limiting process $\mathbf{k}$ covers all real vectors and not just the discrete set where $E_{ij}^L(\mathbf{k},t)$ is defined at a particular value of L.

This same argument needs to be repeated in the case of triple velocity correlations. Thus, define

$$T_{ijn}^L(\mathbf{k},\mathbf{l},t) = \left(\frac{L}{2\pi}\right)^6 \overline{\hat{u}_i(\mathbf{k})\hat{u}_j(\mathbf{l})\hat{u}_n(-\mathbf{k}-\mathbf{l})} \tag{5.144}$$

where, by a generalization of the argument leading to Eq. (2.61), use has been made of the fact that

$$\overline{\hat{u}_i(\mathbf{k},t)\hat{u}_j(\mathbf{l},t)\hat{u}_n(\mathbf{m},t)} = 0 \quad \textit{unless} \quad \mathbf{k}+\mathbf{l}+\mathbf{m} = 0. \tag{5.145}$$

Substituting for the Fourier components using Eq. (2.52) then gives

$$T_{ijn}^L(\mathbf{k},\mathbf{l},t) = \left(\frac{L}{2\pi}\right)^6 \frac{1}{L^9}\int_{\mathcal{V}_L}\int_{\mathcal{V}_L}\int_{\mathcal{V}_L} \overline{u_i(\mathbf{x},t)u_j(\mathbf{y},t)u_n(\mathbf{z},t)}e^{-\imath\mathbf{k}\cdot(\mathbf{x}-\mathbf{z})-\imath\mathbf{l}\cdot(\mathbf{y}-\mathbf{z})}d\mathbf{x}d\mathbf{y}d\mathbf{z}. \tag{5.146}$$

In homogeneous turbulence the triple velocity correlation within the integral depends on just the two vectors $\mathbf{r} = \mathbf{x} - \mathbf{z}$ and $\mathbf{s} = \mathbf{y} - \mathbf{z}$. This justifies the definition

$$S_{ijn}(\mathbf{r}, \mathbf{s}, t) \equiv \overline{u_i(\mathbf{z} + \mathbf{r}, t)u_j(\mathbf{z} + \mathbf{s}, t)u_n(\mathbf{z}, t)}, \tag{5.147}$$

which is to be contrasted with the two-point triple correlation $S_{ij,n}$ in Eq. (2.31). Changing the $\mathbf{x}$ and $\mathbf{y}$ variables in Eq. (5.146) to $\mathbf{r}$ and $\mathbf{s}$, respectively, using Eq. (5.147), and carrying out the $\mathbf{z}$ integration in Eq. (5.146) yields

$$T^L_{ijn}(\mathbf{k}, \mathbf{l}, t) = \left(\frac{1}{2\pi}\right)^6 \int_{\mathcal{V}_L}\int_{\mathcal{V}_L} S_{ijn}(\mathbf{r}, \mathbf{s}, t)e^{-\imath\mathbf{k}\cdot\mathbf{r}-\imath\mathbf{l}\cdot\mathbf{s}}d\mathbf{r}d\mathbf{s}. \tag{5.148}$$

In the limit as $L \to \infty$ this becomes

$$T_{ijn}(\mathbf{k}, \mathbf{l}, t) = \left(\frac{1}{2\pi}\right)^6 \int\int S_{ijn}(\mathbf{r}, \mathbf{s}, t)e^{-\imath\mathbf{k}\cdot\mathbf{r}-\imath\mathbf{l}\cdot\mathbf{s}}d\mathbf{r}d\mathbf{s} \tag{5.149}$$

as the Fourier transform of $S_{ijn}(\mathbf{r}, \mathbf{s}, t)$.

Now we are prepared to consider the limit of Eq. (5.140) as $L \to \infty$. Multiply this equation by $(L/2\pi)^3$ and note through Eqs. (5.141), (5.143), and (4.66) that

$$\lim_{L\to\infty}\left(\frac{L}{2\pi}\right)^3 E_{\mathbf{k}} = \frac{1}{2}\mathcal{E}_{ii}(\mathbf{k}, t) = \frac{E(k, t)}{4\pi k^2}. \tag{5.150}$$

Moreover, it may be calculated using Eq. (5.144) that

$$\begin{aligned}\lim_{L\to\infty}\left(\frac{L}{2\pi}\right)^3 \sum_{\mathbf{l}} \overline{\hat{u}_j(\mathbf{l})\hat{u}_m(\mathbf{k}-\mathbf{l})\hat{u}_i(-\mathbf{k})} &= \lim_{L\to\infty}\sum_{\mathbf{l}}\left(\frac{2\pi}{L}\right)^3 T^L_{jmi}(\mathbf{l}, \mathbf{k}-\mathbf{l}, t)\\ &= \int T_{jmi}(\mathbf{l}, \mathbf{k}-\mathbf{l}, t)d\mathbf{l}\end{aligned} \tag{5.151}$$

where the second equality depends on the fact, implied by Eqs. (5.148) and (5.149), that $\lim_{L\to\infty} T^L_{jmi} = T_{jmi}$, and that $(2\pi/L)^3$ is the volume surrounding each of the wave number vectors in the sum (since $\mathbf{k} = 2\pi\mathbf{n}/L$). The integral appears in the last term in Eq. (5.151) as the limit of the Riemann sums as $L \to \infty$. Collecting these results together, it has been established that

$$\begin{aligned}&\frac{\partial E(k, t)}{\partial t} + 2\nu k^2 E(k, t)\\ &\quad = 2\pi k^2 M_{ijm}(\mathbf{k}) \int (T_{jmi}(\mathbf{l}, \mathbf{k}-\mathbf{l}, t) - T_{jmi}(\mathbf{l}, -\mathbf{k}-\mathbf{l}, t))d\mathbf{l}.\end{aligned} \tag{5.152}$$

Using the homogeneity properties of $T_{jmi}(\mathbf{l}, -\mathbf{k}-\mathbf{l}, t)$ it can be shown that the second term in the integral on the right-hand side of Eq. (5.152) contributes minus that of the first term so that the equation simplifies to

$$\frac{\partial E(k, t)}{\partial t} + 2\nu k^2 E(k, t) = 4\pi k^2 M_{ijm}(\mathbf{k}) \int T_{jmi}(\mathbf{l}, \mathbf{k}-\mathbf{l}, t)d\mathbf{l}. \tag{5.153}$$

The form of the stretching term in Eq. (5.153) should be contrasted with that derived in Eq. (5.122) incorporating the scalar correlation function $k(r)$.

5.8.3 Applications to Turbulence Theory

The spectral analysis of turbulence that has been introduced in this section provides a very different way of developing and applying theoretical ideas about turbulence than is pursued in the engineering analysis of turbulent flows that is considered in Chapters 9 and 10. In the present case, in essence, the focus is on developing mathematical models of the ways in which the Fourier modes interact so as to create energy transfer between scales. The end result of such theories is to develop closure to the energy spectrum equation in situations such as homogeneous turbulence where the complexity of transport phenomena is absent. Conventional closure modeling includes phenomenonological expressions such as were seen in Eq. (5.78) that make no pretense of explaining the detailed physics of how the energy transfer takes place. Rather they are designed to fit in with RANS and LES modeling of general turbulent flows found in applications.

By framing the dynamics of the energy spectrum in the form of Eq. (5.153) it becomes possible to apply models of the statistical behavior of multi-point velocity correlations to effect closure to the governing equations. Thus, in particular, the joint probability distribution for the velocity at three points in a turbulent flow can be applied in the evaluation of the transfer term $T_{jml}(\mathbf{l}, \mathbf{k} - \mathbf{l}, t)$ by way of its definition in Eq. (5.149) in terms of the triple velocity correlation in Eq. (5.147). The joint statistics of the velocity field in turbulent flow are not known rigorously though approximations are possible that enable solution of Eq. (5.153). For example, if $\mathbf{u}(\mathbf{x}, t)$ were a Gaussian random field – meaning that the velocity components at any N locations have a joint probability density function that is Gaussian in form – then it may be shown that all higher order correlations between velocities can be reduced to combinations of second-order correlations, in which case a closed set of equations is derived from Eq. (5.153). Although the turbulent velocity field is not exactly Gaussian it can be argued that it may nonetheless be described as being a perturbation away from Gaussianity. By various techniques for modeling the departure of random fields from Gaussianity it becomes possible to develop closed systems of equations from which the energy spectrum can be computed.

In the theory known as EDQNM (Eddy-damped quasi-normal Markovian) [19] a quasi-normal (QN) model is assumed in which the fourth-order moments of the velocity field are assumed to depend on second-order moments in the same way as they do for Gaussian fields. Since the fourth-order velocity moments appear in an exact equation for the third-order velocity moments (not given here) that may be derived to accompany Eq. (5.153), the QN hypothesis allows closure to be attained. With just the QN assumption the resulting closure yields unphysical solutions due to the appearance of negative energy spectra. The source of the negative spectra lies in the over-prediction of the third-order moments due to the QN hypothesis and a cure for this is the addition of a damping term in the third moment equation as well as an additional Markovian assumption to the effect that events in the near present should be weighted more highly in determining the spectrum than events in the past. With these hypotheses EDQNM makes prediction of the evolution in time of a $-5/3$ Kolmogorov spectrum, among other results.

In the direct interaction approximation (DIA) [20], the solution for the energy spectrum is based on assuming that the turbulent velocity field can be viewed as a perturbation away from a Gaussian ground state. The method considers the solution to the energy spectrum equation coupled to an exact relation for an infinitesimal response tensor that carries information about how the solutions to the Navier–Stokes equation respond to the presence of small perturbations. The closure is developed via perturbation analysis of Eq. (5.138) and a generalization of Eq. (5.153) that considers the velocity field at two separate times. The methodology substitutes expansions of the Fourier velocity coefficients and perturbation tensor away from the Gaussian state into the transfer terms and truncates at the level of "direct" interactions where triads of wave numbers are coupled together, with some wave numbers acting as intermediaries within the interactions. Solutions to the DIA equations provide a detailed look at the history of the transfer function and the energy spectrum that can compare well with experiments.

5.9 Chapter Summary

The aspect of turbulent flow considered in this and the previous chapter is the motion at small scales and its decay, which plausibly has characteristics that can be found in all turbulent flows regardless of how they are created or maintained. In the final period significant evidence for a $K \sim t^{-5/2}$ decay law exists, including a measured Gaussian correlation function. Other possibilities for the decay have some credence also, including a $-3/2$ final period decay law. Some significant experimental evidence reaching to very high Reynolds numbers was seen to support the $t^{-6/5}$ decay law. In this, the predicted decay rate did not show significant change over three orders of magnitude variation in the Reynolds number. An alternative viewpoint with which to view decay at high Reynolds numbers is associated with a t^{-1} decay law that is the necessary outcome of a balancing of vortex stretching and dissipation that emerges after sufficient time during the decay process. Many other observed decay rates seen in experiments and computations can be explained as the consequence of the flow evolving toward the high and low Reynolds number equilibrium states.

The mathematical simplification deriving from isotropy was seen to help develop tractable expressions for the terms in the ϵ equation specialized to isotropic decay. These relations form the basis for the modeling of these terms in general turbulent flow fields. The choice of constants in the modeled expressions are seen to be tied to the predicted decay rate as long as the vortex stretching term in the dissipation equation is assumed to be subsumed by the dissipation term so it has no independent status in the ϵ equation. Allowing for any residual influence of vortex stretching results in a t^{-1} decay law once turbulent production equilibrates with dissipation.

Many of the topics introduced in this and the previous chapter continue to be of great interest within the field of turbulence. Further progress in understanding the physics of small scales can be of significant benefit in improving current techniques for simulating turbulent flows. Some basic results for setting up the decay problem in spectral space were considered, with a brief indication of how this can be used to develop statistical theories at a more fundamental level than is accomplished in conventional turbulence modeling. Though not included here, there are many other important theoretical approaches to the turbulence problem with implications for the physics behind energy decay and production. A number of the most commonly encountered theories may be found in a number of volumes devoted to the theory of turbulent motion, including [21–26].

References

1 Perot, J.B. (2011) Determination of the decay exponent in mechanically stirred isotropic turbulence. *AIP Advances*, 1, 022 104.

2 Batchelor, G.K. and Townsend, A.A. (1948) Decay of isotropic turbulence in the final period. *Proc. Roy. Soc. London A*, 194, 527–543.

3 Batchelor, G.K. and Townsend, A.A. (1948) Decay of isotropic turbulence in the initial period. *Proc. Roy. Soc. London A*, 193, 539–558.

4 Speziale, C.G. and Bernard, P.S. (1992) The energy decay in self-preserving isotropic turbulence revisited. *J. Fluid Mech.*, 241, 645–667.

5 Saffman, P. (1967) The large-scale structure of homogeneous turbulence. *J. Fluid Mech.*, 27, 581–593.

6 Guckenheimer, J. and Holmes, P.J. (1986) *Nonlinear Oscillations, Dynamical Systems and Bifurcations of Vector Fields*, Springer, New York.

7 Stalp, S.R., Skrbek, L., and Donnelly, R.J. (1999) Decay of grid turbulence in a finite channel. *Phys. Rev. Lett.*, 82, 4831–4834.

8 Touil, H., Bertoglio, J.P., and Shao, L. (2002) The decay of turbulence in a bounded domain. *J. Turbulence*, 3, 049.

9 Sinhuber, M., Bodenschatz, E., and Bewley, G.P. (2015) On the relaxation of turbulence at high Reynolds numbers. *Phys. Rev. Lett.*, 114, 034 501.

10 Djenidi, L. and Antonia, R. (2015) A general self-preservation analysis for decaying homogeneous isotropic turbulence. *J. Fluid Mech.*, 773, 345–365.

11 Djenidi, L., Kamruzzaman, M., and Antonia, R. (2015) Power-law exponent in the transition period of decay in grid turbulence. *J. Fluid Mech.*, 779, 544–555.

12 Lesieur, M. (1997) *Turbulence in Fluids*, Kluwer Academic Publishers, Dordrecht, 3rd edn.

13 von Kármán, T. and Howarth, L. (1938) On the statistical theory of isotropic turbulence. *Proc. Roy. Soc. London*, 164A, 192–215.

14 Slater, L.J. (1960) *Confluent Hypergeometric Functions*, Cambridge University Press, Cambridge.

15 Sedov, L.I. (1944) Decay of isotropic turbulent motions of an incompressible fluid. *Dokl. Akad. Nauk. SSSR*, 42, 116–119.

16 Korneyev, A.I. and Sedov, L.I. (1976) Theory of isotropic turbulence and its comparison with experimental data. *Fluid Mech. – Soviet Res.*, 5, 37–48.

17 Bernard, P.S. (1985) Energy and vorticity dynamics in decaying isotropic turbulence. *Intl. J. Eng. Sci.*, 23, 1037–1057.

18 Hildebrand, F.B. (1976) *Advanced Calculus for Applications*, Prentice-Hall, Inc., Englewood Cliffs, New Jersey, 2nd edn.

19 Orszag, S.A. (1970) Analytical theories of turbulence. *J. Fluid Mech.*, 41, 363–386.

20 Kraichnan, R. (1959) The structure of isotropic turbulence at very high Reynolds numbers. *J. Fluid Mech.*, 5, 497–543.

21 McComb, W.D. (2014) *Homogeneous Isotropic Turbulence: Phenomenology, Renormalization and Statistical Closures*, Oxford University Press, Oxford. International Series of Monographs on Physics Vol. 162.

22 Oberlack, M. and Busse, F. (eds) (2014) *Theories of Turbulence*, International Centre for Mechanical Sciences Courses and Lectures No. 442, Springer-Verlag, Wien.

23 Leslie, D.C. (1973) *Developments in the Theory of Turbulence*, Clarendon Press, Oxford.
24 Stanišić, M.M. (1988) *The Mathematical Theory of Turbulence*, Springer, New York, 2nd edn.
25 Frisch, U. (1995) *Turbulence: The Legacy of A. N. Kolmogorov*, Cambridge University Press, Cambridge.
26 Chorin, A.J. (1994) *Vorticity and Turbulence*, Springer-Verlag, New York.
27 Kolmogorov, A.N. (1941) On degeneration of isotropic turbulence in an incompressible viscous fluid. *Dokl. Akad. Nauk. SSSR*, 31, 319–323.
28 Proudman, I. and Reid, W.H. (1954) On the decay of a normally distributed and homogenous turbulent velocity field. *Phil. Trans. Roy. Soc. London A*, 247, 539–558.
29 Batchelor, G.K. and Proudman, I. (1956) The large scale structure of homogeneous turbulence. *Phil. Trans. Roy. Soc. London A*, 248, 369–405.
30 Saffman, P. (1967) Note on decay of homogeneous turbulence. *Phys. Fluids*, 10, 1349–1349.

Problems

5.1 Derive Eq. (5.37) starting from Eq. (5.31).

5.2 Write a code (e.g., in MATLAB) that can numerically solve the coupled equations (5.1) and (5.46) for $K/K(0)$ and $\epsilon/\epsilon(0)$ for self-similar decay. For given initial conditions for R_T and $S^*_{K_0}$, find the solutions for a range of G^*_0 values and make observations of how they are changed by the different G^*_0 regimes.

5.3 Using the code in the previous problem, calculate and plot the time history of the terms in the ϵ equation for small and large values of G^*_0. Interpret the results.

5.4 In an early theory of Kolmogorov [27] it was hypothesized that the Loitsianski integral given by

$$\overline{u^2}(t)\int_0^\infty r^4 f(r/L(t))dr = C, \tag{5.154}$$

where C is a constant and $L(t)$ is a length scale of the energy containing eddies, is an invariant of isotropic decay, that is, remains constant in time (an assumption now not regarded as correct [28, 29]). Show how Eq. (5.154) together with the assumption that the decay rate ϵ depends on $(\overline{u^2})^{3/2}/L$ as in Eq. (4.46), so that

$$\frac{d\overline{u^2}}{dt} = -A\frac{\overline{u^2}^{3/2}}{L} \tag{5.155}$$

means that $\overline{u^2}$ obeys a $t^{-10/7}$ decay law.

5.5 Demonstrate that if the expression on the left-hand side of Eq. (4.78) is an invariant of the decay, and that the integrand has the self-similar form

$$\Psi(r/L(t)) \equiv 3f(r) + r\frac{df}{dr}(r), \tag{5.156}$$

then if Eq. (5.155) also holds it can be shown that $\overline{u^2}$ decays according to $t^{-6/5}$ [30].

5.6 Explain how Eq. (5.88) becomes equivalent to the convection term $\rho\overline{U}_k\frac{\partial R_{ij}}{\partial x_k}$ in Eq. (3.53) as $\mathbf{y} \rightarrow \mathbf{x}$.

6

Turbulent Transport and its Modeling

This chapter considers the phenomenon of turbulent transport that is of essential importance for predicting and understanding turbulent flows. We review some of the classical ideas concerning the modeling of molecular transport and show how these motivate traditional efforts at accounting for turbulent transport. The fundamental differences between molecular and turbulent transport are then made clear with the help of a Lagrangian analysis of backward fluid particle paths that exposes the origin of turbulent transport and the conditions under which traditional models can be justified. The modeling of homogeneous shear flows is then considered, which provides an opportunity to examine the physics of transport in isolation from more complex shear flow phenomena. Some of the traditional turbulent transport models for general flows are examined followed by discussions of the physics and modeling of vorticity transport.

6.1 Molecular Momentum Transport

It was previously suggested that the velocity correlations appearing in the tensor $\rho\overline{\mathbf{u} \otimes \mathbf{u}}$ have the physical interpretation of representing the turbulent flux of momentum. For example, the particular component $\rho(\overline{\mathbf{u} \otimes \mathbf{u}})_{12} = \rho\overline{uv}$ represents a flux of x momentum, as represented by ρU in the y direction caused by turbulent velocity fluctuations v. Some insight into how to model such turbulent transport correlations may be had by considering the seemingly analogous case of molecular momentum transport in non-dense gases associated with the random motion of molecules.

The hydrodynamic velocity field, $\mathbf{U}(\mathbf{x}, t)$, can be interpreted as being the result of averaging the velocities of molecules contained in a small sensing volume centered at $\mathbf{x}$ at time t. If $\mathbf{C}$ denotes the molecular velocities, and $\langle ... \rangle$ denotes the process of averaging over molecules in a sensing volume, then $\mathbf{U} = \langle \mathbf{C} \rangle$. Analogous to Eq. (2.14), a molecular velocity fluctuation vector $\mathbf{c}$ can be introduced so that

$$\mathbf{C} = \mathbf{U} + \mathbf{c}, \tag{6.1}$$

where, by definition, $\langle \mathbf{c} \rangle = 0$. The relevance of this to turbulent transport comes from showing that analogously to the interpretation of $\rho\overline{\mathbf{u} \otimes \mathbf{u}}$ as a turbulent momentum flux, the tensor $\rho\langle \mathbf{c} \otimes \mathbf{c} \rangle$ represents the flux of momentum in a non-dense gas. In this case models of the physics associated with the latter may have some relevance in devising models of the former.

Turbulent Fluid Flow, First Edition. Peter S. Bernard.

Companion website: www.wiley.com/go/Bernard/Turbulent_Fluid_Flow

To see the meaning of $\rho\langle \mathbf{c} \otimes \mathbf{c} \rangle$ consider the particular case of $\rho\langle c_1 c_2 \rangle = \rho\langle C_1 c_2 \rangle$ accounting for the flux of x momentum ρC_1 in the y direction for a unidirectional flow in which $\mathbf{U} = (U(y), 0, 0)$. Let $p(c_1, c_2)$ be the joint probability density function for the c_1 and c_2 molecular velocity components so that

$$p(c_1, c_2)dc_1 dc_2 \tag{6.2}$$

is the probability that the molecular velocity fluctuation in the x direction is between c_1 and $c_1 + dc_1$ and the molecular velocity in the y direction is between c_2 and $c_2 + dc_2$.

Consider molecules next to an interior surface of area dA that is oriented with normal in the y direction, as shown in Figure 6.1. If n is the number density of molecules per unit volume then $np(c_1, c_2)dc_1 dc_2$ is the number density of particles having velocities within the velocity range $c_1 \to c_1 + dc_1, c_2 \to c_2 + dc_2$. All molecules of this group that lie within the fluid volume of size $c_2\ dA\ dt$ abutting the surface will cross it in time dt. Consequently $nc_2\ dA\ dt\ p(c_1, c_2)dc_1 dc_2$ is the number of molecules crossing the surface within the specified velocity range in time dt. If M is the mass of a molecule then the total x momentum component crossing the surface dA in time dt by these particular molecules is

$$nMC_1 c_2 dA\ dt\ p(c_1, c_2)dc_1 dc_2 = \rho C_1 c_2 dA\ dt\ p(c_1, c_2)dc_1 dc_2 \tag{6.3}$$

since $\rho = Mn$. Dividing Eq. (6.3) by $dA\ dt$ and integrating over all the possible molecular velocities, the total momentum flux (momentum/area-sec) is found to be

$$\int_{-\infty}^{\infty} dc_1 \int_{-\infty}^{\infty} dc_2 \rho C_1 c_2 p(c_1, c_2) = \rho\langle c_1 c_2 \rangle, \tag{6.4}$$

by definition of the joint pdf $p(c_1, c_2)$.

The preceding discussion can be generalized to show that $\rho\langle \mathbf{c} \otimes \mathbf{c} \rangle \mathbf{n} = \rho(\mathbf{c} \cdot \mathbf{n})\mathbf{c}$ represents the molecular flux of momentum through a surface oriented in the $\mathbf{n}$ direction moving at the speed of the local velocity field. Moreover, by considering the balance of momentum for an arbitrary material fluid element in a non-dense gas [1], it may be shown that the momentum flux tensor is identical to the stress tensor that appears in the momentum equation for any fluid. Consequently, it is the case that for non-dense gases

$$\sigma = -\rho\langle \mathbf{c} \otimes \mathbf{c} \rangle, \tag{6.5}$$

which may be contrasted with Eq. (2.21), linking the Reynolds stress tensor to the turbulent momentum flux.

Returning to the consideration of $\rho\langle c_1 c_2 \rangle$, according to Eqs. (3.2) and (6.5)

$$\sigma_{12} = \mu \frac{dU}{dy} = -\rho\langle c_1 c_2 \rangle, \tag{6.6}$$

which shows that the molecular momentum flux can be expressed as a gradient momentum diffusion process. This is the key idea that is generalized to encompass momentum transport modeling in turbulent flow. Further guidance in developing turbulent transport models can be obtained by adapting a classical derivation of an expression for μ for non-dense gases [2] to the particular circumstances of turbulent flow.

To see how μ can be modeled, note that the molecules in a non-dense gas can be expected to travel a distance on the order of the mean free path, λ, between collisions,

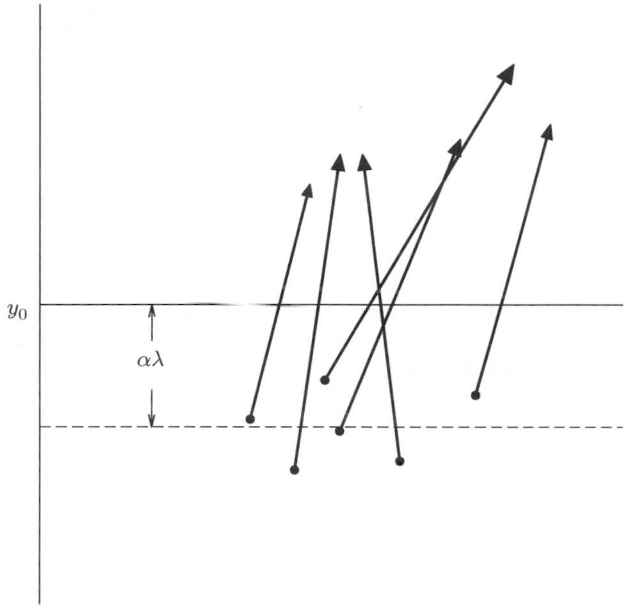

Figure 6.1 Molecules crossing the surface $y = y_0$ in the positive y direction during the mixing time, on average, carry unchanged the momentum from the location $y_o - \alpha\lambda$.

without a change in momentum. Moreover, it can be hypothesized that due to the high molecular speed and the frequency of collisions, a given molecule rapidly shares its momentum with neighboring molecules. Thus, the time between collisions may be taken as representative of a mixing time over which molecules successfully exchange their momentum with the surrounding molecules.

The rate at which molecules cross a surface in units of molecules/area-sec is given by $\frac{1}{4}nc$, where $c \equiv \langle \mathbf{c} \cdot \mathbf{c} \rangle^{1/2}$ is the magnitude of the molecular velocity fluctuation. The molecules crossing a surface $y = y_0$ in the positive y direction, as shown in Figure 6.1, on average carry momentum from the location $y_0 - \alpha\lambda$ across the surface. Here, α is an empirical constant that turns out to be smaller than 1. In the presence of a velocity field $U(y)$ the typical molecule crossing the surface carries, unchanged, the momentum $MU(y_0 - \alpha\lambda)$ that it acquires after its last collision. After crossing the surface, the momentum is exchanged with the fluid on the other side via collisions. By this reckoning, the rate at which momentum is carried upwards across the surface is $\frac{1}{4}\rho c U(y_0 - \alpha\lambda)$. Conversely, molecules traveling across the surface from above to below carry streamwise momentum at the rate $\frac{1}{4}\rho c U(y_0 + \alpha\lambda)$. The net flux of momentum/area-sec is the difference between the momentum going up and that going down so that

$$\rho\langle c_1 c_2 \rangle \approx \frac{1}{4}\rho c(U(y_0 - \alpha\lambda) - U(y_0 + \alpha\lambda)). \tag{6.7}$$

Since $\alpha\lambda$ is very small, in all but the most exceptional circumstances (e.g., in a shock wave where gas properties can change dramatically on the scale of λ), the linear approximation

$$U(y_0 \pm \alpha\lambda) = U(y_0) \pm \alpha\lambda \frac{dU}{dy}(y_0) \tag{6.8}$$

is very well justified. In this case, Eq. (6.7) gives

$$\rho\langle c_1 c_2\rangle \approx -\frac{1}{2}\alpha\lambda\rho c\frac{dU}{dy}, \tag{6.9}$$

showing that the momentum flux obeys a gradient diffusion law. In view of Eq. (6.6) it has thus been deduced that the viscosity is given by

$$\mu = \frac{1}{2}\alpha\lambda\rho c. \tag{6.10}$$

This shows that the viscosity is proportional to the density, mean magnitude of the molecular velocity fluctuation, and the mean free path. Note, as well, that Eq. (6.9) provides additional justification for the linear stress/rate-of-strain law in Eq. (3.2) since the phenomenological connection between σ and the velocity field should be independent of the molecular composition of the fluid.

In summary, it may be concluded that the validity of Eq. (6.9), and by extension Eq. (6.10), depends on the confluence of three essential properties of molecular momentum transport: (i) mixing occurs after a well defined mixing time, (ii) momentum is preserved between collisions, and (iii) a purely linear variation of U over the mixing length can be justified.

6.2 Modeling Turbulent Transport by Analogy to Molecular Transport

Since the form of σ in a Newtonian fluid is given in Eq. (3.2), it is evident that the constitutive model for σ can also be interpreted as a model for the molecular momentum flux. Consequently, by assuming an analogy between molecular and turbulent momentum fluxes, the constitutive model for σ can be appropriately adapted as a representation of the Reynolds stress tensor. Proceeding formally in this case leads to the general 3D constitutive law for the Reynolds stress tensor given by

$$\sigma_t = -\frac{2}{3}\rho K I + \mu_t\left(\nabla\overline{U} + \nabla\overline{U}^t\right), \tag{6.11}$$

which may be contrasted with Eq. (3.11) for the mean viscous stress tensor in incompressible flow. In place of the mean pressure appearing in Eq. (3.11) it is appropriate to have $\frac{2}{3}\rho K$ in Eq. (6.11) in order to guarantee consistency under a contraction of indices. The scalar parameter μ_t is referred to as the *eddy viscosity*, which remains to be determined and may be contrasted with the molecular viscosity μ appearing in Eq. (3.11).

The tensor form in Eq. (6.11) is predicated on the isotropy assumption appropriate to Newtonian fluids where there is no underlying directional bias in the molecular motions causing transport. By adopting the form in Eq. (6.11) for the Reynolds stress tensor, an implicit isotropy assumption is made that is not necessarily compatible with the physics of turbulent momentum transport. In other words, there are reasonable grounds for believing that the random eddying motion in the turbulent flow that causes transport has directional biases originating in the particular 3D distribution of the mean velocity field. This is unlike the situation as regards the Navier–Stokes equation. Below it will be seen that even in simple shear flows next to a boundary the anisotropic form taken by the Reynolds stress tensor on the left-hand side of Eq. (6.11) conflicts with that of the mean rate-of-strain tensor appearing on the right-hand side. Some discussion of alternatives to Eq. (6.11) that accommodate anisotropy will be considered in Chapter 10.

Despite the issue of implied isotropy, the use of Eq. (6.11) is traditional in most RANS modeling where the main focus is on developing theories for predicting μ_t. For transport in a mean velocity field $\overline{U}(y)$, the analogous form of Eq. (6.6) for the turbulent momentum flux derived from Eq. (6.11) is

$$\frac{\sigma_{T_{12}}}{\rho} = -\overline{uv} = \nu_t \frac{d\overline{U}}{dy}, \tag{6.12}$$

where $\nu_t = \mu_t/\rho$ and there is a long history of efforts devoted to finding expressions for ν_t in this case. A general consideration is that if Eq. (6.12) arises from a comparable mechanism as led to Eq. (6.6), then it is reasonable to assume that

$$\mu_t = C_\mu \rho\, \mathcal{V} \mathcal{L} \tag{6.13}$$

where $\mathcal{V}$ and $\mathcal{L}$ are appropriate velocity and length scales that take on the roles played by c and λ, respectively, and C_μ is an empirical constant. How to pick values for the turbulent velocity and length scales remains to be determined, and additional information about the flow is needed in order to select them.

Two of the most common directions that have been pursued in the historical development of transport modeling are the Prandtl mixing length theory [3] and that based on the use of the local values of K and ϵ [4]. For the mixing length theory, which will be considered in Section 9.2.1, the scale $\mathcal{L}$ is presumed to be determined by geometrical considerations such as the distance from a solid boundary. The velocity scale, assuming a mean velocity field $\overline{U}(y)$, is taken to be of the form

$$\mathcal{V} = \left| \mathcal{L} \frac{d\overline{U}}{dy} \right|, \tag{6.14}$$

which represents a rough estimate of how much the local $\overline{U}$ at fluid particle positions might change over the distance that particles move during the mixing time. With these choices of scales, the Prandtl mixing length model takes the form

$$\mu_t = C_\mu \rho \mathcal{L}^2 \left| \frac{d\overline{U}}{dy} \right|, \tag{6.15}$$

where additional considerations are necessary to select a value of $\mathcal{L}$.

In contrast to the mixing layer approach, the K–ϵ model, which is the subject of Section 9.2.2, assumes that $\mathcal{V} = \sqrt{K}$, a choice that is similar in principle to the appearance of c in the molecular transport case. The appropriate length scale is taken to be that associated with the eddy turnover time as defined in Eq. (4.46) and is given by $K^{3/2}/\epsilon$. Consequently,

$$\mu_t = C_\mu \rho \frac{K^2}{\epsilon}, \tag{6.16}$$

where now additional equations are necessary to allow for the calculation of K and ϵ.

Justification for models such as Eqs. (6.15) and (6.16) depends on seeing to what extent the three conditions that form the basis for the molecular transport model in Eq. (6.10) are satisfied. A formal means of answering these questions lies in an analysis based on tracing back through time to see what dynamical processes in the flow create the transport correlation. This Lagrangian analysis of the motion that is considered in the next section explains why and how there is a net transport of momentum in the flow field. It will also bring into focus some of the main issues surrounding when and if the models of the form in Eqs. (6.11) and (6.12) can be justified.

6.3 Lagrangian Analysis of Turbulent Transport

The essence of the gradient transport law that underlies the physical model in Figure 6.1 is that the mixing length is much smaller than the extent of the region over which a linear approximation to the mean velocity field is valid. In the case of turbulent transport, the mixing length – as determined by the action of turbulent eddies in the flow – is surely of a much larger scale than the exceedingly small mean free path that is relevant to the viscous stress tensor. Examination of the typical relationship between a mean velocity field and its local linear approximation, as shown in Figure 6.2, suggests that there is not much leeway in the motion of particles during the mixing time before they travel distances over which a linear approximation to the mean velocity is not acceptable.

Though local linear approximations to $\overline{U}(y)$ may not be legitimate over the mixing length in turbulent flow, the basic concept of turbulent mixing in which fluid particles "carry" momentum from initial points to final points over a mixing time, to thus cause a net momentum transport, need not necessarily be inappropriate. In other words, the use

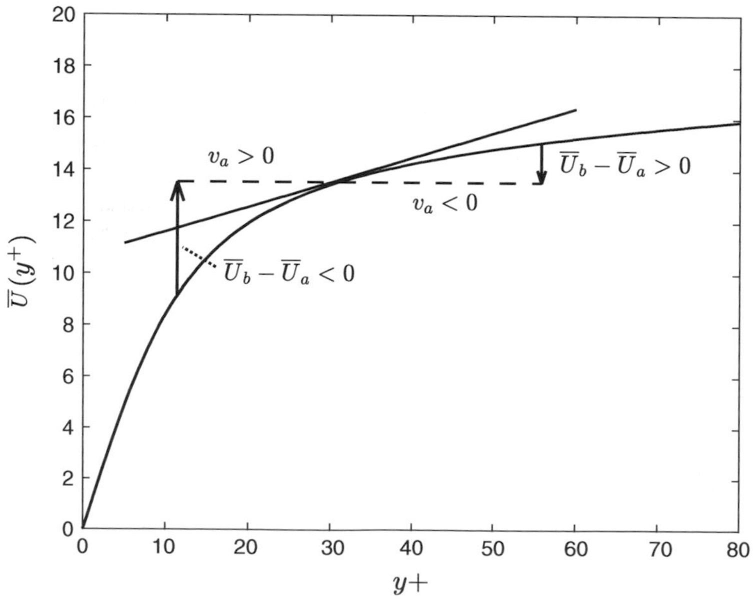

Figure 6.2 A local linear approximation to the mean velocity field $\overline{U}$ at a point **a** in a channel flow is inappropriate for fluid particles traveling significant distances during the mixing time. Fluid particles traveling toward the wall located at $y^+ = 0$ have $v_a < 0$, $\overline{U}_b - \overline{U}_a > 0$, and vice versa for particles traveling away from the wall.

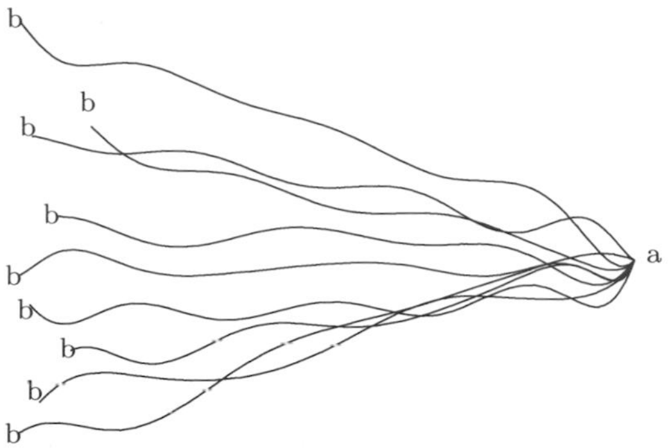

Figure 6.3 Ensemble of paths, each with a different initial position **b**, arriving at **a**.

of a gradient model to express the mean velocity changes may be unjustified, even if the mixing concept itself is realistic. To see this, consider the set of fluid particles arriving at a point **a** at time t for a set of realizations of the random flow field, as shown in Figure 6.3. Paths differ between realizations except for having a common end point. Denote the paths by $\mathbf{X}(s)$, with s being time, so that by definition $\mathbf{X}(t) = \mathbf{a}$. Time $s < t$ denotes motion prior to arriving at **a** while $s > t$ is future motion. For a positive time $\tau > 0$, denote the position of the fluid particle at the earlier time $t - \tau$ as **b** so that $\mathbf{X}(t - \tau) = \mathbf{b}$. In contrast to the point **a**, which is the common destination point for the ensemble of paths, **b** is a random position that varies from one realization of the flow to the next, as illustrated in Figure 6.3.

The velocity of the fluid particle having path $\mathbf{X}(s)$, say $\mathbf{V}(s) \equiv \frac{d\mathbf{X}}{ds}(s)$, must everywhere equal the local velocity of the fluid at the location of the fluid particle. Consequently,

$$\frac{d\mathbf{X}}{ds}(s) = \mathbf{U}(\mathbf{X}(s), s), \tag{6.17}$$

which links the velocity from the Lagrangian viewpoint on the left-hand side to the Eulerian velocity field $\mathbf{U}$ evaluated at the location of the fluid particle $\mathbf{X}(s)$ on the right-hand side.

The quantity $\mathbf{U}(\mathbf{X}(s), s)$ is random both through the velocity field $\mathbf{U}$ – which varies from realization to realization of a turbulent flow – and from the fact that $\mathbf{X}(s)$ changes from one realization to the next. If it is imagined that the ensemble averaging needed in computing the non-random mean field $\overline{\mathbf{U}}$ has been done a priori, then the velocity of a fluid particle can be decomposed according to

$$\mathbf{U}(\mathbf{X}(s), s) = \overline{\mathbf{U}}(\mathbf{X}(s), s) + \mathbf{u}(\mathbf{X}(s), s). \tag{6.18}$$

Note that $\overline{\mathbf{U}}(\mathbf{X}(s), s)$ in this expression is random only because the deterministic mean velocity field is being evaluated at the random point $\mathbf{X}(\mathbf{s})$. At time t Eq. (6.18) gives

$$\mathbf{U}_a = \overline{\mathbf{U}}_a + \mathbf{u}_a \tag{6.19}$$

at the destination point **a**. At $t - \tau$ a similar decomposition using Eq. (6.18) gives

$$\mathbf{U}_b = \overline{\mathbf{U}}_b + \mathbf{u}_b \tag{6.20}$$

with the subscript b referring to the spatial point **b**. Integration of Eq. (6.17) between $t - \tau$ and t gives

$$\mathbf{a} - \mathbf{b} = \mathbf{L} \tag{6.21}$$

where

$$\mathbf{L} = \int_{t-\tau}^{t} \mathbf{U}(\mathbf{X}(s), s)ds \tag{6.22}$$

is the change in position of fluid particles from **b** to **a** in time τ. Using the previous definitions the identity

$$u_a = u_b + (\overline{U}_b - \overline{U}_a) + (U_a - U_b) \tag{6.23}$$

can be written, which is motivated by the desire to express the velocity fluctuation u_a at position **a** in terms of its value earlier in time on the path, namely u_b, plus the factors which have caused it to change since then. Equation (6.23) shows that u_a differs from u_b because of $\overline{U}_b - \overline{U}_a$ representing the difference in the local mean velocity field between the beginning and end points of the path in question, and $U_a - U_b$, which is the change in the fluid particle velocity. $\overline{U}_b - \overline{U}_a$ expresses how the local velocity fluctuation changes merely because the fluid particle has changed position in a non-uniform mean field, while $U_a - U_b$ reflects changes due to acceleration or deceleration of the fluid particle. Thus, even for non-accelerating particles for which $U_a = U_b$, the local velocity fluctuation components can change due to changes in the local mean velocity field experienced along the particle path.

Substituting Eq. (6.23) into $\overline{u_a v_a}$ yields the identity

$$\overline{u_a v_a} = \overline{u_b v_a} + \overline{v_a(\overline{U}_b - \overline{U}_a)} + \overline{v_a(U_a - U_b)}. \tag{6.24}$$

Among the three terms on the right-hand side, the first has the property of converging to $\overline{u_a v_a}$ for small τ and becoming zero as τ is increased. In essence, the time when $\overline{u_b v_a}$ reaches zero gives an upper limit to the mixing time. Within this time period the correlation in $\overline{u_a v_a}$ is created by events transpiring in the flow field. For times larger than this there is no correlation between u_b and v_a.

The second term on the right-hand side of Eq. (6.24), henceforth referred to as the displacement transport term $\Phi_D = \overline{v_a(\overline{U}_b - \overline{U}_a)}$, is a formal mathematical statement of the idea behind the classical turbulent mixing argument that has eddies carrying momentum over a mixing time to then share their momentum with the surrounding fluid. If a local linear approximation to the mean velocity field is assumed to be valid for the movement of fluid particles over the mixing time, as will be considered in the next section, then Φ_D will yield a gradient model. However, this is a separate step that does not have to be taken. To see how Φ_D comes about, consider the circumstances in Figure 6.2. A particle with $v_a < 0$ has most likely descended from above so that $\overline{U}_b - \overline{U}_a > 0$. The opposite will be the case for $v_a > 0$ so that fluid particles ejecting outwards from below will have $\overline{U}_b - \overline{U}_a < 0$. Either way the displacement transport term is negative, and hence on average it is not zero.

The last term in Eq. (6.24) contains contributions to transport specifically from changes in the momentum of fluid particles during the mixing time that have come about due to their acceleration or deceleration by the action of viscous and pressure forces. Phenomena such as this are absent in the physics of the molecular model leading to Eqs. (6.12), since molecules are assumed to retain their momentum over the mixing time.

Numerical evaluation of the terms in Eq. (6.24) using backward particle paths collected from a DNS of the flow between two parallel plates [5] (i.e., channel flow, to be considered in the next chapter) provides a means of discerning the relative importance

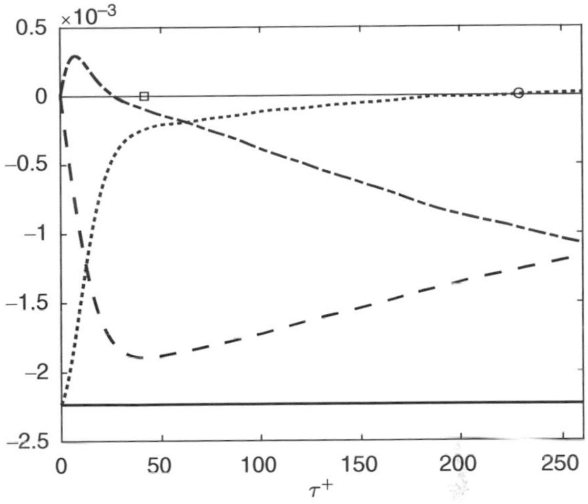

Figure 6.4 Decomposition in Eq. (6.24) at $y^+ = 54.8$. $\cdots$, $\overline{u_b v_a}$ with zero point, τ_m, denoted by a circle; $--$, $\overline{v_a(\overline{U}_b - \overline{U}_a)}$ with minimum denoted by a square; $-\cdot-$, $\overline{v_a(U_a - U_b)}$; —, $\overline{u_a v_a}$.

of the different physical processes responsible for the Reynolds stress. A typical result, which is similar to that occurring at any point in the channel, is shown in Figure 6.4, where Eq. (6.24) is evaluated for the particular location $y^+ = 54.8$, where $y^+ = yU_\tau/\nu$ and $U_\tau = \sqrt{\nu d\overline{U}/dy(0)}$ is referred to as the *friction velocity*. As expected, $\overline{u_b v_a}$ goes to zero with increasing time delay. Most significantly, the displacement term is seen to reach a minimum that is of the same magnitude as $\overline{u_a v_a}$. The time delay at which this occurs, say τ_D, is a natural choice for the mixing time since it reflects the strongest manifestation of the mixing idea.

The dominance of displacement transport in accounting for the Reynolds stress is illustrated in Figure 6.5, where the terms in Eq. (6.24) are evaluated across the lower half of a channel covering the region $0 \leq y^+ \leq 1000$. The negativity of $\overline{uv}$ in this half of a channel reflects the fact that it represents transport of streamwise momentum toward the lower boundary (i.e., in the negative y direction). In the figure, the remaining terms in Eq. (6.24), which have a small effect, are plotted together. To more fully explain the transport process that leads to the result in Figure 6.5 it is necessary to consider how the trends in these curves originate in specific events occurring in the turbulent field. The next section considers this question by examining some aspects of the ensembles of fluid particle paths that have been used in evaluating the terms in Eq. (6.24).

6.4 Transport Producing Motions

In order to focus on the particular kinds of motions in a turbulent flow that produce displacement transport it is helpful to distinguish those paths that are most responsible for the existence of the displacement term. Thus, consider a correlation such as $\overline{u_a v_a}$ that is negative in the lower half of the channel. Events for which u_a and v_a have opposite signs must take precedence over events where they have the same sign, if the correlation is to be negative. Consider the N paths that contribute to each of the correlations in Eq. (6.24) at a particular y^+ value ordered from the smallest to the largest contributors. When the term in question is negative, the ranking is from most positive to the most negative, as shown in Figure 6.6(a) for the data for $\overline{uv}$ at $y^+ = 84.8$. By then calculating partial sums $\sum_{i=1}^{n} u^i v^i$, $n = 1, 2, \ldots N$, as seen in Figure 6.6b, it can be seen that a point is reached,

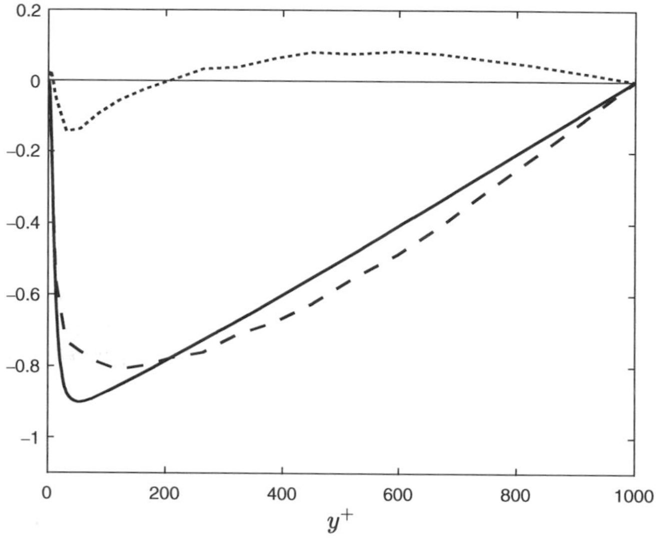

Figure 6.5 Evaluation of Eq. (6.24) at τ_D computed across the channel. —, $\overline{u_a v_a}$; – –, $\overline{(\overline{U}_b - \overline{U}_a)v_a}$; · · · , $\overline{(U_a - U_b)v_a} + \overline{u_b v_a}$.

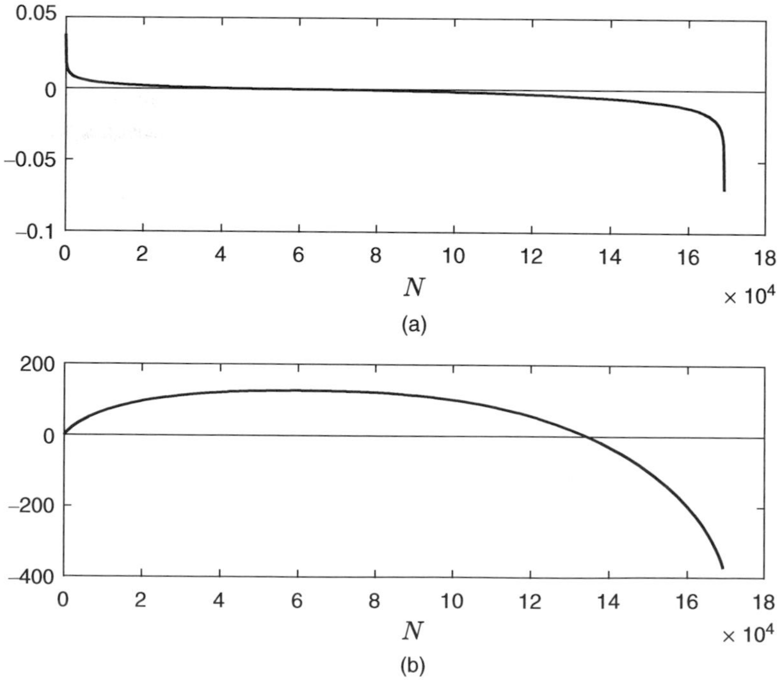

Figure 6.6 Contributions to $\overline{uv}$ at $y^+ = 84.8$ from a data set consisting of 169,344 points in the lower channel half. (a) Individual contributions ranked from largest to smallest. (b) Cumulative sum of contributions in (a) showing zero crossing at $N_0 = 134{,}543$.

say $n = N_0$, where the sums change sign from positive to negative. The collection of contributions for $n \geq N_0$ may be viewed as the reason why the correlation is negative, since the other contributions cancel out between plus and minus. The fraction of events that are responsible for the correlation given by $(N - N_0)/N$ reveals useful information about how the Reynolds shear stress comes about.

Figure 6.7 is a plot for each of the terms in Eq. (6.24) of the fraction of events at each y^+ position whose contribution is not canceled by events with the opposite sign. It is seen in Figure 6.7 that for a large portion of the channel, Φ_D and $\overline{uv}$ follow a very similar trend in which approximately 20% of events are responsible for their occurrence. The percentage for each of these rises toward 30% at $y^+ = 30$ near the wall. Toward the center line the terms drop toward zero, as does $\overline{uv}$ itself, as the natural cancellation of plus and minus events must take over in a symmetric flow field. Figure 6.7 also shows that the contributions of $\overline{u_b v_a}$ and $\overline{v_a(U_a - U_b)}$ are the result of relatively rare events in the flow field. On the whole there is a great cancellation of plus and minus events for these terms so that, in fact, for virtually the entire channel considerably less than 1% of events in the samples explain why these are non-zero.

By examining in detail the flow field in the neighborhood of some of the individual events making significant contributions to the Reynolds shear stress, it becomes possible to gain an understanding of the kinds of motion that produce transport and to see what causes these motions. Two examples of paths that are among the largest contributors to Φ_D in a particular simulation are shown in Figures 6.8 and 6.9 together with velocity quiver plots in the cross section of the channel at the particle locations. In both figures the velocity plot reveals the presence of vortical objects with a significant streamwise orientation. The particle paths that accompany them are driven by the action of the vortices. In the case of Figure 6.8 a high-speed particle is being swept toward the wall in what is referred to as a *sweep* motion, which tends to dominate the events leading to the

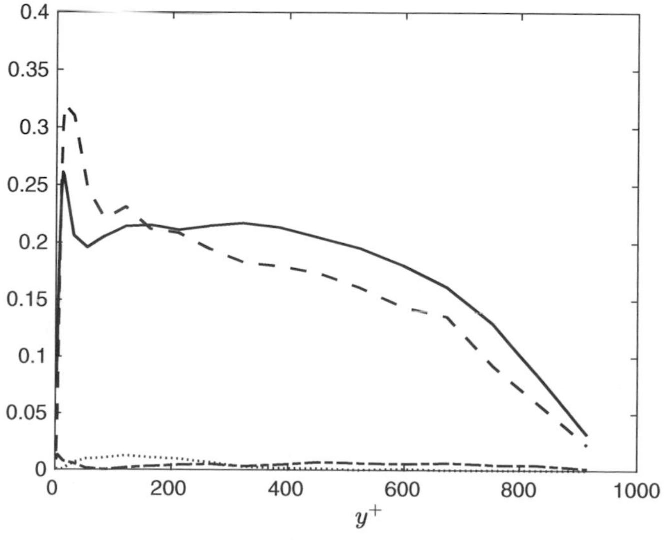

Figure 6.7 Fraction of points in the data ensembles that account for the local computed values of the terms in Eq. (6.24). —, $\overline{u_a v_a}$; −−, $(\overline{U}_b - \overline{U}_a)v_a$; − · −, $\overline{(U_a - U_b)v_a}$; · · · , $\overline{u_b v_a}$.

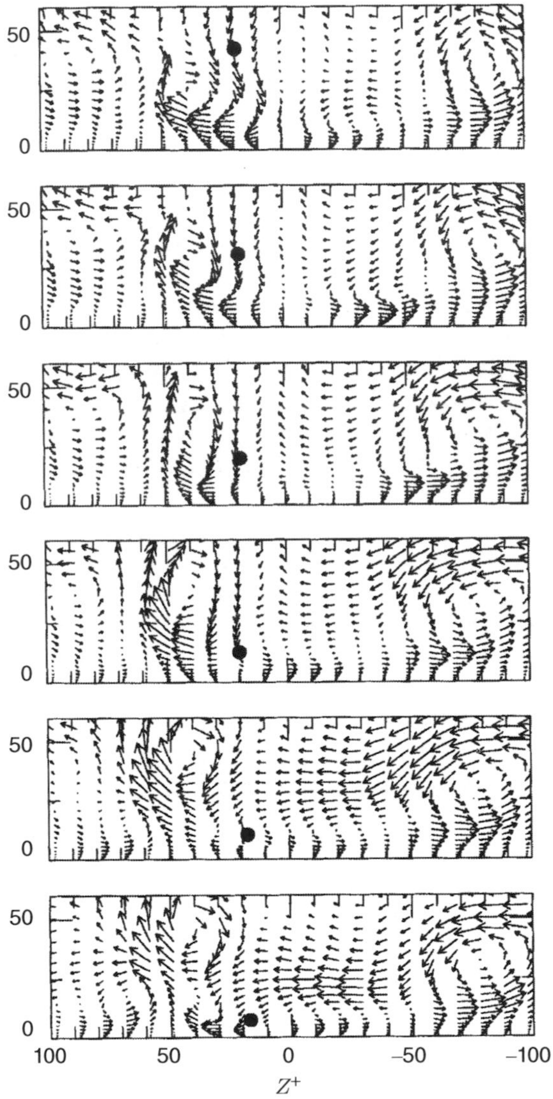

Figure 6.8 Fluid particle arriving at $y^+ = 7.3$ [7] due to a sweep event. Time increases moving from top to bottom image. Reprinted with the permission of Cambridge University Press.

Reynolds shear stress inside the buffer layer [6]. Figure 6.9 shows a low-speed particle being carried away from the wall in an example of the *ejection* events that are most common just outside the buffer layer. The spatial scale over which the paths extend in these figures is determined by the coherent vortices that control the motion of fluid particles over the mixing time. Conversely, the mixing time may be thought of as the time over which coherent vortices can exert influence over the motions of fluid particles. During this time the net momentum flux associated with the Reynolds stress is created. If transport is linked to the presence of vortices in the turbulent flow, then understanding the nature of vortical structures is an important part of explaining the physics of turbulence. The nature and dynamics of the vortical structure in turbulence near walls will be addressed in Chapter 8.

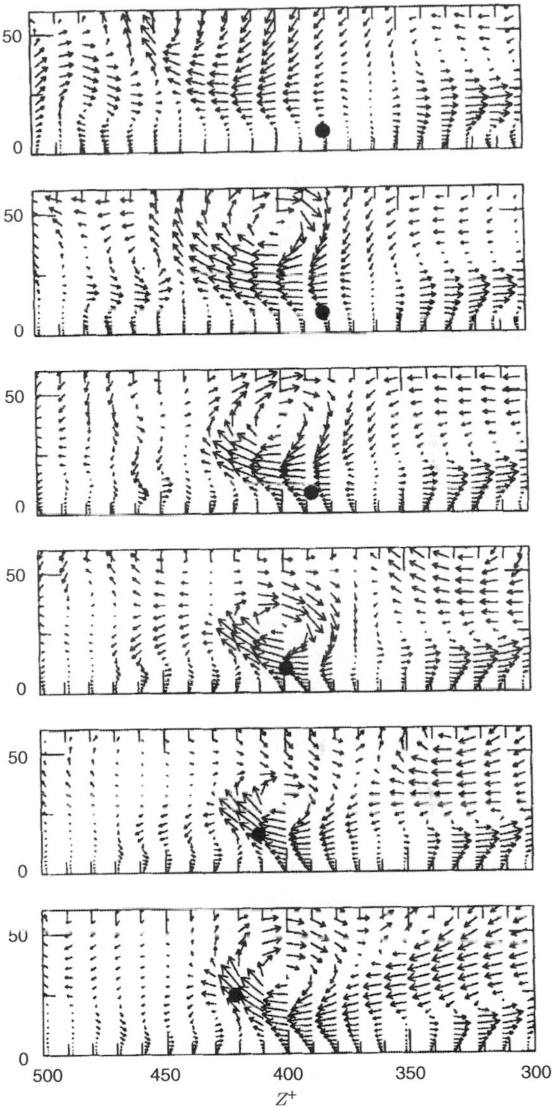

Figure 6.9 Fluid particle arriving at $y^+ = 24.6$ [7] due to an ejection event. Time increases moving from top to bottom image. Reprinted with the permission of Cambridge University Press.

6.5 Gradient Transport

Since gradient transport modeling, as in Eq. (6.6), is the most common way of modeling the Reynolds stress tensor, it is of interest to see how such a law fits in with the formal transport analysis given in the previous section. If gradient transport is valid then it should arise from the displacement transport mechanism under the further hypothesis that the change in the local mean velocity field along particle paths during the mixing time can be approximated by a linear variation in $\overline{U}$. In the latter case, according to Eq. (6.21)

$$\overline{U}_b = \overline{U}(\mathbf{a} - \mathbf{L}) \tag{6.25}$$

and expanding the right-hand side in a Taylor series

$$\overline{U}_b \approx \overline{U}_a - L_2 \frac{d\overline{U}}{dy} \tag{6.26}$$

where higher order terms have been omitted. Substituting Eq. (6.26) into the displacement transport term yields

$$\overline{v(\overline{U}_b - \overline{U}_a)} = -\overline{vL_2}\frac{d\overline{U}}{dy} + \dots . \tag{6.27}$$

Consequently, if a gradient transport is valid, it would arise in the displacement transport term and would have an eddy viscosity, as given in Eq. (6.12) in the form

$$\nu_t = \overline{vL_2}. \tag{6.28}$$

Defining a Lagrangian auto-correlation function

$$f_{vv}(s) = \frac{\overline{v(\mathbf{X}(t), t)v(\mathbf{X}(t+s), t+s)}}{\overline{v(\mathbf{X}(t), t)^2}}, \tag{6.29}$$

and substituting for L_2 using Eq. (6.22), (6.28) may be written as

$$\nu_t = \overline{v^2}\mathcal{T}_{22}, \tag{6.30}$$

where $\mathcal{T}_{22}$ is a Lagrangian integral scale defined by

$$\mathcal{T}_{22} = \int_{-\infty}^{0} f_{vv}(s)ds, \tag{6.31}$$

where it is safe to assume that $f_{vv}(s) = 0$ for $|s|$ large enough.

If gradient transport were physically accurate then the eddy viscosity in Eq. (6.30) should well approximate the model eddy viscosity

$$\nu_t = \frac{-\overline{uv}}{d\overline{U}/dy} \tag{6.32}$$

that is needed to guarantee an accurate calculation of $\overline{U}$. As seen in Figure 6.10 where these are compared, there are some significant discrepancies between Eqs. (6.30) and (6.32) that imply that non-gradient physics may have some role in determining the Reynolds shear stress. Conflicts in the eddy viscosity are seen to occur beyond $y^+ = 500$ where the physical eddy viscosity is constant and the modeled version decreases, and near the wall where there are small quantitative and qualitative differences between the exact and modeled eddy viscosity.

If the Reynolds shear stress predicted by a gradient law using the physical eddy viscosity given in Eq. (6.30) is compared to the exact $\overline{uv}$, as is done in Figure 6.11, it is seen that, in fact, gradient physics is not a viable description of turbulent transport. In particular, the small differences in eddy viscosities near the wall in Figure 6.10 translate into very large differences in Figure 6.11. Moreover, despite the much larger discrepancy in viscosities far from the wall seen in Figure 6.10, by $y^+ = 400$ it can be said that gradient physics is not unreasonable. This may reflect the greater suitability of a linear model to the mean velocity field in the central region of channel flow. The errors in the gradient model are most severe in the near-wall region where the variation in the mean velocity

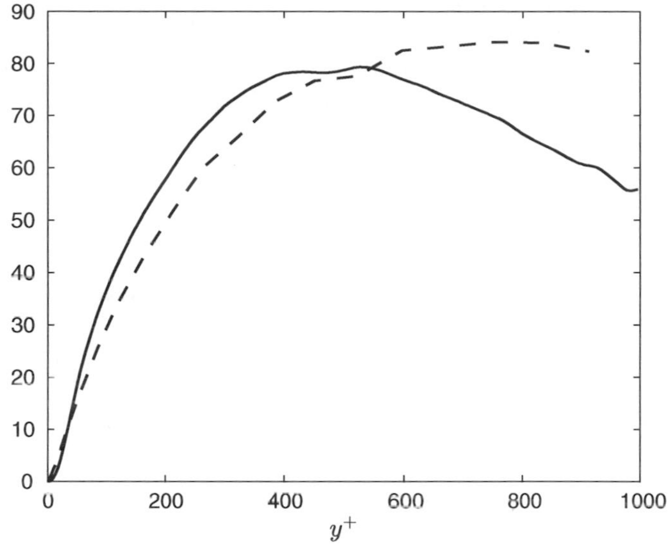

Figure 6.10 Eddy viscosity in channel flow: ––, $\mathcal{T}_{22}^+\overline{v^2}^+$; —, ν_t^+.

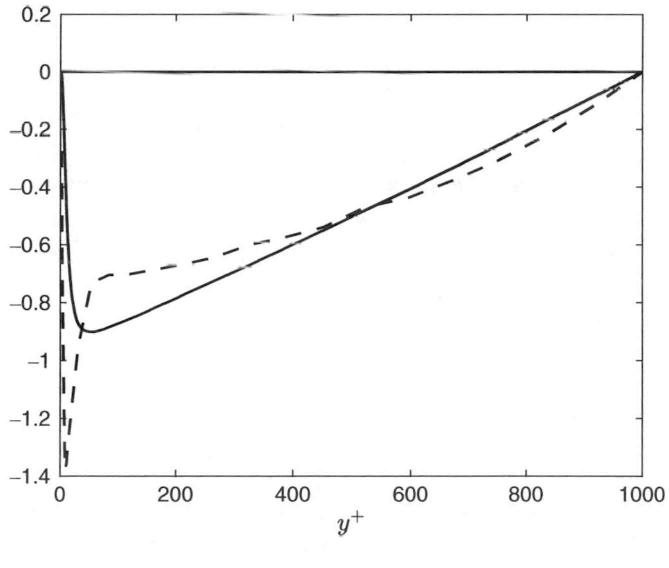

Figure 6.11 Inadequacy of gradient transport physics: —, $\overline{uv}^+$; ––, $-\mathcal{T}_{22}^+\overline{v^2}^+\, d\overline{U}^+/dy^+$.

field is more rapid and the particles travel much further than the distances over which a linear approximation to $\overline{U}$ is valid.

Despite the shortcomings of Eq. 6.12 that have been revealed in Figure 6.11, the eddy viscosity model seemingly has the virtue that a positive ν_t is computed everywhere in the domain. This occurs because ν_t is everywhere positive as a result of the fact that $\overline{uv}$ and $d\overline{U}/dy$ are zero at the channel centerline where they both change sign. This result enables the use of a gradient law despite the departures of ν_t from its physical value,

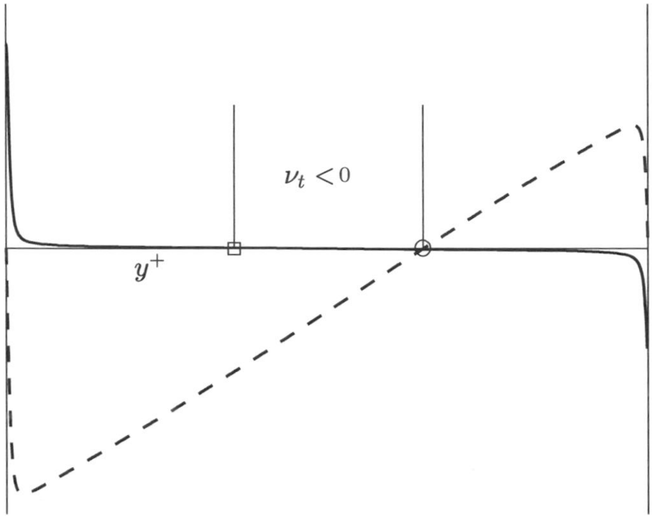

Figure 6.12 A region of negative eddy viscosity can be expected in a channel flow with one rough wall (on the left) and one smooth wall (on the right). —, $d\overline{U}/dy$; ––, $\overline{uv}$; □, zero crossing of $d\overline{U}/dy$; ○, zero crossing of $\overline{uv}$.

a circumstance that is largely fortuitous and helps explain the difficulties of gradient transport modeling that will be observed in later chapters. To see why Eq. (6.12) is fragile, consider the well-known case of the flow in a channel with one smooth and one rough wall [8, 9]. Rough walls are associated with higher shearing so the mean velocity change is steeper on the rough wall than the smooth wall. This means that $\overline{U}$ reaches a peak value shifted from the channel centerline to a point closer to the rough wall. Consequently, the zero crossing in the mean velocity gradient is closer to the rough than the smooth wall, as illustrated in Figure 6.12. On the side of the channel with higher shear, where the wall is rough, the Reynolds shear stress $\overline{uv}$ also must increase in magnitude. In view of the linear variation in $\overline{uv}$ throughout the center of the channel, its zero crossing must thus move toward the smooth wall as also shown in the figure. This creates a zone in the middle of the channel where $\nu_t < 0$. In fact, at the zero crossing in $d\overline{U}/dy$ the eddy viscosity is infinite. Such behavior of the eddy viscosity fits no physical model of diffusion. Moreover, numerical schemes tend to become unstable if the eddy viscosity is negative.

An important implication of this discussion is that there is likely to be many instances where the assumed gradient law in Eq. (6.11) imposes a positive ν_t in places where a strict evaluation of its terms would require $\nu_t < 0$. This is in addition to computing any errors in the magnitude of ν_t or the likelihood that ν_t cannot have the same values for all components of the $\overline{u_i u_j}$ tensor.

6.6 Homogeneous Shear Flow

The idealized flow consisting of a constant, uniform mean shear $S \equiv d\overline{U}/dy > 0$ superimposed on homogeneous, isotropic turbulence offers a convenient vantage point with which to assess the consequences of the gradient Reynolds stress model in a situation where it is likely to be a very good, if not exact, model of momentum transport. For this flow Eq. (6.27) has no truncated terms since there are no higher derivatives to $\overline{U}$. At the same time there is no compelling reason why the remaining terms in Eq. (6.24) should be significant, particularly in view of the symmetry of the homogeneous shearing. Consequently, it is expected that the relation

$$\overline{uv} = -\mathcal{T}_{22}\overline{v^2}S \tag{6.33}$$

is well justified.

The K equation (3.37) in homogeneous shear flow reduces to

$$\frac{dK}{dt} = \mathcal{P} - \epsilon, \tag{6.34}$$

where the production term

$$\mathcal{P} = -\overline{uv}\frac{d\overline{U}}{dy} \tag{6.35}$$

is non-zero owing to the presence of non-zero $\overline{uv}$. This is the fundamental difference between homogeneous shear flow and the previously considered case of isotropic decay. Note that $\mathcal{P} > 0$, since $\overline{uv} < 0$ for transport associated with a gradient $S > 0$.

The presence of a mean shear also causes the production terms $P_\epsilon^1, P_\epsilon^2$ in the ϵ equation (3.42) to be non-zero and in this case the ϵ equation takes the form

$$\frac{d\epsilon}{dt} = P_\epsilon^1 + P_\epsilon^2 + P_\epsilon^4 - \Upsilon_\epsilon. \tag{6.36}$$

The insistence on isotropy in homogeneous shear flow means that the terms P_ϵ^4 and Υ_ϵ will not be different than they were in the previous consideration of isotropic decay.

In homogeneous shear flow, the definitions in Eqs. (3.43) and (3.44) can be combined and simplified to yield

$$P_\epsilon^1 + P_\epsilon^2 = -2\nu\overline{\omega_1\omega_2}\, S, \tag{6.37}$$

in which the isotropic relation to the effect that

$$\overline{\frac{\partial u_i}{\partial x_m}\frac{\partial u_j}{\partial x_n}} = \overline{\frac{\partial u_i}{\partial x_n}\frac{\partial u_j}{\partial x_m}} \tag{6.38}$$

is used.

To bring Eq. (6.37) into a useful form a means of evaluating the vorticity covariance $\overline{\omega_1\omega_2}$ is required. In fact, it has been observed that in simple shear flows the approximation

$$\frac{\overline{\omega_1\omega_2}}{\zeta} \sim \frac{\overline{uv}}{2K} \tag{6.39}$$

is excellent. For example, Figure 6.13 compares the ratio of the two sides of the equation in a channel flow and is seen to be approximately constant. Taking the constant to be C_{ϵ_1} and noting that $\epsilon = \nu\zeta$ leads to the model

$$P_\epsilon^1 + P_\epsilon^2 = C_{\epsilon_1}\mathcal{P}\frac{\epsilon}{K} \tag{6.40}$$

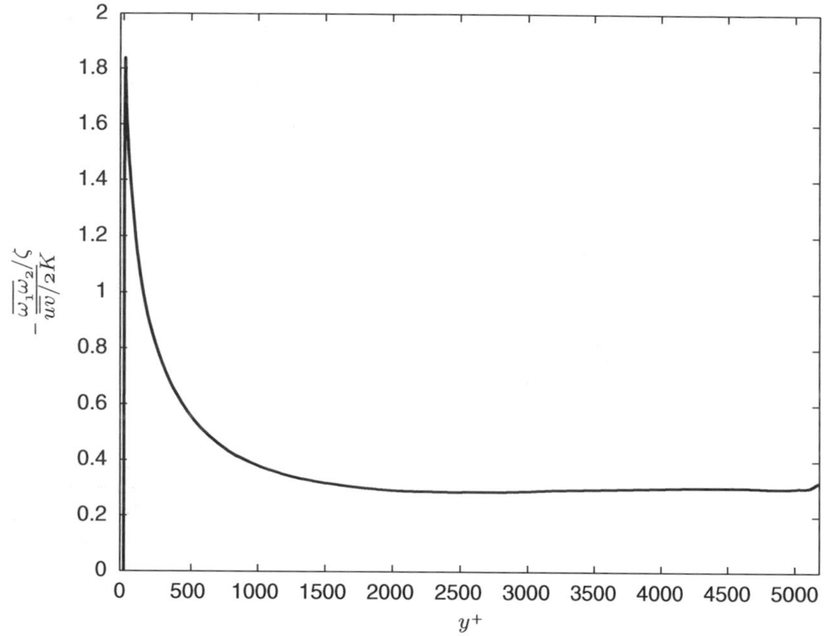

Figure 6.13 Demonstration of the near constancy of $(\overline{\omega_1\omega_2}/\zeta)/(\overline{uv}/2K)$ in the central region of channel flow for $R_\tau = 5186$. Data taken from [10].

For the terms P_ϵ^4 and Υ_ϵ in Eq. (6.36), the expressions (5.71) and (5.72), respectively, are also available for the present application, so that

$$P_\epsilon^4 - \Upsilon_\epsilon = S_K^* R_T^{\frac{1}{2}} \frac{\epsilon^2}{K} - G^* \frac{\epsilon^2}{K}. \tag{6.41}$$

Allowing for the possibility that vortex stretching does not get preempted by dissipation, as was done in Section 5.4, it is again assumed that

$$G^* = (S_K^* - C_{\epsilon_3})\sqrt{R_T} + C_{\epsilon_2} \tag{6.42}$$

where $C_{\epsilon_3} = 0$ produces the standard model that is traditionally used in RANS modeling. Substituting Eqs. (6.40)–(6.42) into (6.36) gives the ϵ equation in the form

$$\frac{d\epsilon}{dt} = C_{\epsilon_1} \mathcal{P} \frac{\epsilon}{K} + C_{\epsilon_3} R_T^{\frac{1}{2}} \frac{\epsilon^2}{K} - C_{\epsilon_2} \frac{\epsilon^2}{K}, \tag{6.43}$$

which is to be solved in conjunction with (6.34) once an appropriate model of P is introduced.

Before considering the nature of solutions to the coupled equations (6.34) and (6.43), it is helpful to consider evidence from experimental and theoretical studies concerning the behavior of K and ϵ in homogeneous shear flow. Note that even though it is unlikely that homogeneous shear flow can be created in the real world, nonetheless an approximation to such flow can be achieved in physical experiments in which turbulence is generated by an appropriately designed grid placed in a wind tunnel [11, 12]. It is also possible to approximate homogeneous shear flow numerically [13]. For the most part these studies have observed exponential growth in K and ϵ at approximately the same rate over the

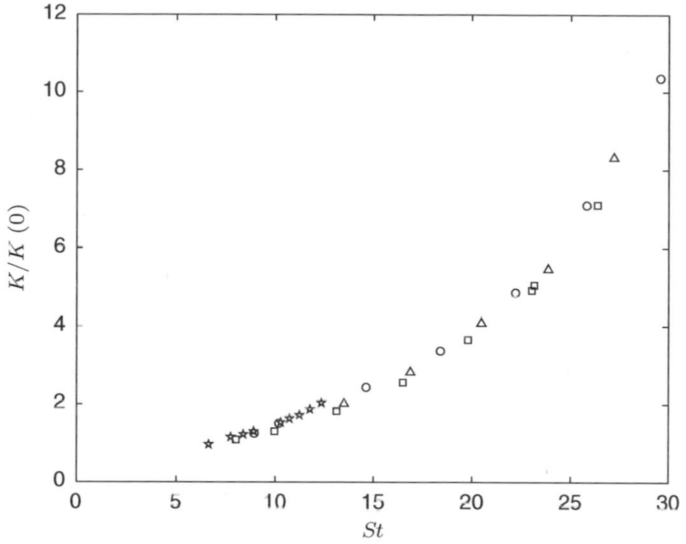

Figure 6.14 Measured $K/K(0)$ in homogeneous shear flow for $St < 30$ from [15]. Reprinted with permission of Cambridge University Press.

times which have been examined, generally $St < 30$, as shown in Figure 6.14. If K and ϵ grow at the same rate, then their ratio should become constant, and indeed such a trend has been seen in experiments. Generally, the following asymptotic relations are observed:

$$\frac{SK}{\epsilon} \approx 6.0 \tag{6.44}$$

and

$$\frac{\mathcal{P}}{\epsilon} \approx 1.80, \tag{6.45}$$

though in some calculations using LES (e.g., [14]) even by time $St = 30$ these terms have not equilibrated.

Introducing the non-dimensional time $t^* = St$ and the function $K^*(t^*)$ satisfying

$$K^*(St) = K(t)/K(0), \tag{6.46}$$

then Eq. (6.34) may be transformed to

$$\frac{dK^*}{dt^*} = \frac{\epsilon}{SK}\left(\frac{\mathcal{P}}{\epsilon} - 1\right)K^* \tag{6.47}$$

so that at times for which Eqs. (6.44) and (6.45) hold, Eq. (6.47) may be solved, yielding

$$K^*(t^*) = e^{0.13t^*}. \tag{6.48}$$

Consistent with the exponential growth seen in Figure 6.14 a similar analysis can be applied to the ϵ equation with the traditional assumption that $C_{\epsilon_3} = 0$, resulting in an equation for $\epsilon^*(t^*)$ that also has exponential growth as a solution (see Problem 6.2).

The long-time behavior of homogeneous shear flow thus far remains beyond the purview of either physical experiments or simulation. Two principal scenarios have been put forward as to what ultimately happens at longer times. Townsend [16] hypothesized that, given enough time, a production-equals-dissipation equilibrium results in which the terms on the right-hand sides of Eqs. (6.34) and (6.43) achieve balances so that K and ϵ asymptote to constant finite values. It is clear from Eq. (6.47) that for this to happen Eq. (6.45) must be valid only for relatively short times. An alternative viewpoint is that the exponential growth in K and ϵ observed for relatively short times continues unabated for long times as well. Such a scenario can only pertain to the ideal case since, clearly, unlimited growth in K is not physical.

To explore the nature of the solutions to Eqs. (6.34) and (6.43) for long times it is necessary to numerically solve the coupled equations once a model for $\overline{uv}$ is applied so that the set of equations forms a closed system. For this purpose, Eq. (6.33) can be taken as an exact result, and invoking the isotropy assumption it follows that $\overline{v^2} = 2K/3$. To model $\mathcal{T}_{22}$, one possibility is to make the plausible assumption that $\mathcal{T}_{22}$ is proportional to the eddy turnover time K/ϵ, which is, in effect, the approach followed in the K–ϵ closure to be discussed in Chapter 9. Specifically, assuming that

$$\mathcal{T}_{22} = \frac{3}{2} C_\mu \frac{K}{\epsilon} \tag{6.49}$$

in this case yields a coupled set of equations for K and ϵ for homogeneous shear flow in the form

$$\frac{dK}{dt} = C_\mu \frac{K^2}{\epsilon} S^2 - \epsilon \tag{6.50}$$

and

$$\frac{d\epsilon}{dt} = C_{\epsilon_1} C_\mu K S^2 + C_{\epsilon_3} R_T^{\frac{1}{2}} \frac{\epsilon^2}{K} - C_{\epsilon_2} \frac{\epsilon^2}{K}. \tag{6.51}$$

Solutions to Eqs. (6.50) and (6.51) can be used to illustrate the physics of homogeneous shear flow at long times. For this purpose the choice $C_{\epsilon_1} = 1.45$ is made in conformity to the traditional value used to obtain a desired decay rate in the case of isotropic turbulence decay, as discussed in Section 5.3. The standard values $C_\mu = 0.09$ and $C_{\epsilon_2} = 1.9$ are traditionally obtained by arguments having their origin in near-wall turbulent boundary layers and will be discussed in Chapter 8. For the classical analysis of homogeneous shear flow $C_{\epsilon_3} = 0$ has the effect of eliminating vortex stretching as an independent process in the same way it was done for isotropic turbulence in the previous chapter. A solution with $C_{\epsilon_3} = 0.1$ is also considered here so as to get insight into the effect of keeping vortex stretching in the homogeneous shear flow case.

The numerical solutions for K and ϵ for times comparable to those seen in typical data sets are shown in Figure 6.15. The computed growth rate in K and ϵ is similar in nature and magnitude to that seen in a LES simulation [17]. If the same solutions are extended in time, as shown in Figure 6.16, it is seen that the traditional K–ϵ solution with $C_{\epsilon_3} = 0$ grows exponentially for all time. In contrast, the presence of an unrestricted role for vortex stretching by having $C_{\epsilon_3} \neq 0$ leads to an eventual curtailment of the exponential growth in K and ϵ, and the appearance of a production-equals-dissipation equilibrium along the lines predicted by Townsend [16].

The asymptotic values of K and ϵ can be obtained by setting the right-hand sides of Eqs. (6.50) and (6.51) to zero. Solving these coupled equations yields the

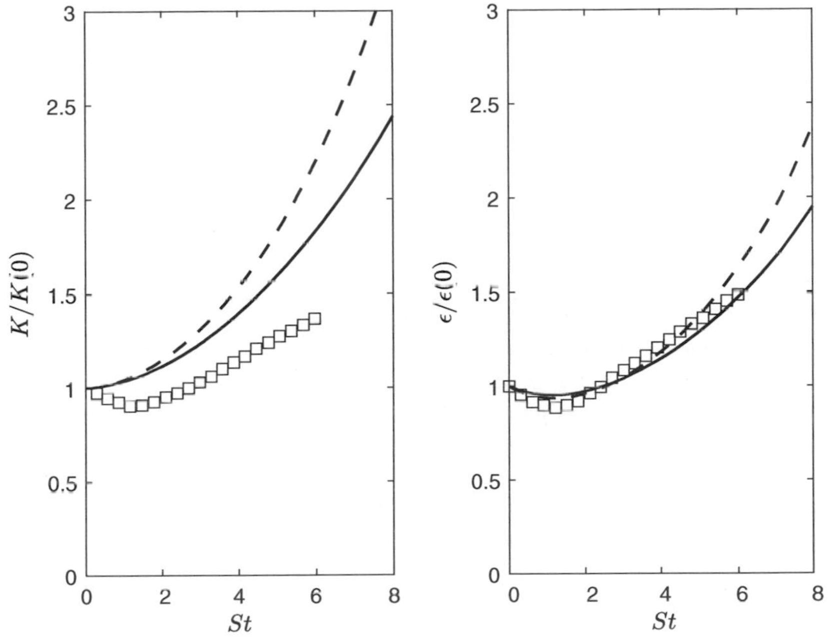

Figure 6.15 Computed solution for $K/K(0)$ (left) and $\epsilon/\epsilon(0)$ (right) in homogeneous shear flow: —, with vortex stretching; – –, without vortex stretching; ○, LES calculation [17].

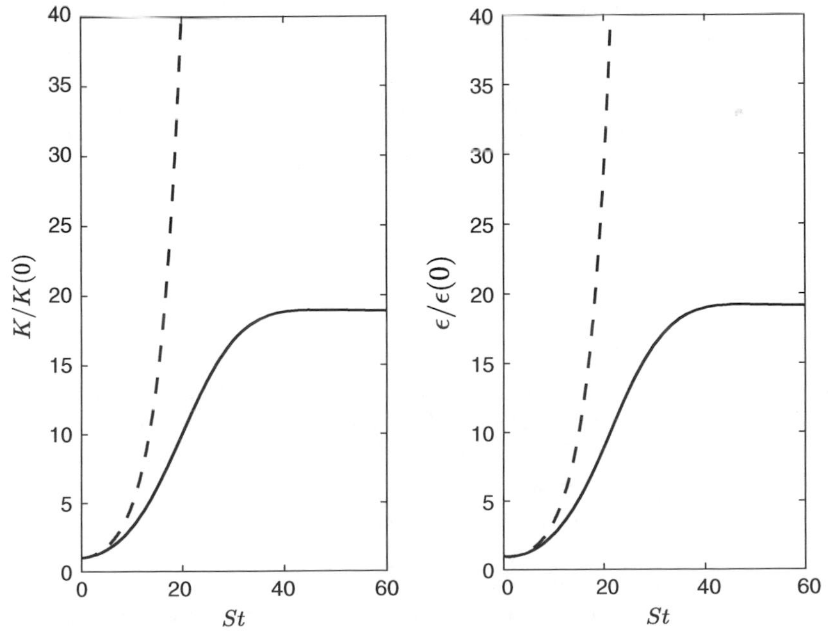

Figure 6.16 Computed solutions for $K/K(0)$ (left) and $\epsilon/\epsilon(0)$ (right) in homogeneous shear flow: —, with vortex stretching; – –, without vortex stretching.

asymptotic solutions

$$K_\infty = \frac{\sqrt{C_\mu}(C_{\epsilon_2} - C_{\epsilon_1})^2}{C_{\epsilon_3}^2} \nu S \tag{6.52}$$

and

$$\epsilon_\infty = \frac{C_\mu(C_{\epsilon_2} - C_{\epsilon_1})^2}{C_{\epsilon_3}^2} \nu S^2, \tag{6.53}$$

respectively. The magnitudes of the asymptotic values in these equations increase rapidly with the inverse of the square of C_{ϵ_3}. When this parameter is small and vortex stretching is relatively weak, the solution is likely to pass through an initial growth phase which for all intents and purposes has the appearance of unbounded exponential growth. Eventually, however, the growth in dissipation caused by vortex stretching leads to a saturation in K which then causes an equilibration in the dissipation rate. The model highlights the essential role of vortex stretching acting as an additional source of dissipation that ultimately controls what kind of equilibrium takes place.

While it appears that current experimental and numerical simulations of homogeneous shear flow, which are limited to $St < 30$, do not necessarily contradict either of the alternative models, it may be argued that the production-equals-dissipation scenario is the more likely of the two outcomes. In particular, despite the fact that model equations were used in the analysis, it is apparent that the issue is equivalent to the question as to whether or not vortex stretching retains an independent role with respect to dissipation in the ϵ (or ζ) equation at high Reynolds numbers. An affirmative answer in the case of isotropic decay in Section 5.3 led to a richer understanding of the decay process. A similar conclusion holds here too, particularly when it is recognized that the alternative depends on the exact annihilation by dissipation of vortex stretching as an independent physical process. The stretching term with $C_{\epsilon_3} \neq 0$ has also been shown [18] to provide a means of curing the stagnation point anomaly to the effect that the coupled K–ϵ equations have a tendency to overpredict turbulent energy at stagnation points in the flow field.

6.7 Vorticity Transport

The complexities of modeling momentum transport, including the likelihood that pressure and viscous forces affect the momentum of fluid particles during the mixing time, led Taylor [19, 20] to develop a vorticity transport model for use in Eq. (3.58). In this way the need to model the Reynolds stress was avoided. While the vorticity in 2D flows will stay constant on fluid particle paths apart from viscous effects, such is not the case in 3D flow due to the action of vortex stretching during the mixing time. Thus, it is not a priori evident that vorticity transport modeling necessarily offers benefit over Reynolds stress modeling by, for example, having the transport quantity stay fixed on particle paths prior to mixing with the surrounding fluid. Nevertheless it is instructive to consider how vorticity transport might be modeled by applying the backward particle path analysis scheme used in Section 6.3.

To adapt the methodology previously applied to momentum transport to the case of vorticity transport [21] begin with the exact decomposition

$$\omega_j^a = \omega_j^b + \left(\overline{\Omega}_j^b - \overline{\Omega}_j^a\right) + \left(\Omega_j^a - \Omega_j^b\right), \tag{6.54}$$

whose terms mimic the velocity decomposition in Eq. (6.23). Integration of Eq. (3.65) along a fluid particle path gives for the last term

$$\Omega_j^a - \Omega_j^b = \int_{t-\tau}^{t} \Omega_k(s)\frac{\partial U_j}{\partial x_k}(s)ds + \int_{t-\tau}^{t} \nu\nabla^2\Omega_j(s)ds. \tag{6.55}$$

The first integral in Eq. (6.55) accounts for changes in vorticity due to vortex stretching and the second from viscous effects. Substituting Eqs. (6.54) and (6.55) into the flux correlation yields

$$\overline{u_i^a\omega_j^a} = \overline{u_i^a(\overline{\Omega}_j^b - \overline{\Omega}_j^a)} + \int_{t-\tau}^{t} \overline{u_i\Omega_k(s)\frac{\partial U_j}{\partial x_k}(s)}ds + \int_{t-\tau}^{t} \overline{\nu u_i\nabla^2\Omega_j(s)}ds \tag{6.56}$$

where it has been assumed that τ is sufficiently large so that the mixing condition $\overline{u_i^a\omega_j^b} = 0$ is satisfied. Vortex stretching terms are accommodated in Eq. (6.56) by the second term on the right-hand side.

The first term on the right-hand side of Eq. (6.56), representing displacement transport, can be developed in the same way as was done for momentum transport in Eq. (6.27). Thus, substituting a Taylor series expansion for $\overline{\Omega}_j^b$ yields the gradient model

$$\overline{u_i(\overline{\Omega}_j^b - \overline{\Omega}_j)} = -\int_{t-\tau}^{t} \overline{u_i u_k(s)}ds\,\frac{\partial\overline{\Omega}_j}{\partial x_k}\ \ldots \tag{6.57}$$

The stretching term in Eq. (6.56) can be approximated in such a way as to extract linear terms in the mean vorticity consistent with the modeling that results in Eq. (6.57). In particular, substituting mean and fluctuating quantities for $\Omega_i(s)$ and $\partial U_i/\partial x_j(s)$ in the stretching term leads to the model

$$\int_{t-\tau}^{t} \overline{u_i\Omega_k(s)\frac{\partial U_j}{\partial x_k}(s)}ds = \int_{t-\tau}^{t} \overline{u_i\frac{\partial u_j}{\partial x_k}(s)}ds\,\overline{\Omega}_k + \ldots \tag{6.58}$$

where among the truncated terms are those that contain only fluctuating quantities and are non-linear in the mean field. Also omitted is a term containing $\partial\overline{U}_i/\partial x_j$ which may be seen to be identically zero in the case of a simple shear flow. For simplicity in what follows the viscous term in Eq. (6.56) is also omitted. Putting the results together yields the vorticity transport law

$$\overline{u_i\omega_j} = -\int_{t-\tau}^{t} \overline{u_i u_k(s)}ds\,\frac{\partial\overline{\Omega}_j}{\partial x_k} + \int_{t-\tau}^{t} \overline{u_i\frac{\partial u_j}{\partial x_k}(s)}ds\,\overline{\Omega}_k, \tag{6.59}$$

where a gradient term is complemented by a vortex stretching term that is also first order in τ.

6.7.1 Vorticity Transport in Channel Flow

Channel flow, which is considered in detail in the next chapter, offers an opportunity to consider the physical appropriateness of Eq. (6.59) and especially discover the role that the vortex stretching terms might have in the overall flux of vorticity. For unidirectional shear flows with mean velocity $\overline{U}(y)$, $\overline{\Omega}_3 = -d\overline{U}/dy$ is the only non-zero mean vorticity component. Specializing Eq. (6.59) to this case it is found that five of the nine vorticity flux components are predicted to be identically zero, specifically $\overline{u\omega_1}$, $\overline{v\omega_2}$, $\overline{w\omega_3}$, $\overline{u\omega_2}$, and $\overline{v\omega_1}$. In fact, this result is consistent with the implications of the symmetry and homogeneity conditions in channel flow. The remaining correlations given by Eq. (6.59) are non-zero and are given by

$$\overline{w\omega_1} = \mathcal{Q}_{313}\overline{w\frac{\partial u}{\partial z}}\overline{\Omega}_3 \tag{6.60}$$

$$\overline{w\omega_2} = \mathcal{Q}_{323}\overline{w\frac{\partial v}{\partial z}}\overline{\Omega}_3 \tag{6.61}$$

$$\overline{v\omega_3} = -\mathcal{T}_{22}\overline{v^2}\frac{d\overline{\Omega}}{dy} + \mathcal{Q}_{233}\overline{v\frac{\partial w}{\partial z}}\overline{\Omega}_3 \tag{6.62}$$

and

$$\overline{u\omega_3} = -\mathcal{T}_{12}\overline{uv}\frac{d\overline{\Omega}_3}{dy} + \mathcal{Q}_{133}\overline{u\frac{\partial w}{\partial z}}\overline{\Omega}_3. \tag{6.63}$$

In these relations additional Lagrangian integral scales are defined similar to the way that $\mathcal{T}_{22}$ is defined in Eq. (6.31). Thus,

$$\overline{uv}\ \mathcal{T}_{12}(\tau) = \int_{t-\tau}^{t} \overline{u(\mathbf{x},t)v(\mathbf{X}(s),s)}ds \tag{6.64}$$

and

$$\overline{w\frac{\partial u}{\partial z}}\ \mathcal{Q}_{313}(\tau) = \int_{t-\tau}^{t} \overline{w(\mathbf{x},t)\frac{\partial u}{\partial z}(\mathbf{X}(s),s)}ds, \tag{6.65}$$

with similar definitions for the scales $\mathcal{Q}_{323}$, $\mathcal{Q}_{233}$, and $\mathcal{Q}_{133}$. Among the four correlations in Eqs. (6.60)–(6.63) that are not identically zero, $\overline{w\omega_1}$ and $\overline{w\omega_2}$ do not originate in gradient physics and would be predicted to be zero if the influence of vortex stretching were neglected.

A partial test of the validity of Eqs. (6.60)–(6.63) in channel flow of width $2h$ can be made with the help of DNS data to evaluate all quantities except for the time scales, which are expected to be non-negative. Thus the question is whether or not positive, constant values of the time scales can be found to cause the predicted fluxes on the right-hand sides of Eqs. (6.60) to (6.63) agree with their DNS values on the left-hand side. The result of such an exercise is shown in Figures (6.17–6.20) using the values (in wall units) $\mathcal{T}_{22}^+ = 4.8$, $\mathcal{T}_{12}^+ = 12.3$, $\mathcal{Q}_{233}^+ = 5.5$, $\mathcal{Q}_{323}^+ = 9.5$, $\mathcal{Q}_{133}^+ = 16.3$, and $\mathcal{Q}_{313}^+ = 0.95$. Evidently, the figures suggest that by including the first-order vortex stretching model, the essentials of the turbulent vorticity flux can be accounted for. In particular, the gradient terms in Eqs. (6.62) and (6.63) capture most of the transport away from the boundary, while near the wall the stretching terms successfully account for non-gradient transport.

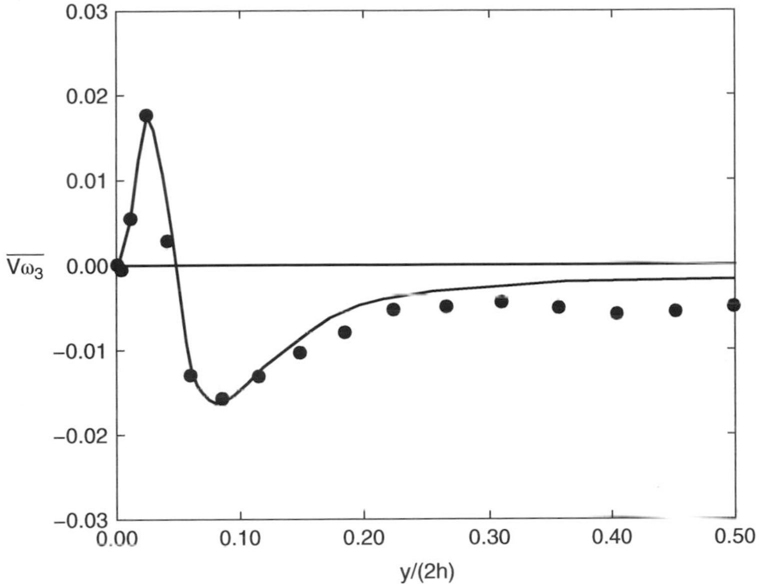

Figure 6.17 $\overline{v\omega_3}$: •, DNS results; —, prediction from Eq. (6.62). From [21]. Copyright ©Springer-Verlag.

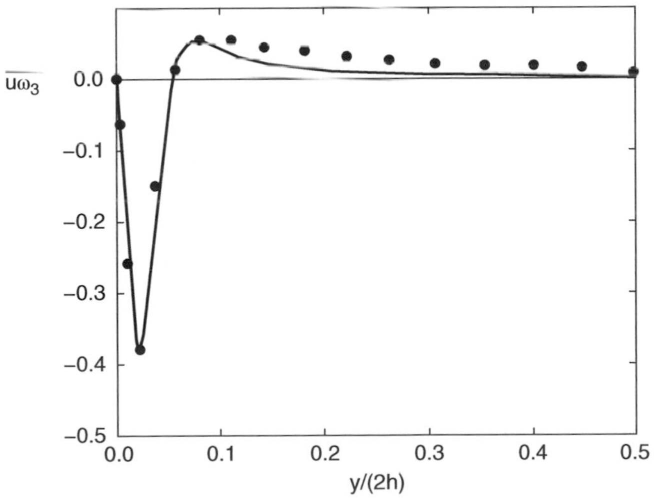

Figure 6.18 $\overline{u\omega_3}$: •, DNS results; —, prediction from Eq. (6.63). From [21]. Copyright ©Springer-Verlag.

Similarly, the computed forms of $\overline{w\omega_1}$ and $\overline{w\omega_2}$ in Eqs. (6.60) and (6.61) are satisfactorily explained by the trends in the stretching correlations. In particular, in all four cases the stretching terms have the correct sign and functional behavior near the wall. These results suggest that Eq. (6.59) may provide a means of reasonably capturing the physics of vorticity transport in general 3D flows.

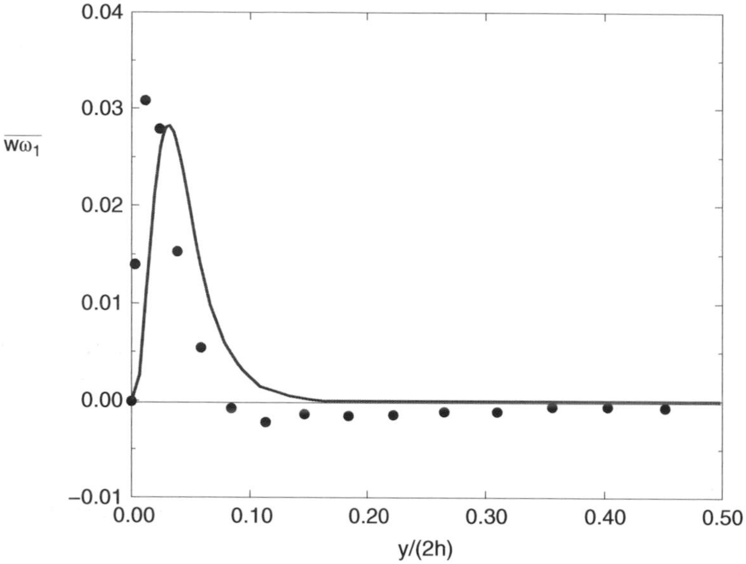

Figure 6.19 $\overline{w\omega_1}$: •, DNS results; —, prediction from Eq. (6.60). From [21]. Copyright ©Springer-Verlag.

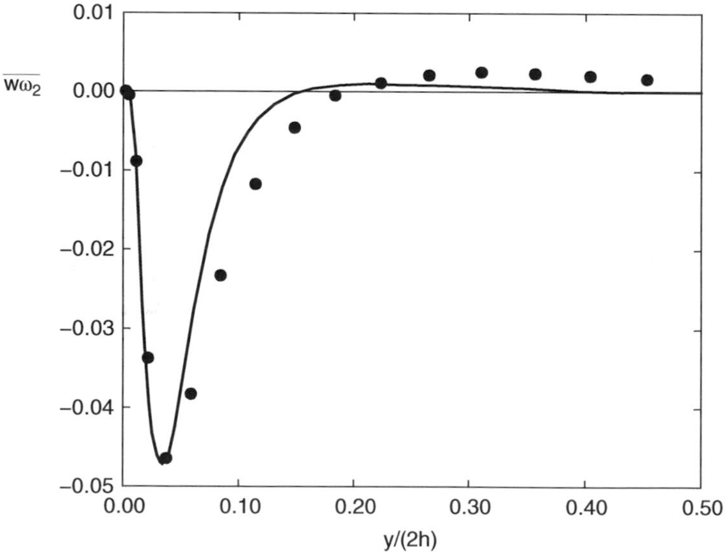

Figure 6.20 $\overline{w\omega_2}$: •, DNS results; —, prediction from Eq. (6.61). From [21]. Copyright ©Springer-Verlag.

6.8 Chapter Summary

The focus of this chapter has been on the physics of turbulent transport, the property of turbulent flows that is likely the most important as far as engineering interest is concerned. It is not surprising then that prediction of the rate of turbulent momentum transport as represented by the Reynolds stresses forms a major part of the effort to derive accurate turbulent flow prediction schemes. Here it was shown that the requirements necessary to justify a gradient Reynolds stress model are typically not satisfied in turbulent flow. The problem stems from the large distances covered by fluid particles in the mixing time, which invalidates the local linear approximation to $\overline{U}$ that underlies the gradient law. In some flows, the gradient model can be used successfully so long as the eddy viscosity can be chosen empirically. Similar considerations for vorticity transport show the need to include a term accounting for the effect of vortex stretching on the transport process if a viable model is to be developed.

A dynamical link was described between the Reynolds shear stress on the one hand and the presence of vortical eddying motions in turbulent flow on the other. The presence of a net Reynolds shear stress appears to be the result of a vortex -driven process that includes approximately one-fifth of the flow field at any given time. This provides an incentive to pursue approaches such as LES where much of the essential vortical activity can be captured. The discussion of homogeneous shear flow in this chapter showed that the standard modeling of the ϵ equation may be improved by the addition of an independent vortex stretching effect. This is seen to enrich the physical analysis of both isotropic decay and homogeneous shear flow.

References

1 Bernard, P.S. (2015) *Fluid Dynamics*, Cambridge University Press, Cambridge.

2 Chapman, S. and Cowling, T.G. (1952) *The Mathematical Theory of Non-Uniform Gases*, Cambridge University Press, London, 2nd edn.

3 Prandtl, L. (1932) Zur turbulenten ströhren und längs platten. *Ergebn. Aerodyn. Versuchanstalt*, 4, 18–29.

4 Jones, W.P. and Launder, B.E. (1972) The prediction of laminarization with a two-equation model of turbulence. *Intl. J. Heat and Mass Transfer*, 15, 301–314.

5 Bernard, P.S. and Erinin, M.A. (2018) Fluid particle dynamics and the non-local origin of Reynolds shear stress. *J. Fluid Mech.*, 847, 520–551.

6 Wallace, J.M., Eckelmann, H., and Brodkey, R.S. (1972) The wall region in turbulent shear flow. *J. Fluid Mech.*, 54, 39–48.

7 Bernard, P.S., Thomas, J.M., and Handler, R.A. (1993) Vortex dynamics and the production of Reynolds stress. *J. Fluid Mech.*, 253, 385–419.

8 Maubach, K. and Rehme, K. (1972) Negative eddy diffusivities for asymmetric turbulent velocity profiles? *Int. J. Heat Mass Transfer*, 15, 425–432.

9 Hanjalić, K. and Launder, B.E. (1972) Fully developed asymmetric flow in a plane channel. *J. Fluid Mech.*, 51, 301–335.

10 Lee, M. and Moser, R.D. (2015) Direct numerical simulation of channel flow up to $Re_\tau \approx 5200$. *J. Fluid Mech.*, 774, 395–415.

11 Tavoularis, S. and Corrsin, S. (1981) Experiments in nearly homogeneous turbulent shear flows with a uniform mean temperature gradient, part 1. *J. Fluid Mech.*, 104, 311–347.

12 Tavoularis, S. and Karnik, U. (1989) Further experiments on the evolution of turbulent stresses and scales in uniformly sheared turbulence. *J. Fluid Mech.*, 204, 457–478.

13 Sukheswalla, P., Vathianathan, T., and Collins, L.R. (2013) Simulation of homogeneous turbulent shear flows at higher Reynolds numbers: numerical challenges and a remedy. *J. Turbulence*, 14, 60–97.
14 Rogers, M.M., Moin, P., and Reynolds, W.C. (1986) *The structure and modeling of the hydrodynamic and passive scalar fields in homogeneous turbulent shear flow, Tech. Rep. TF-25*, Stanford University.
15 Rohr, J.J., Itsweire, E.C., Helland, K.N., and Van Atta, C.W. (1988) An investigation of the growth of turbulence in a uniform mean shear flow. *J. Fluid Mech.*, 187, 1–33.
16 Townsend, A.A. (1956) *The Structure of Turbulent Shear Flow*, Cambridge University Press, New York.
17 Bardina, J., Ferziger, J.H., and Reynolds, W.C. (1983) *Improved turbulence models based on large-eddy simulation of homogeneous, incompressible turbulent flows, Tech. Rep. TF-19*, Stanford University.
18 Abid, R. and Speziale, C. (1996) The freestream matching condition for stagnation point turbulent flows: An alternative formulation. *J. Appl. Mech.*, 63, 95–100.
19 Taylor, G.I. (1915) Eddy motion in the atmosphere. *Phil. Trans. Roy. Soc. London*, 215, 1–26.
20 Taylor, G.I. (1932) The transport of vorticity and heat through fluids in turbulent motion. *Proc. Roy. Soc.*, 135A, 685–705.
21 Bernard, P.S. (1990) Turbulent vorticity transport in three dimensions. *Theor. Comp. Fluid Dyn.*, 2, 165–183.

Problems

6.1 For a simple shear flow $\overline{U}(y)$ examine the predicted models for the normal Reynolds stresses as given by Eq. (6.11).

6.2 Investigate the behavior of ϵ as given by Eq. (6.43) using the same strategy as was done in the case of the K equation leading to Eqs. (6.47) and (6.48). For this assume that $C_{\epsilon_3} = 0$ and apply the asymptotic relations (6.44) and (6.45).

6.3 Show that the equilibrium states in homogeneous shear flow as predicted by the K–ϵ closure keeping C_{ϵ_3} not zero are equal to Eqs. (6.52) and (6.53).

6.4 Compute the solution to the K–ϵ equations (6.50) and (6.51) for a range of parameter values using an ODE solver such as ode45 in MATLAB. In particular, investigate how the solution changes when C_{ϵ_3} is and is not zero. Compute the asymptotic values of the quantities in Eqs. (6.44) and (6.45).

6.5 Reformulate Eqns. (6.50) and (6.51) based on having $\mathcal{T}_{22}$ constant and show that this does not affect the properties of the long-term behavior of the solutions in so far as whether or not C_{ϵ_3} is zero or not zero.

7

Channel and Pipe Flows

Channel, pipe, and boundary layer flows have the common property of containing flow past a geometrically simple solid boundary, with many of their differences concentrated in the region outside the immediate wall vicinity. Note that though the pipe wall is not flat, nonetheless for high Reynolds number flow it may be assumed that the velocity in the viscous region near the wall is not strongly affected by curvature. Comparisons between the channel, pipe, and boundary layer flows [1–3] show that while their mean statistics have much in common near the solid boundary, there are some differences that likely reflect the effects of the various outer flows on the near- wall structure.

Investigations of channel, pipe, and boundary layer flows have tended to concentrate on those particular aspects of each flow that are most readily studied. For example, channel flow is substantially easier to simulate via DNS than a pipe or boundary layer so this has been computed more extensively and at higher Reynolds numbers. On the other hand, reproducing channel flow in a laboratory setting is difficult since it requires a large spanwise physical extent in order to eliminate end effects at the sides of the flow field [4, 5]. Placing the walls closer together can alleviate end effects at the expense of making measurements difficult in the narrow space between walls. Such problems grow more difficult with Reynolds number.

Pipe flow is substantially easier to study in a laboratory than channel flow and so there is considerable information available about turbulence in pipes taken from physical experiments [6]. Turbulent boundary layers forming on a flat plate provide an ideal environment with which to study vortical structures in turbulent flow because of the ease of access and the capability they provide for studying the evolution of structure from transition into the fully turbulent region. Boundary layers can be created in wind tunnels or found in nature. In recent years they can be simulated profitably by DNS. Our discussions of the channel, pipe, and boundary layer flows will play to their separate strengths. In this chapter channel and pipe flows are considered, providing a basic background in the physics of turbulence from a statistical point of view. The next chapter considers boundary layers and there, after noting some statistical properties, the structural features of wall turbulence are considered in some detail.

7.1 Channel Flow

The flow between two parallel plates driven by a constant pressure gradient is illustrated in Figure 7.1. Under laminar flow conditions, known as Poiseulle flow, apart from an

Turbulent Fluid Flow, First Edition. Peter S. Bernard.

Companion website: www.wiley.com/go/Bernard/Turbulent_Fluid_Flow

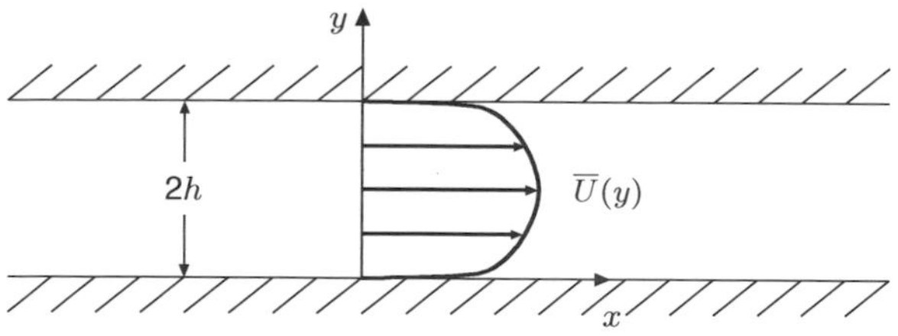

Figure 7.1 Geometry of channel flow.

entrance region where the velocity field will develop with downstream distance, the fluid velocity is fixed in the x direction in a fully developed state. The velocity is parabolic and given by

$$U(y) = \frac{-1}{2\mu}\frac{\partial P}{\partial x}y(2h - y) \tag{7.1}$$

(see [7]). Here, y is the coordinate normal to the plates, located at $y = 0$ and $2h$, x is in the flow direction, and z is the spanwise direction. In an experimental setting Poiseulle flow is generally maintained for Reynolds numbers $R_e = hU_m/\nu$ less than approximately 1000 [8, 9], where

$$U_m = \frac{1}{2h}\int_0^{2h} \overline{U}(y)dy \tag{7.2}$$

is the mean bulk velocity. The actual critical Reynolds number where instabilities cause the appearance of turbulent flow varies from one facility to the next because of local flow conditions. When the channel flow is turbulent the velocity field is 3D, though the mean velocity is given by $\mathbf{U}(y) = (\overline{U}(y), 0, 0)$.

The symmetry of fully developed channel flow is such that its numerical simulation can employ periodic boundary conditions in the spanwise and streamwise directions. An important requirement for this to be accurate is that the spatial extent of the computational domain be large enough so that the flow becomes fully randomized within one period of the computational domain. For example, two-point longitudinal velocity correlation functions such as Eq. (2.38) should vanish before the end of the domain. If not, the turbulent flow structures will not be able to develop naturally to their full extent, so that the downstream end of a structure will affect its upstream end. In this case the dynamics of the flow field is affected by having structures that are, in effect, infinitely long.

The use of periodic boundary conditions means that highly efficient and accurate spectral methods can be applied to the simulations [10]. Consequently, many investigations of the flow physics in shear flows have been based on DNS calculations of channel flow. Channel flow simulations are often categorized using the Reynolds number

$$R_\tau = \frac{U_\tau h}{\nu} \tag{7.3}$$

based on the friction velocity

$$U_\tau \equiv \sqrt{\frac{\tau_w}{\rho}} \tag{7.4}$$

where

$$\tau_w = \mu \frac{d\overline{U}}{dy}(0) \tag{7.5}$$

is the wall shear stress. An attractive feature of this choice of Reynolds number is that it may be thought of as the ratio of the length scale, h, expressing the size of the channel, and ν/U_τ, which is a measure of the size of flow features in the near-wall viscous region. Thus, the ratio of these lengths as expressed by R_τ gives an idea of how many multiples of ν/U_τ there are between the wall and the center of the channel. When this is large there is a clear separation between the viscous wall flow and flow in the interior of the channel.

From the first successful simulation at $R_\tau = 180$ [11] there has been a steady progression of simulations to higher and higher Reynolds numbers. A recent simulation [10] has $R_\tau = 5186$. Interest in increasing R_τ stems from the expectation that some special properties of turbulent flow do not become manifest until a sufficiently high Reynolds number regime where there is a large separation of scales between the smallest and largest.

It is interesting to compare the attributes of channel flow simulations over the almost 30-year period between the $R_\tau = 180$ simulation [11] and the $R_\tau = 5186$ simulation [10]. The former has $R_e = U_m h/\nu = 3300$, the latter has $R_e = 125{,}000$ so that the Reynolds number has increased by a factor of approximately 38. The mesh size has increased from $192 \times 129 \times 160 = 3{,}962{,}880$ mesh points covering a region $4\pi h \times 2h \times 2\pi h$ to a grid with $10240 \times 1536 \times 7680 = 120{,}795{,}955{,}200$ mesh points covering the volume $8\pi h \times 2h \times 3\pi h$, which is three times larger than that of the earlier simulation. The number of mesh points has grown by the factor of 10,161 taking into account the different domain sizes. According to the estimate in Section 4.3 it is expected that for the factor of 38 gain in Reynolds number the mesh would have to grow by the factor $38^{9/4} = 3585$ which is a factor of approximately 2.8 less than the actual increase in mesh points.

As for the speed of the computations, the lower Reynolds number simulation was performed on a Cray XMP supercomputer with theoretical peak speed of 400 megaflops. The simulation required 250 CPU hours and simulated the flow over the dimensionless time interval $TU_\tau/h = 10$. The more recent simulation at a higher Reynolds number was performed on Mira, which is an IBM Blue Gene/Q supercomputer with peak performance of approximately 8.6 petaflops. Under this measure, Mira is approximately 20M times faster than the Cray XMP. This simulation ran for $TU_\tau/h = 7.8$, required 20M corehours that were executed over a 9-month period [12]. Comparison between the simulations separated by 30 years gives a useful perspective with which to appreciate the realities of the ways in which DNS is developing and will continue to develop into the future. In particular, it appears from these results that the previous estimates of how grid and computational cost are likely to rise with Reynolds number, discussed in Section 4.3, are somewhat conservative.

The wealth of information provided by channel flow simulations affords an opportunity to examine many interesting properties of turbulent shear flow without models or missing information. Particularly insightful is the balance of forces that determine the mean velocity as well as the balances of processes that create the Reynolds stresses. Here we consider these and other statistics taken from the $R_\tau = 5186$ [10] and related simulations.

7.1.1 Reynolds Stress and Force Balance

For a fully developed mean flow $\overline{U}(y)$, simplification of Eq. (3.8) for the special properties of channel flow yields

$$0 = -\frac{\partial \overline{P}}{\partial x} + \frac{d}{dy}\left(\mu \frac{d\overline{U}}{dy} - \rho\overline{uv}\right) \tag{7.6}$$

$$0 = -\frac{\partial \overline{P}}{\partial y} - \rho\frac{d\overline{v^2}}{dy} \tag{7.7}$$

for the x and y momentum equations, respectively. The mean spanwise momentum equation has no content since all its terms are zero. Since $\overline{U}, \overline{u^2}, \overline{v^2}$, and $\overline{uv}$ are functions of y only, taking an x derivative of both Eqs. (7.6) and (7.7) shows that $\partial\overline{P}/\partial x$ is independent of x and y so that it is constant everywhere in fully developed channel flow.

Integration of Eq. (7.7) outward from the wall at a fixed x shows that

$$\overline{P}(x, y) = \overline{P}(x, 0) - \rho\overline{v^2}(y) \tag{7.8}$$

since $\overline{v^2} = 0$ at a fixed solid boundary. Consequently, at any arbitrary streamwise locations in fully developed channel flow, $\overline{P}(x, y)$ is a minimum where $\overline{v^2}(y)$ is a maximum. This may be contrasted with the laminar case where the pressure remains constant on every cross-section. Since the streamwise pressure gradient is constant everywhere, the pressure difference between two downstream locations, namely, $\overline{P}(x, y) - \overline{P}(x + L, y)$, is also independent of y.

Integration of (7.6) over the area $0 \leq x \leq L, 0 \leq y \leq 2h$ yields the force balance

$$\Delta\overline{P}\ 2h - 2\tau_w L = 0, \tag{7.9}$$

where

$$\Delta\overline{P} = -L\frac{\partial \overline{P}}{\partial x} = \overline{P}(x, 0) - \overline{P}(x + L, 0) \tag{7.10}$$

is the pressure drop between x locations. In deriving Eq. (7.9) the symmetry of channel flow is used to conclude that $d\overline{U}/dy(0) = -d\overline{U}/dy(2h)$. According to Eq. (7.9) the net pressure force on a finite streamwise segment of the channel flow is balanced by the mean viscous shear acting on the fluid at the wall surface. Even though the Reynolds stress, $\overline{uv}$, does not appear explicitly in Eq. (7.9), it nonetheless exerts an influence on the force balance through its effect on τ_w.

Figure 7.2 compares the mean velocity normalized by the mean centerline velocity for a turbulent channel flow at $R_\tau = 5186$ with the similarly scaled parabolic velocity profile in Poiseulle flow. The turbulent flow profile is distinguished from the Poiseulle flow by its very different near-wall behavior. In effect, high momentum fluid in the central region of turbulent channel flow is much better able to penetrate to points close to the solid surface than is high momentum fluid in the central region of a laminar channel flow. The result is the steep velocity gradients near the wall in the turbulent flow case.

Differentiating Eq. (7.1) and using Eq. (7.5) gives

$$\mu\frac{dU}{dy} = \tau_w\left(1 - \frac{y}{h}\right), \tag{7.11}$$

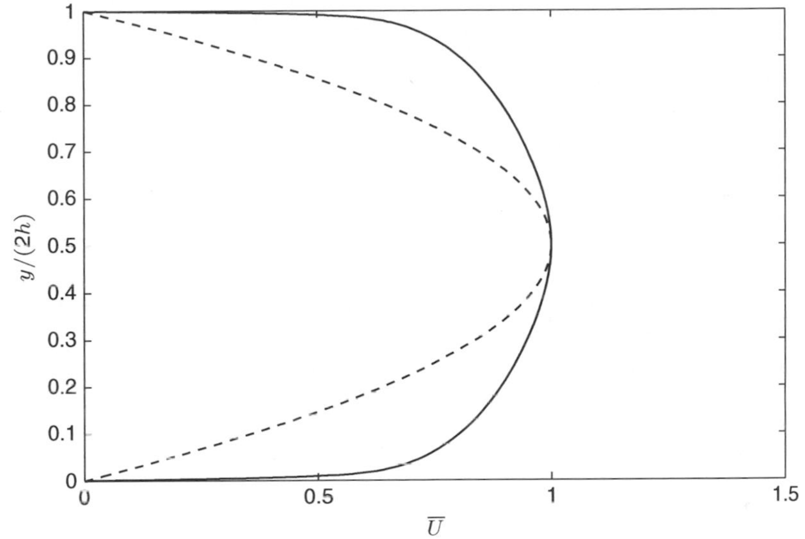

Figure 7.2 Average velocity in channel flow of width $2h$ scaled by mean centerline velocity, U_{cl}: —, turbulent flow; ––, laminar flow.

showing that there is a linear variation in the shear stress across Poiseulle flow in a channel. This means that there is a constant diffusive momentum flux traveling toward each boundary from the fluid interior. The net viscous force on a fluid layer is the same across the channel independent of position, and this force is balanced by an equal and opposite net pressure force.

In the case of turbulent flow, Eqs. (7.9) and (7.10) give

$$-\frac{\partial \overline{P}}{\partial x} = \frac{\tau_w}{h}. \tag{7.12}$$

Substituting this into Eq. (7.6), integrating from 0 to y, and rearranging terms gives

$$\mu \frac{d\overline{U}}{dy} - \rho \overline{uv} = \tau_w \left(1 - \frac{y}{h}\right), \tag{7.13}$$

which generalizes the equivalent laminar result in Eq. (7.11). It is seen that in this case it is the sum of the mean viscous diffusion plus the turbulent diffusion that varies linearly across the channel.

The decomposition of the total viscous plus turbulent stress is illustrated in Figure 7.3 where for convenience the terms in Eq. (7.13) have been scaled by the friction velocity in which case

$$\frac{d\overline{U}^+}{dy^+} - \overline{uv}^+ = 1 - \frac{y^+}{R_\tau}, \tag{7.14}$$

where

$$y^+ = \frac{U_\tau y}{\nu} \tag{7.15}$$

represents lengths scaled in wall units, $\overline{U}^+ = \overline{U}/U_\tau$, and $\overline{uv}^+ = \overline{uv}/U_\tau^2$. Figure 7.3 shows that the mean viscous momentum diffusion is confined to a relatively thin layer next to

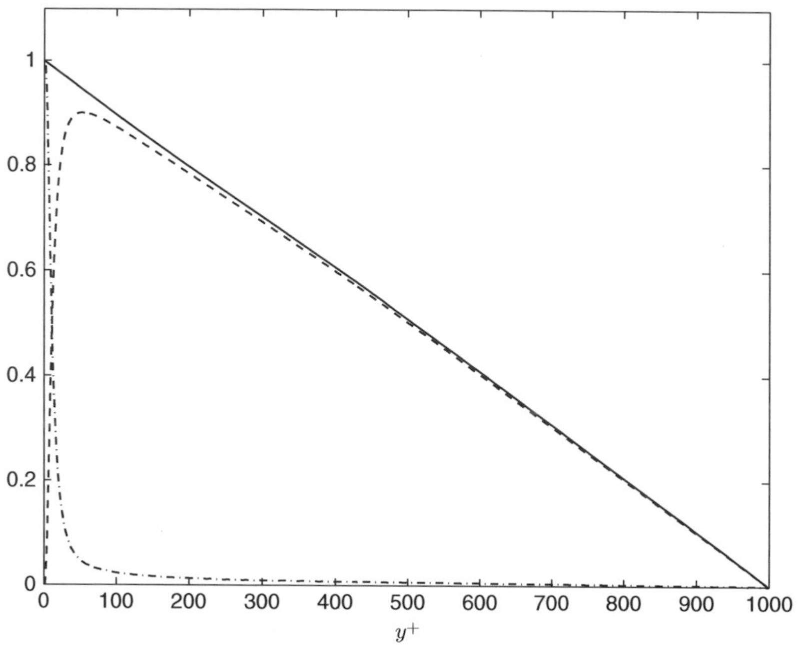

Figure 7.3 Decomposition of the total stress as given by Eq. (7.14) in turbulent channel flow: – · –, $d\overline{U}^+/dy^+$; – –, $-\overline{uv}^+$; —, $1 - y^+/R_\tau$. Data taken from [13].

the wall. Compensating for the drop off in molecular transport is the turbulent momentum transport as represented by the Reynolds shear stress $\overline{uv}^+$, which is anti-symmetric across the channel. In the lower half $\overline{uv}^+ < 0$ so the flux is toward the lower wall and $\overline{uv}^+ > 0$ on the upper half so it is toward the upper boundary. The location of the peak turbulent momentum flux is at $y^+ \approx 53$. $\overline{uv}^+$ varies almost exactly linearly throughout the central region of the channel where the mean shear stress is negligible.

Another useful viewpoint with which to examine the channel flow is through the balance of forces in the mean momentum equation. After applying a scaling to the terms in Eq. (7.6) this takes the form

$$0 = \frac{1}{R_\tau} + \frac{d^2\overline{U}^+}{dy^{+2}} - \frac{d\overline{uv}^+}{dy^+}, \tag{7.16}$$

in which Eq. (7.12) has been used and $1/R_\tau$ represents the non-dimensional constant pressure force. A numerical evaluation of Eq. (7.16) is given in Figure 7.4. Close to the wall the dominant and counterbalancing forces consist of large gains in momentum by its transport in the direction of the surface and momentum loss associated with molecular diffusion to the wall surface. Away from the immediate wall vicinity there is a balance between the pressure force and a slight momentum loss driven by its net transport toward the wall. The figure shows that by a wide margin the most significant mean forces at work in the channel flow are confined to a relatively thin layer within $y^+ = 70$ of the wall surface. Here, the complexity of the near-wall flow arises out of the interaction between viscous forces and momentum transport. This is dissimilar to the physics of the central region.

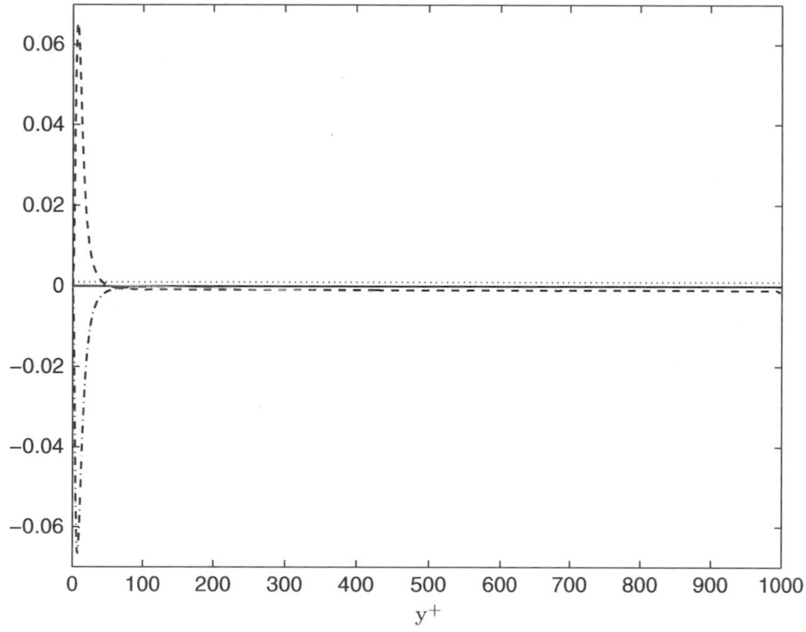

Figure 7.4 Decomposition of the mean momentum equation (7.16) in turbulent channel flow: – · –, viscous force; ––, turbulent transport; · · ·, pressure force. Data taken from [13].

7.1.2 Mean Flow Similarity

The mean velocity field in turbulent channel flow, as illustrated in Figure 7.2, suggests that there is a near-wall flow region in the form of a boundary layer and a very much different, almost uniform flow in the central region of the channel. In fact, the demarcation of the flow into separate zones with specially identifiable scaling behavior has proven to be a useful way of gaining a richer understanding of the motion of wall-bounded flows. A depiction of the various regions as they are traditionally given in analyzing boundary layer flow is given in the next chapter in Figure 8.3. Flow in the region closest to the wall is referred to as the *viscous sublayer*, a thin region where the influence of viscosity in shaping the flow is of critical importance. The center of the channel is referred to as the *outer* or *core* region where the flow is largely independent of the direct influence of viscosity. Between these is the *intermediate region*, which is also referred to as the *fully turbulent layer* due to the maxima in the magnitudes of the Reynolds stresses occurring in this region. The intermediate layer is also referred to as the *overlap layer* since it should scale in such a way as to accommodate both the near- and far-wall behaviors.

The existence of an intermediate layer, and particularly one with well-defined attributes, depends to some extent on the magnitude of the Reynolds number. In fact, it is generally believed that the conditions necessary for the flow to settle into a distinct similarity form in the overlap layer requires high Reynolds numbers so that the near- and far-field flows are widely separated from each other. This then allows an intermediate zone to appear that has properties that may have a universal form for all large Reynolds numbers. The outer portion of the viscous region located at the beginning of the intermediate zone is often referred to as the *buffer layer*. It is here where strong viscous effects and the turbulent motions associated with turbulence production and maximum momentum transport coexist, as was observed in Figures 7.3 and 7.4.

7.1.3 Viscous Sublayer

The viscous sublayer is distinguished by the essential role that viscosity plays in the flow adjacent to solid boundaries. Some indication of how this influences the trend in $\overline{U}$ near the surface can be derived by evaluating Eq. (7.6) at $y = 0$ and using Eq. (7.12) to obtain

$$\frac{d^2\overline{U}}{dy^2}(0) = -\frac{1}{h}\frac{d\overline{U}}{dy}(0), \tag{7.17}$$

where the fact that $\partial\overline{uv}/\partial y = 0$ at the boundary follows from the identity

$$\frac{\partial\overline{uv}}{\partial y} = \overline{\frac{\partial u}{\partial y}v} + \overline{u\frac{\partial v}{\partial y}}. \tag{7.18}$$

Differentiating Eq. (7.6) with respect to y and using the fact that

$$\frac{\partial^2\overline{uv}}{\partial y^2}(0) = 0 \tag{7.19}$$

(see Problem 7.1), it is found that

$$\frac{d^3\overline{U}}{dy^3}(0) = 0. \tag{7.20}$$

Substituting Eqs. (7.17) and (7.20) into the Taylor series expansion

$$\overline{U}(y) = \sum_{n=0}^{\infty} \frac{y^n}{n!}\frac{d^n\overline{U}}{dy^n}(0) \tag{7.21}$$

gives

$$\overline{U}(y) = h\frac{d\overline{U}}{dy}(0)\left(\frac{y}{h} - \frac{y^2}{2h^2}\right) + O\left(\frac{y}{h}\right)^4, \tag{7.22}$$

where the expression in the last term, which is of order of magnitude proportional to $(y/h)^4$, represents the remaining terms in the expansion. After scaling via U_τ, Eq. (7.22) becomes

$$\overline{U}^+(y^+) = y^+ - \frac{(y^+)^2}{2R_\tau} + \dots \tag{7.23}$$

where the first omitted term in this series is proportional to $(y^+)^4$. Equation (7.23) makes clear that $\overline{U}^+$ is linear in y^+ close to the wall, satisfying

$$\overline{U}^+(y^+) = y^+ \tag{7.24}$$

for some region near the surface. Experiments [14] and simulation show that Eq. (7.24) is accurate until $y^+ \approx 5$, and beyond this point the higher order terms in Eq. (7.23) begin to contribute, leading to a deviation from linearity.

The result in Eq. (7.24) is an example of the kind of general result that is sought in a similarity analysis of turbulent flow. In particular, the functional dependence between

$\overline{U}^+$ and y^+ is consistent with a similarity hypothesis to the effect that the flow adjacent to a wall should depend at most on flow variables via the relation

$$\overline{U} = f(y, \tau_w, \nu, \rho), \tag{7.25}$$

where the function f is to be determined. Because the only two quantities in Eq. (7.25) depending on mass are τ_w and ρ, they must combine to form U_τ. Similarly, only U_τ and ν depend on time so they must combine together to form the length scale ν/U_τ. It then follows from Eq. (7.25) that

$$\frac{\overline{U}}{U_\tau} = f(y^+), \tag{7.26}$$

which is known as the *law of the wall*. Equation (7.24) supplies the specific information that f has a linear form very close to the surface.

Since Eq. (7.24) holds independently of the Reynolds number, it is an example of what is termed *complete similarity*. *Partial similarity* would occur if an explicit Reynolds number dependence persisted in the similarity relation, as it does in Eq. (7.23).

7.1.4 Intermediate Layer

A central result of measurements and classical analyses of flow in the intermediate layer is the expectation that the mean streamwise velocity can be described via a logarithmic law [15, 16]. While there are many studies that purport to show log-law behavior in the channel flow, a more definitive reckoning of the validity of a log law and other similarity laws can be had from recently available simulations of channel flow at substantially higher Reynolds numbers than had previously been feasible. Before considering such results, it is instructive to consider some of the traditional motivations for why the occurrence of a log law in the mean velocity field is expected.

Thus, consider the flow in the neighborhood of $y^+ = 50$, where Figure 7.3 shows that $\overline{uv}$ has a minimum. A crude, yet reasonable approximation to $\overline{uv}$ in this region is to assume the existence of a constant stress layer in the vicinity of this location where according to Eq. (7.14) $\overline{uv}^+ \approx -1$, so that

$$|\overline{uv}| \approx \tau_w/\rho. \tag{7.27}$$

If Eq. (7.27) is legitimate, and assuming that the distance to the wall y is a relevant length scale, then a simple dimensional argument suggests that

$$\frac{d\overline{U}}{dy} = f(y, \tau_w, \rho), \tag{7.28}$$

where this relation is given in terms of $\frac{d\overline{U}}{dy}$, instead of $\overline{U}$, to avoid constraints imposed by the non-slip condition at the boundary. As was noted previously, τ_w and ρ combine to give U_τ, and it may be concluded from Eq. (7.28) that

$$\frac{d\overline{U}}{dy} \sim \frac{U_\tau}{y}. \tag{7.29}$$

Introducing a dimensionless constant of proportionality, κ, known as the Kármán constant, Eq. (7.29) becomes

$$\frac{d\overline{U}}{dy} = \frac{U_\tau}{\kappa y}. \tag{7.30}$$

Expressing Eq. (7.30) in wall units and integrating gives

$$\overline{U}(y^+) = \frac{1}{\kappa} \log y^+ + B, \tag{7.31}$$

where B is a constant. The implication of Eq. (7.31) is that the law of the wall as implied by Eq. (7.26) transitions from the linear form in Eq. (7.24) into a logarithmic form outside the viscous sublayer.

While the derivation of Eq. (7.31) is not rigorous, such a mathematical form is somewhat consistent with the measured properties of $\overline{U}$ when viewed in a semi-logarithmic plot, as shown in Figure 7.5. It is seen that the three mean velocity plots spanning $R_\tau = 542$ to 5186 overlay each other for $y^+ \leq 100$ and then individually diverge by rising above the apparent log-law form. The higher the Reynolds number the greater the y^+ extent that the curve appears to stay within the linear form.

A more precise determination of whether a log law is present can be had by examining the trends in the quantity

$$\beta = y^+ \frac{d\overline{U}^+}{dy+} \tag{7.32}$$

which will be constant, and equal to $1/\kappa$ according to Eq. 7.31, in log-law regions if they exist. Figure 7.6 shows the computed β for the same velocity profiles displayed in Figure 7.5. The figure reveals that true log-law behavior has not yet materialized for the solutions with $R_\tau \leq 1000$ despite their appearances in Figure 7.4. For the highest R_τ

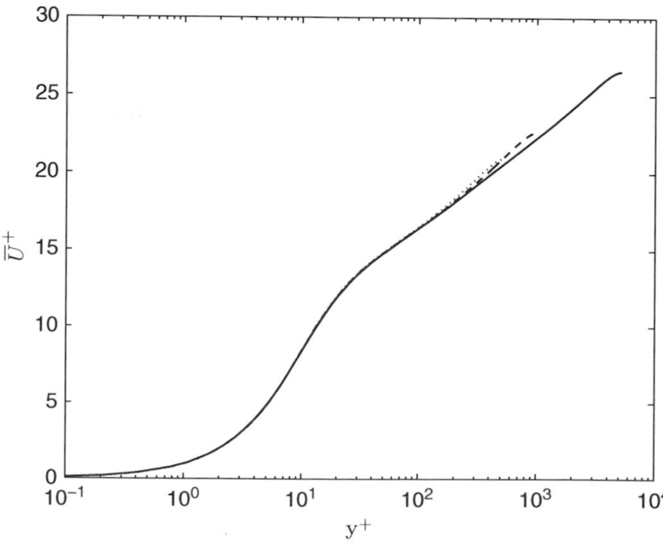

Figure 7.5 Semi-log plot of $\overline{U}^+$ showing an approximate log-law behavior. $\cdots$, $R_\tau = 541$; ––, $R_\tau = 1000$; —, $R_\tau = 5186$. Data from [10, 13].

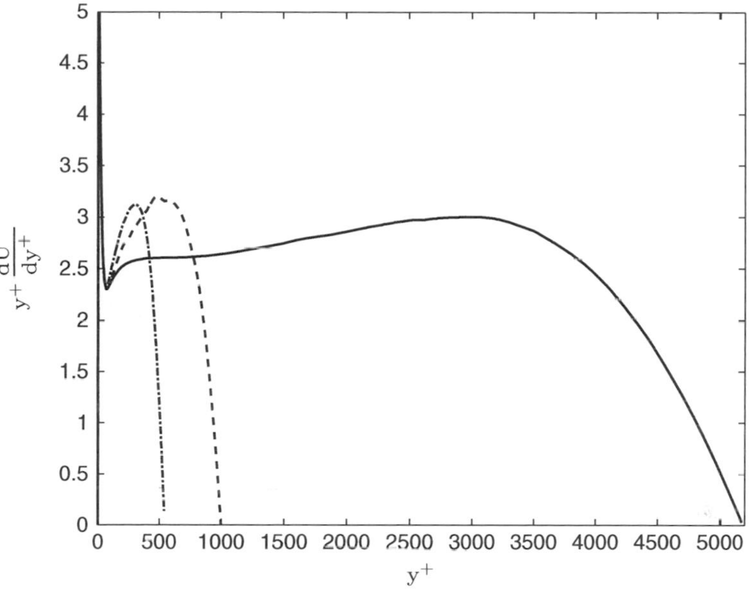

Figure 7.6 β as defined in Eq. (7.32) for the mean velocities in Figure 7.5. — · —, $R_\tau = 541$; – –, $R_\tau = 1000$; —, $R_\tau = 5186$. Data from [10, 13].

solution it does appear that for the approximate region $350 \leq y^+ \leq 830$, β is for the most part constant and equal to 2.604, implying that $\overline{U}$ satisfies a log law with $\kappa = 0.384$. Since $y^+ = 830$ corresponds to $y = 0.16h$ for this simulation, it may be hypothesized that the log law, if it exists, begins at $y^+ = 350$ and ends by $y = 0.16h$. If the log law must begin at $y^+ = 350$ for all R_τ, then a calculation reveals that for $R_\tau < 2200$ there is no y^+ range where a log law can be expected to hold, in agreement with Figure 7.6. One final point is that if the plot of β is examined more closely in the region where it appears that a log law holds, it will be seen that the curve is not exactly constant but has a small but definite slope. However, by modifying the derivation of the log law to include higher order terms in an expansion in $1/R_\tau$, it is possible to account for the small difference from exact log-law behavior [17, 18].

7.1.5 Velocity Moments

Some indication of the anisotropy brought into turbulent flow by the presence of a solid wall is evident in the behavior of the three normal components of the Reynolds stress tensor. Their values computed in a DNS of channel flow at $R_\tau = 1000$ and at $R_\tau = 5186$ [10] are shown in Figure 7.7 covering the half channel, and in Figure 7.8 they are shown in the near-wall region plotted with respect to y^+. The effect of Reynolds number is negligible in the core region of the channel. Near the wall, for the larger R_τ, the rapid changes in the correlations are shifted closer to the surface than the lower R_τ and display more rapid variations.

The effect of the wall in suppressing velocity fluctuations in the wall-normal direction is evident in the plots of $\overline{v^2}$. In the core region of the channel there is a semblance of isotropy between $\overline{v^2}$ and $\overline{w^2}$, but not with $\overline{u^2}$. The particular way in which anisotropy

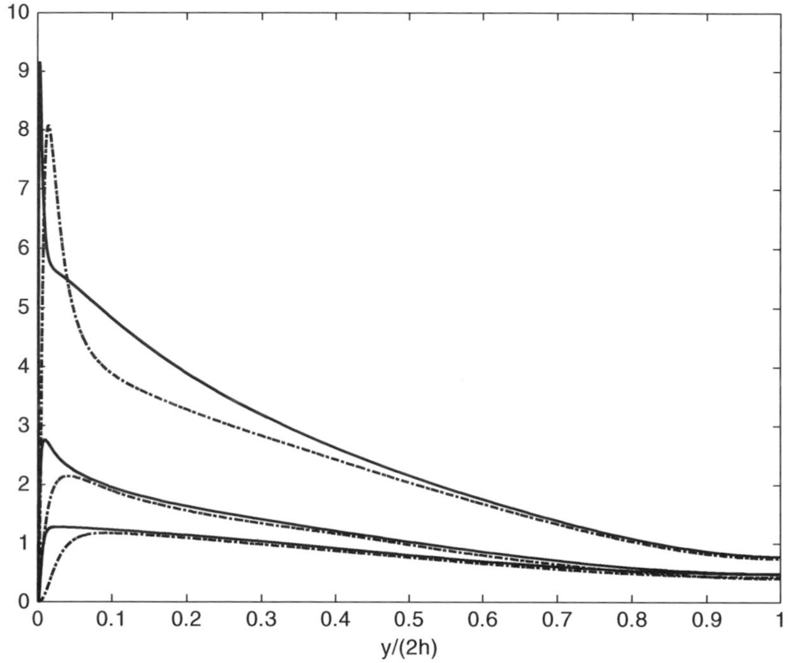

Figure 7.7 Normal Reynolds stresses in channel flow. —, $R_\tau = 5186$; – · –, $R_\tau = 1000$. Top curves are $\overline{u^2}$, middle curves are $\overline{w^2}$, lower curves are $\overline{v^2}$. Data taken from [10, 13].

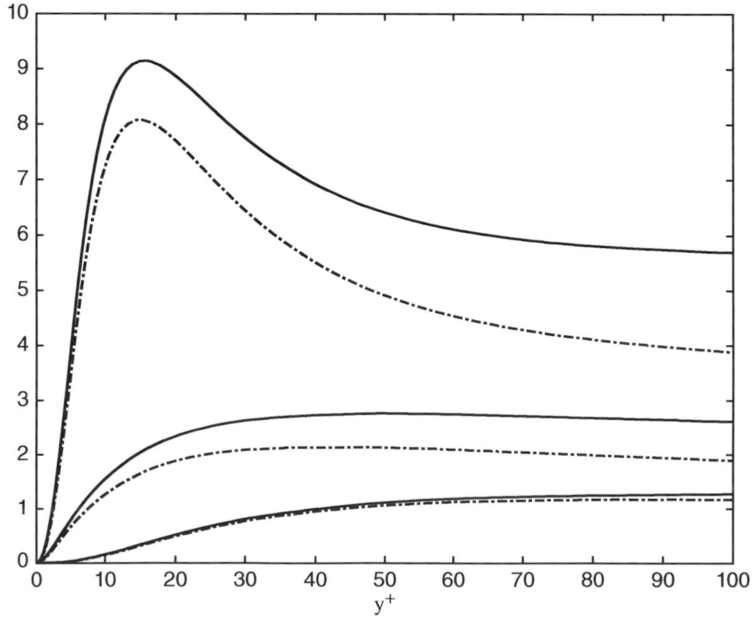

Figure 7.8 Normal Reynolds stresses in channel flow plotted with respect to y^+. —, $R_\tau = 5186$; – · –, $R_\tau = 1000$. Top curves are $\overline{u^2}$, middle curves are $\overline{w^2}$, lower curves are $\overline{v^2}$. Data taken from [10, 13].

is expressed near the wall, including a large spike in $\overline{u^2}$ at $y^+ \approx 15.5$ for both Reynolds numbers, must have its origin in the structural aspects of the flow. It reflects the idiosyncrasies of streamwise momentum transport in the presence of a steep mean velocity gradient whose presence means that small movements of fluid can produce large local u velocity fluctuations. The processes working to create the normal stress anisotropy are not entirely understood at the present time, though consideration of near-wall vorticity structure in the next chapter sheds some light on the issue.

Near the wall, despite the scaling in wall units, Figure 7.8 shows that there are some significant Reynolds number effects in the streamwise and spanwise normal stresses and much less so in the wall-normal stress, which appears to be independent of R_τ. Only in the very near-wall region, within approximately $y^+ = 2.5$ of the wall, do all three normal stresses appear to scale independent of the Reynolds number. However, a least squares fit of a quadratic polynomial to $u^+_{rms} \equiv \sqrt{\overline{u^2}^+}$ for $y^+ \leq 10$ for the $R_\tau = 5186$ solution in Figure 7.8 gives approximately

$$\frac{du^+_{rms}}{dy^+}(0) = 0.5 \tag{7.33}$$

and

$$\frac{d^2u^+_{rms}}{dy^{+^2}}(0) = -0.038. \tag{7.34}$$

These values change to 0.47 and −0.032 for the $R_\tau = 1000$ solution. Evidently, there are some persistent Reynolds number effects for these simulations.

As noted previously, $\overline{U}$ is linear near the boundary out to approximately $y^+ = 5$ and it is of interest to see what can be said about the normal stresses next to the wall. In this regard, consider a Taylor series expansion of u_{rms} about $y = 0$. Using boundary layer scaling this is

$$u^+_{rms} = \frac{du^+_{rms}}{dy^+}(0)y^+ + \frac{1}{2}\frac{d^2u^+_{rms}}{dy^{+^2}}(0)y^{+2} + O((y^+)^3) \tag{7.35}$$

The distance from the boundary over which u^+_{rms} can be modeled as linear can be analyzed by considering the ratio

$$\frac{u^+_{rms}}{\overline{U}^+} = \frac{du^+_{rms}}{dy^+}(0) + y^+\left(\frac{1}{R_\tau}\frac{du^+_{rms}}{dy^+}(0) + \frac{1}{2}\frac{d^2u^+_{rms}}{dy^{+^2}}(0)\right) + O((y^+)^2), \tag{7.36}$$

which is derived from Eqs. (7.23) and (7.35). Where u^+_{rms} behaves linearly near the wall – within the region where $\overline{U}^+$ is linear – will show up in the constancy of the ratio $u^+_{rms}/\overline{U}^+$. Using the results in Eqs. (7.33) and (7.34) it follows that near the wall

$$\frac{u^+_{rms}}{\overline{U}^+} = 0.5 + y^+\left(\frac{0.25}{R_\tau} - 0.019\right) + \dots. \tag{7.37}$$

This suggests that linearity of u^+_{rms} is maintained until $y^+ \approx 2$.

The spanwise rms fluctuation satisfies a similar relation as Eq. (7.35) in the form

$$w^+_{rms} = \frac{dw^+_{rms}}{dy^+}(0)y^+ + \frac{1}{2}\frac{d^2w^+_{rms}}{dy^{+^2}}(0)y^{+2} + O((y^+)^3) \tag{7.38}$$

and computations show that at the wall this has the approximate numerical value

$$w^+_{rms} = 0.25y^+ + \dots. \tag{7.39}$$

Using the continuity equation, it may be shown that

$$\frac{dv^+_{rms}}{dy^+} = 0, \tag{7.40}$$

and, consequently, near the wall

$$v^+_{rms} = \frac{1}{2}\frac{d^2v^+_{rms}}{dy^{+2}}(0)\ y^{+2} + O((y^+)^3). \tag{7.41}$$

Computations show that this is given numerically as approximately

$$v^+_{rms} = 0.006y^{+2} + O((y^+)^3). \tag{7.42}$$

To the extent that Eqs. (7.37), (7.39), and (7.42) provide good approximations to the normal Reynolds stresses within a short distance of the wall, they may serve as useful boundary conditions in enabling the development of models that include anisotropy of the near-wall flow.

In addition to the logarithmic trend in $\overline{U}$ in the intermediate layer, there has been some analysis, notably the attached eddy model of Townsend [19], that has suggested that the streamwise normal Reynolds stress $\overline{u^2}$ should decrease according to a log law in the intermediate layer. While there has been some experimental verification of this result, DNS simulations [10] have shown that at least for R_τ up to 5186 the log behavior is not observed. On the other hand, though lacking a theoretical explanation as of yet, log behavior is found for the spanwise normal Reynolds stress $\overline{w^2}$.

7.1.6 Turbulent Kinetic Energy and Dissipation Rate Budgets

Simplification of the kinetic energy equation (3.37) to the form appropriate to channel flow yields

$$0 = -\overline{uv}\frac{d\overline{U}}{dy} - \epsilon - \frac{1}{\rho}\frac{d\overline{pv}}{dy} + \nu\frac{d^2K}{dy^2} - \frac{1}{2}\frac{d\overline{vu_j^2}}{dy}, \tag{7.43}$$

where the terms on the right-hand side account for kinetic energy production, dissipation, pressure work, viscous diffusion, and turbulent transport, respectively. How these effects balance near the boundary is shown in Figure 7.9 for the simulation at $R_\tau = 5186$ [10], where quantities are given in wall units. It is interesting to note that for $y^+ > 30$ up to the channel centerline the turbulent kinetic energy distribution is maintained almost exclusively by a balance between production and dissipation. The largest production rate occurs at $y^+ \approx 12$, which is close to the peak in K itself, while dissipation is largest at the wall surface and has a local plateau off the boundary.

Turbulent transport of kinetic energy is important mainly near the wall. It is negative in the range $8 < y^+ < 30$ and positive within $y^+ \approx 8$, suggesting that much of the turbulent energy produced in the peak Reynolds stress zone around $y^+ = 10$ is transferred toward the boundary. At the wall surface, the rate of viscous diffusion of kinetic energy is balanced by its viscous dissipation since the other terms in Eq. (7.43) are identically zero. In other words, in this location molecular diffusion brings energy toward

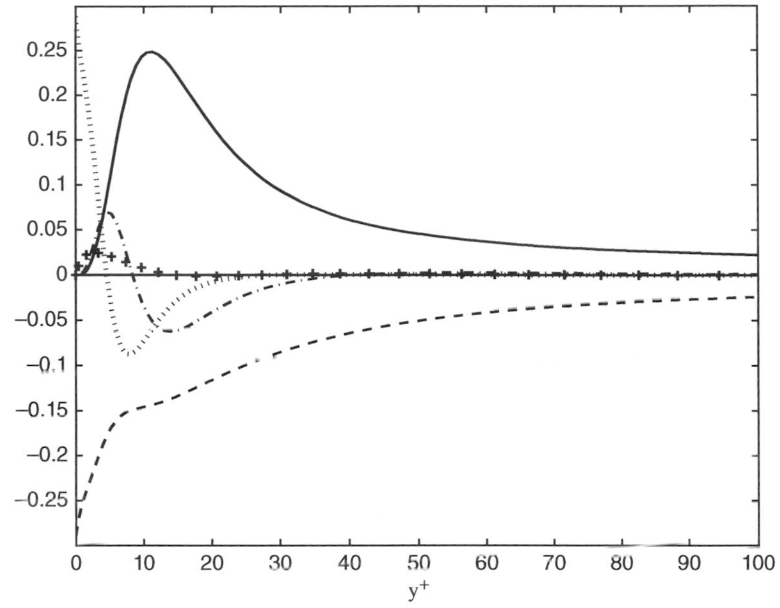

Figure 7.9 Turbulent kinetic energy budget in channel flow $R_\tau = 5186$ [10] scaled with ν and u_τ: —, production; ––, dissipation; +, pressure work; · · ·, viscous diffusion; – · –, turbulent transport.

the surface where it is dissipated. In the buffer region outside the viscous sublayer, the energy balance is more complex and involves transfer, production, dissipation, and pressure work.

The location where the energy production is at a peak can be estimated by first noting that the production term can be written using Eq. (7.14) as

$$-\overline{uv}^+ \frac{d\overline{U}^+}{dy^+} = \left(1 - \frac{y^+}{R_\tau} - \frac{d\overline{U}^+}{dy^+}\right)\frac{d\overline{U}^+}{dy^+}. \tag{7.44}$$

The maximum of the production term occurs where the y^+ derivative of (7.44) vanishes and a calculation shows that this occurs approximately where

$$\left(1 - 2\frac{d\overline{U}^+}{dy^+}\right)\frac{d^2\overline{U}^+}{dy^{+2}} = 0 \tag{7.45}$$

after terms of $O(R_\tau^{-1})$ are dropped. Since $d^2\overline{U}^+/dy^{+2} \neq 0$ in the region where the peak production occurs, as is clear from the viscous diffusion term plotted in Figure 7.9, peak production occurs approximately when

$$\frac{d\overline{U}^+}{dy^+} = \frac{1}{2} = -\overline{uv}^+, \tag{7.46}$$

in which the last equality in Eq. (7.46) comes from Eq. (7.14) assuming that $y^+ << R_\tau$. The point where (7.46) is satisfied is visible in Figure 7.3 at $y^+ \approx 12$ and agrees with Figure 7.9 as well.

Some additional insights into the energy balance come indirectly from the ϵ equation balance shown in Figure 7.10 for $R_\tau = 590$ [20]. Despite the simplifications inherent in

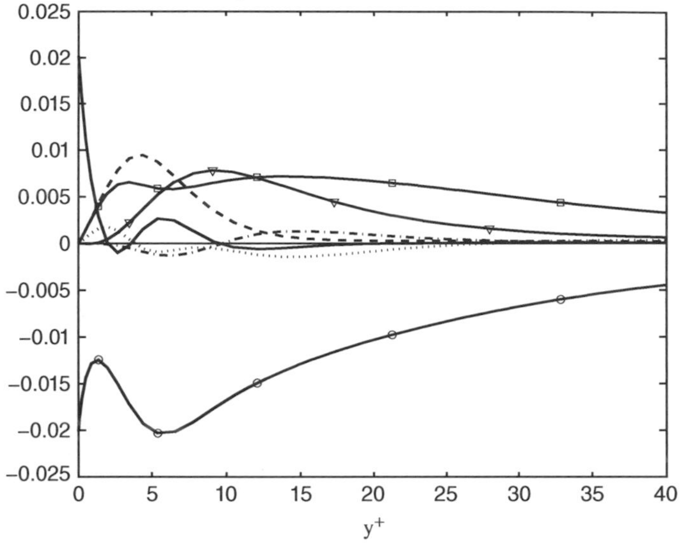

Figure 7.10 ϵ equation budget in channel flow at $R_\tau = 590$ [20] scaled with ν and u_τ. $--$, P^1_ϵ; ∇, P^2_ϵ; $-\cdot-$, P^3_ϵ; $\square$, P^4_ϵ; $\circ$, $-\Upsilon_\epsilon$; —, D_ϵ; $\cdots$, $\Pi_\epsilon + T_\epsilon$.

channel flow, all the terms in Eq. (3.42) for ϵ are populated here. From the figure it is seen that the turbulent vortex stretching term, P^4_ϵ, balances the dissipation rate, Υ_ϵ, away from the channel walls similar to the case of homogeneous shear flow considered in Section 6.6. The production terms P^1_ϵ and P^2_ϵ are significant for $y^+ \leq 25$, while of somewhat less importance is the production term P^3_ϵ and the sum of the pressure and transport terms, $\Pi_\epsilon + T_\epsilon$. Near the wall $-\Upsilon_\epsilon$ has a local minimum off the surface, and at the boundary $-\Upsilon_\epsilon$ and D_ϵ are in balance. It is evident that the factors affecting ϵ near the solid wall are complicated and a very great challenge to model.

7.1.7 Reynolds Stress Budget

The turbulent kinetic energy budget considered above represents the sum of the separate budgets of the normal components of the Reynolds stress. Considering the factors affecting the separate components of the Reynolds stresses by examining the individual terms in their governing equations gives some insight into the physics surrounding the anisotropy of turbulence next to the boundary. The equations determining the normal Reynolds stresses, $\overline{u^2}$, $\overline{v^2}$, and $\overline{w^2}$ for channel flow are derived from Eq. (3.53) in the form

$$0 = -2\overline{uv}\frac{d\overline{U}}{dy} - \epsilon_{11} - \frac{d\overline{u^2v}}{dy} + \Pi_{11} + \nu\frac{d\overline{u^2}}{dy^2} \tag{7.47}$$

$$0 = -\epsilon_{22} - \frac{d\overline{v^3}}{dy} + \Pi_{22} + \nu\frac{d^2\overline{v^2}}{dy^2} - \frac{2}{\rho}\frac{d\overline{pv}}{dy} \tag{7.48}$$

$$0 = -\epsilon_{33} - \frac{d\overline{w^2v}}{dy} + \Pi_{33} + \nu\frac{d^2\overline{w^2}}{dy^2} \tag{7.49}$$

where the dissipation rate ϵ_{ij} is defined in Eq. (3.52) and the pressure-strain term Π_{ij} is defined in Eq. (3.54). Figures 7.11–7.13 taken from DNS data at $R_\tau = 5186$ [10] display the balance of physical effects indicated in Eqs. (7.47)–(7.49), respectively.

It may be noticed that among the three normal stresses only the equation for $\overline{u^2}$ has a direct connection to the mean flow via the production term in which the mean velocity field appears explicitly. The significant action of production in the $\overline{u^2}$ balance throughout most of the channel is seen in Figure 7.11. Since $\overline{v^2}$ and $\overline{w^2}$ lack production via a direct connection to the mean velocity field, they depend on an alternative source of energy. According to Figures 7.12 and 7.13 the pressure-strain term supplies the necessary production for these normal stresses. In fact, it was noted in Eq. (3.56) in reference to the K equation that the sum of the normal components of the pressure-strain correlation are zero, so that if any one of Π_{11}, Π_{22} or Π_{33} is non-zero then at least one of them must be positive and one negative. Figure 7.11 shows that $\Pi_{11} < 0$ everywhere except very close to the wall so that pressure strain acts as a drain on the streamwise normal stress, with this energy being given to the other two components. Thus, energy is produced directly in $\overline{u^2}$ from the mean shearing and is then redirected to the other normal Reynolds stress components through the action of the pressure force.

The precise nature of energy redistribution via the pressure-strain balance is summarized in Figure 7.14 showing a plot of the pressure-strain term in each of the normal Reynolds stress equations. The sum of these terms is identically zero. Near the surface, energy redistribution is for the most part entirely into the spanwise direction originating from both the streamwise and wall-normal directions.

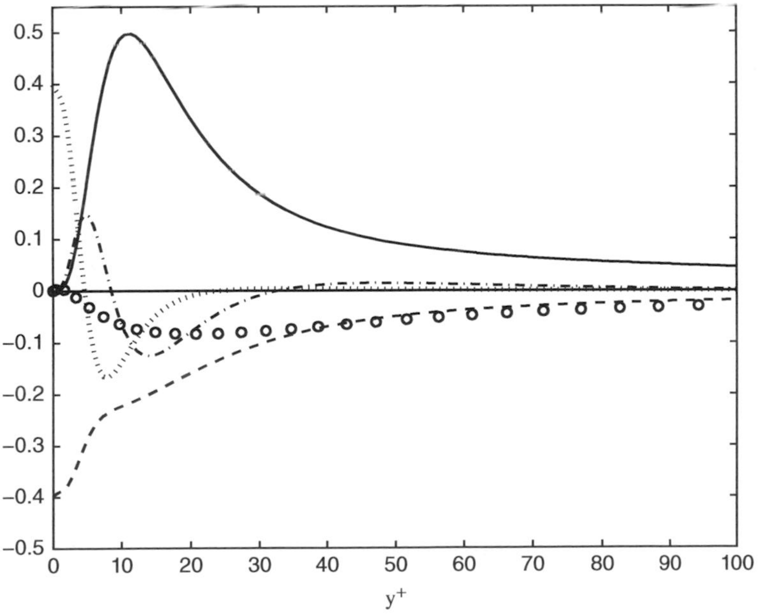

Figure 7.11 $\overline{u^2}$ budget in channel flow for $R_\tau = 5186$ [10]. —, production; – –, dissipation; o, pressure strain; · · ·, viscous diffusion; – · –, turbulent transport.

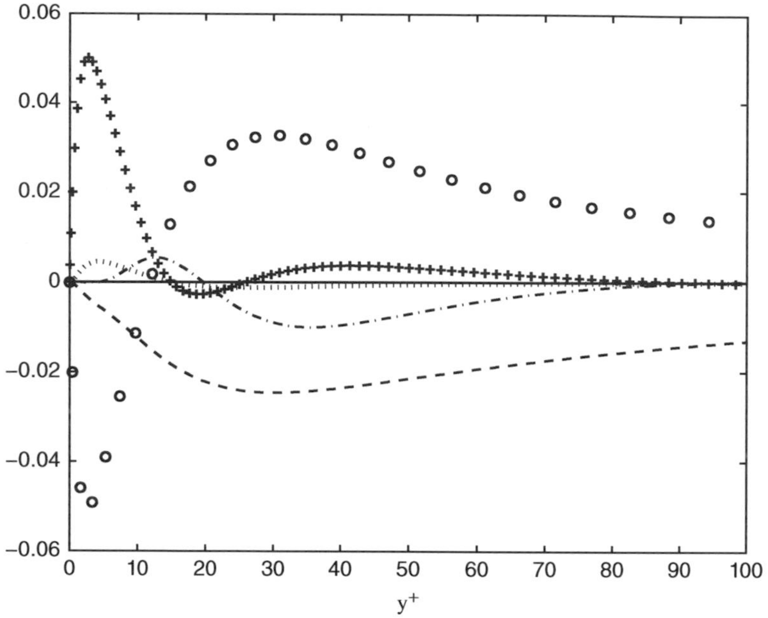

Figure 7.12 $\overline{v^2}$ budget in channel flow for $R_\tau = 5186$ [10]. ––, dissipation; o, pressure strain; +, pressure work; · · ·, viscous diffusion; – · –, turbulent transport.

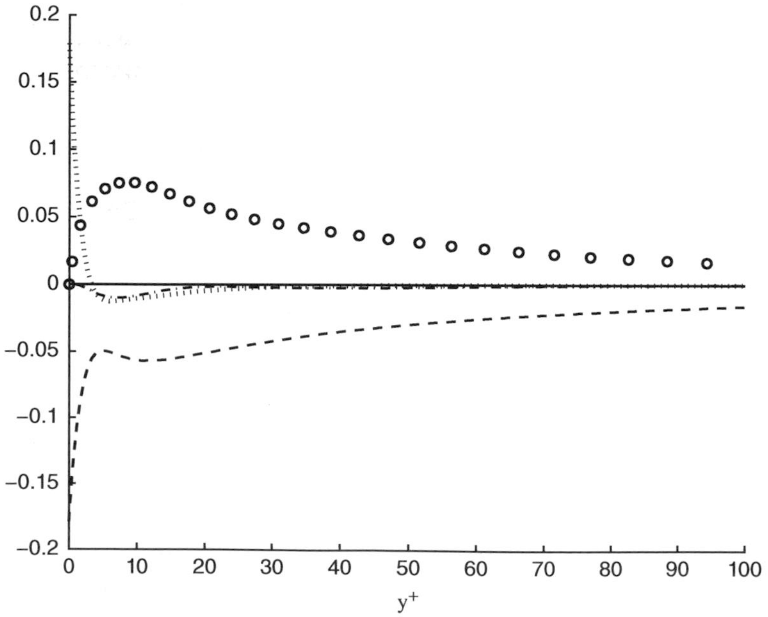

Figure 7.13 $\overline{w^2}$ budget in channel flow for $R_\tau = 5186$ [10]. ––, dissipation; o, pressure strain; · · ·, viscous diffusion; – · –, turbulent transport.

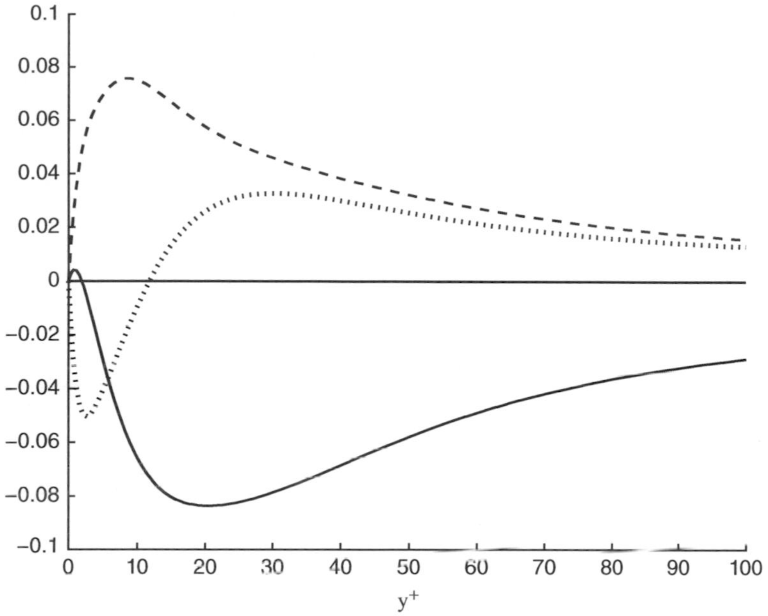

Figure 7.14 Pressure-strain term in normal Reynolds stress equations at $R_\tau = 5186$ [10]: —, Π_{11}; · · ·, Π_{22}; – –, Π_{33}.

In the far field, the balance determining $\overline{u^2}$ is between production on the one hand and viscous dissipation and pressure strain on the other. For $\overline{v^2}$ and $\overline{w^2}$ the far field balance is only between a loss by viscous dissipation and a gain by the pressure-strain term. In all cases the physics is much more complicated within $y^+ = 40$ of the surface. For $\overline{u^2}$ viscous dissipation matches diffusion at the wall surface as it also does in the $\overline{w^2}$ equation, while all the terms in the $\overline{v^2}$ equation are zero at the surface. In the case of the balances of $\overline{v^2}$ and $\overline{w^2}$ shown in Figures 7.12 and 7.13 the pressure work term provides production of $\overline{v^2}$ near the wall that is redirected to the spanwise energy via the pressure-strain term. However, the net effect of the two pressure terms in the $\overline{v^2}$ equation is relatively small, since they mostly cancel, and at the surface they exactly balance. For $\overline{w^2}$, the losses due to dissipation near the wall are balanced mainly by the pressure-strain term which has taken on the role of primary production term. In all this the viscous and turbulent transport terms in each equation account for the spatial redistribution of energy without its production.

Now considering the Reynolds shear stress balance in channel flow, Eq. (3.53) yields

$$0 = -\overline{v^2}\frac{d\overline{U}}{dy} - \epsilon_{12} + \Pi_{12} - \frac{d\overline{uv^2}}{dy} - \frac{1}{\rho}\frac{d\overline{pu}}{dy} + \nu\frac{d^2\overline{uv}}{dy^2} \tag{7.50}$$

where the terms on the right-hand side account for production, dissipation, pressure strain, turbulent transport, pressure work, and viscous diffusion, respectively. The computed Reynolds shear stress budget for channel flow, as shown in Figure 7.15, indicates that $\overline{uv}$, which is negative near the lower channel wall, is produced by the mean

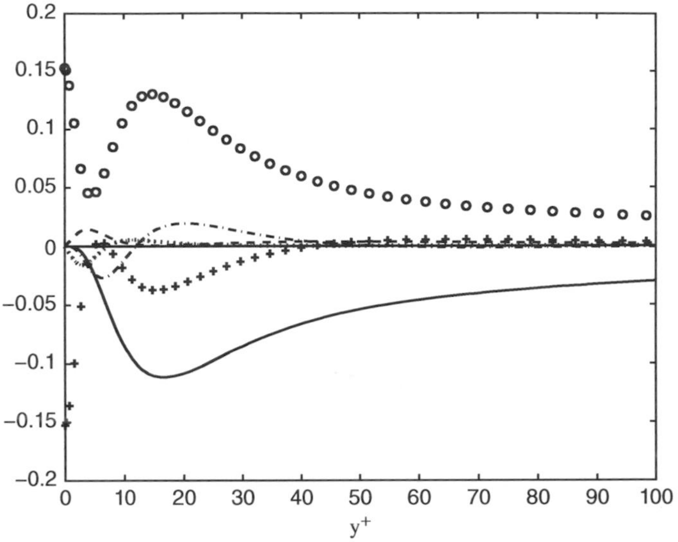

Figure 7.15 $\overline{uv}$ budget in channel flow for $R_\tau = 5186$ [10]. —, production; – –, dissipation; o, pressure strain; +, pressure work; · · ·, viscous diffusion; – · –, turbulent transport.

shear contained in the first term on the right-hand side of Eq. (7.50). Unlike the normal stresses, where production is balanced by dissipation, here the dissipation rate is relatively insignificant so it is the pressure-strain term that balances production. In fact, it may be noted [21] that the combination of ϵ_{12} and the viscous diffusion term in Eq. (7.50) is small throughout the channel, suggesting that the balance of effects leading to changes in $\overline{uv}$ is not strongly dependent on viscosity. This helps make clear the importance of correctly modeling the pressure-strain term if Eq. (7.50) is to be used as the basis for developing a predictive scheme. Similar to what was observed in the $\overline{v^2}$ balance, the sum of the two pressure terms in Eq. (7.50) nearly cancels in the vicinity of the wall. Consequently, incorporating the combined pressure term, instead of its separation into the total pressure work and pressure-strain terms, may be a useful modeling strategy [21].

7.1.8 Enstrophy and its Budget

The distribution across the channel of the variances of the vorticity components $\overline{\omega_1^2}$, $\overline{\omega_2^2}$, and $\overline{\omega_3^2}$ that collectively comprise the enstrophy $\zeta = \overline{\omega_1^2} + \overline{\omega_2^2} + \overline{\omega_3^2}$ is shown in Figure 7.16 for the DNS channel flow data at $R_\tau = 5186$ [10]. To a somewhat greater extent than occurs with the normal Reynolds stresses shown in Figure 7.7, the mean-squared vorticity components converge toward each other away from the wall, though a close examination of the curves reveals that an exact isotropic state is not achieved. In view of the identities,

$$\overline{\omega_1^2}(0) = \overline{(\partial w/\partial y)^2}(0), \tag{7.51}$$

$$\overline{\omega_2^2}(0) = 0, \tag{7.52}$$

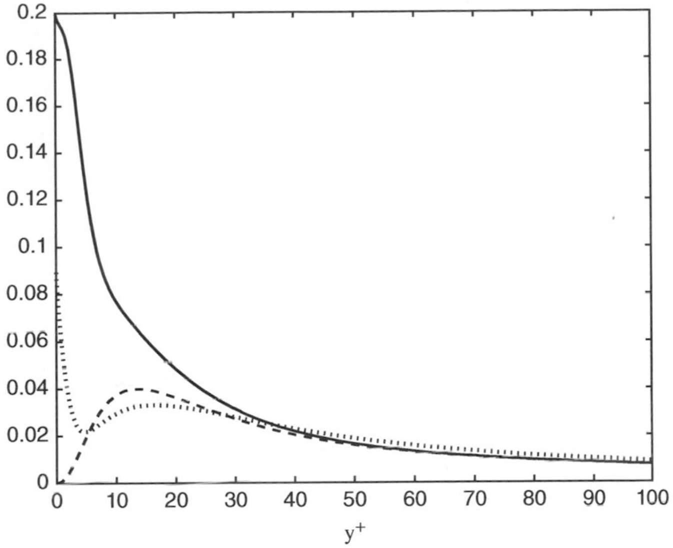

Figure 7.16 Comparison of the enstrophy components in channel flow at $R_\tau = 5186$ [10]. $\cdots, \overline{\omega_1^2}^+$; $--, \overline{\omega_2^2}^+$; $—, \overline{\omega_3^2}^+$.

and

$$\overline{\omega_3^2}(0) = \overline{(\partial u/\partial y)^2}(0), \tag{7.53}$$

the dominance of $\overline{\omega_3^2}$ compared to $\overline{\omega_1^2}$ at the wall surface reflects the fact that there is greater shearing in the streamwise direction at this location than there is in the transverse direction. The trend in $\overline{\omega_3^2}$ can be associated with the large magnitude of $\overline{\Omega}_3 \equiv -\partial \overline{U}/\partial y$ near the wall, as seen in Figure 7.3. Less obvious is the large peak in $\overline{\omega_1^2}(0)$ deriving from the presence of motions in the spanwise direction near the wall surface.

The anisotropic pattern of the enstrophy components in the buffer layer $10 \leq y^+ \leq 30$ must reflect the underlying structure in the wall region flow. Particularly in the case of the local maxima in $\overline{\omega_1^2}$ and $\overline{\omega_2^2}$, whose corresponding mean vorticities are zero in a channel, one can assume that these trends have to do with the properties of vortices (e.g., as in their orientation) in the boundary region. By this reckoning it appears that there is a mechanism at work in the flow causing an increase in streamwise vorticity beyond its local minimum at the upper edge of the viscous sublayer. In the same location there is a noticeable rise in wall-normal vorticity. These observations are explained to some extent by the analysis of the boundary layer structure to be given in Section 8.5.

As seen in Eq. (3.81) the enstrophy has a close relationship to ϵ. In homogeneous turbulence this was exactly $\epsilon/\nu = \zeta$, as discussed in Section 3.6. In channel flow the exact relationship in Eq. (3.81) simplifies to

$$\frac{\epsilon}{\nu} = \zeta + \frac{d^2\overline{v^2}}{dy^2}. \tag{7.54}$$

A plot of the three quantities in Eq. (7.54), scaled in wall coordinates and taken from the channel flow DNS at $R_\tau = 5186$ [10], is shown in Figure 7.17. Despite the presence of

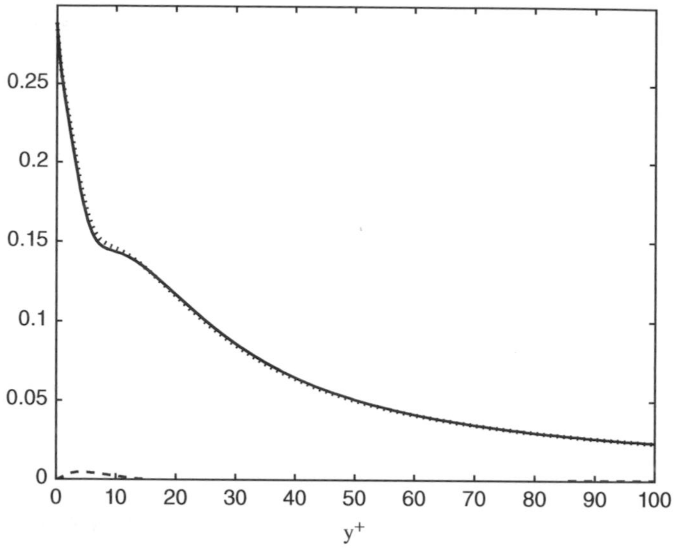

Figure 7.17 Evaluation of the terms in Eq. (7.54) in channel flow with $R_\tau = 5186$ [10]: $\cdots$, $-\epsilon^+$; —, ζ^+; – –, $d^2\overline{v^2}^+/dy^{+^2}$.

significant shearing in the channel flow, as against isotropic turbulence where there is none, ϵ/ν and ζ are virtually identical. This suggests there may be many circumstances wherein $\nu\zeta$ can be used interchangeably with ϵ in representing the dissipation rate.

7.2 Pipe Flow

In view of the widespread commercial importance of fluid transport in pipes, there is considerable need to understand the properties of such flows. Our interest here lies in the behavior of pipe flow in the fully developed turbulent flow region that appears after entrance effects have subsided and the flow has transitioned to turbulence. For Reynolds number based on the pipe diameter $D = 2R_o$ where R_o is the radius, $U_m = Q/A$ is the mean flow velocity, Q is the volume flux in the pipe, and $A = \pi R_o^2$ is the cross-sectional area, transition generally happens for $R_e \approx 2000$. As in the case of channel flow, the precise value of the critical Reynolds number for any particular facility depends on local conditions. How transition in pipe flow occurs, particuarly in regards to the development of vortical structure, is a subject of continued interest [22]. When the flow is fully developed it is unidirectional with velocity $\overline{U}(r)$. Under these circumstances, the averaged streamwise momentum equation (3.8) in cylindrical coordinates is

$$0 = -\frac{\partial \overline{P}}{\partial x} + \frac{1}{r}\frac{d}{dr}\left(\mu r \frac{d\overline{U}}{dr} - \rho r \overline{u v_r}\right) \tag{7.55}$$

where r is the outward radial coordinate with $r = 0$ at the center of the pipe and $r = R_o$ is the wall. The velocity fluctuation vector $\mathbf{u} = (u, v_r, v_\theta)$ has components in the axial, radial, and azimuthal directions, respectively. As in the case of channel flow, the mean axial pressure gradient is constant everywhere.

It is helpful in the following discussion to introduce a coordinate $y \equiv R_o - r$ which is normal to the pipe surface and points toward the pipe center. In terms of y, the mean velocity is given by $\overline{U}^*(y) = \overline{U}(R_o - y)$. However, since it should always be clear from the context which of $\overline{U}$ or $\overline{U}^*$ is being referred to, henceforth, for simplicity, the symbol $\overline{U}$ is used in all cases.

In terms of y, the wall shear stress is

$$\tau_w = \mu \frac{d\overline{U}}{dy}(0). \tag{7.56}$$

Performing an area integration of Eq. (7.55) over the pipe cross-section yields

$$-\frac{\partial \overline{P}}{\partial x}\pi R_o^2 = 2\pi R_o \tau_w, \tag{7.57}$$

where the fact that

$$\frac{d\overline{U}}{dr}(R_o) = -\frac{d\overline{U}}{dy}(0) \tag{7.58}$$

is used. Equation (7.57) is the pipe flow equivalent of Eq. (7.12) for channel flow. It is seen from this that if τ_w is known from having measured $\overline{U}$ then Eq. (7.57) gives the pressure drop. At the same time, the volumetric flux of fluid through the pipe,

$$Q = 2\pi \int_0^R \overline{U}(r) r dr, \tag{7.59}$$

can be determined from $\overline{U}$, and thus so too the average mass flow velocity. For any given distribution of $\overline{U}$, the Reynolds numbers $R_e \equiv U_m D/\nu$ and $R_\tau = U_\tau D/\nu$ can be determined.

Through their common dependence on the mean velocity field, there is an implied functional relationship between $d\overline{P}/dx$, R_τ, and R_e that is useful for determining forces and flow rates in pipes. In practice, the friction factor for pipe flow f defined by

$$f = \frac{\Delta \overline{P}}{\Delta x}\frac{2D}{\rho U_m^2} \tag{7.60}$$

for a pressure drop of ΔP over the prescribed length Δx is used to connect these quantities. Taking advantage of Eq. (7.57) to replace $\Delta \overline{P}$ since the pressure gradient is constant, Eq. (7.60) becomes

$$f = 8\frac{R_\tau^2}{R_e^2}. \tag{7.61}$$

The functional form of f is generally found from physical experiments over a wide range of Reynolds numbers and for both smooth and rough wall pipes. The result is codified in the standard Moody diagram used to find f in engineering design work [23, 24]. Analytic expressions can be developed for limited ranges of Reynolds number. For example, for $R_e < 100{,}000$ in a smooth pipe this consists of the Blasius law [25] to the effect that

$$f = 0.266 R_e^{-1/4}. \tag{7.62}$$

Equating Eq. (7.62) with Eq. (7.61) then yields

$$R_\tau = 0.182 R_e^{7/8} \tag{7.63}$$

for this particular case.

7.2.1 Mean Velocity

In recent years considerable effort has been expended in developing and instrumenting pipe flow at very high Reynolds numbers [3, 6]. This utilizes a pressurized system in order to increase density and hence the Reynolds number. For high Reynolds number flow care must be taken to ensure the smoothness of the pipe walls to prevent the inadvertent occurrence of a rough-wall regime.

Measurements of pipe flow aim to establish the validity of similarity laws in the same way they have been analyzed in channel flow. With velocity and length scaling based on U_τ and ν the flow near the wall within the viscous sublayer is found to obey

$$\overline{U}^+ = y^+, \tag{7.64}$$

as in Eq. (7.24) for channel flow. Further from the wall the same classical arguments that hold for a channel also suggest that a log law may be valid in the intermediate layer of pipe flow.

For high Reynolds number pipe flow such as has been realized in physical experiments it is expected that a substantial core region develops for which the viscosity will not be appropriate for scaling the mean velocity field. In this case a classical argument predicts similarity in the form of the *velocity defect law* [26]

$$\frac{\overline{U}_{cl} - \overline{U}(y)}{U_\tau} = g(\xi), \tag{7.65}$$

where $\overline{U}_{cl}$ is the mean centerline velocity and $\xi \equiv y/R_o$ is a similarity variable. In practice, the validity of Eq. (7.65) is found to extend well beyond the core region to encompass much of the intermediate region as well. Moreover, the defect law Eq. (7.65) should also be relevant to the central region of channel flow and the outer region of boundary layers.

If Eq. (7.65) applies in the overlap region between the viscous sublayer and the core region, then it may be argued [27] that $\overline{U}/U_\tau$ from Eqs. (7.26) and (7.65) may be equated to each other so that

$$f(y^+) = \frac{U_{cl}}{U_\tau} - g(\xi). \tag{7.66}$$

Differentiating this with respect to y gives

$$\frac{df}{dy^+}(y^+)\frac{U_\tau}{\nu} = -\frac{dg}{d\xi}(\xi)\frac{1}{R_o} \tag{7.67}$$

and multiplying both sides of the equation by y gives

$$y^+\frac{df}{dy^+}(y^+) = -\xi\frac{dg}{d\xi}(\xi). \tag{7.68}$$

In order for Eq. (7.68) to be self-consistent for arbitrary values of the similarity variables y^+ and ξ it is necessary that each side of the equation be constant. In other words, this prevents multi-valuedness in the sense that for a fixed value of ξ the left-hand side of Eq. (7.68) should not change for different values of R_τ, and hence y^+. Setting the constant to $1/\kappa$ it follows from Eq. (7.68) that

$$y^+\frac{df}{dy^+}(y^+) = \frac{1}{\kappa} \tag{7.69}$$

which once again, after integration, gives the log law Eq. (7.31). The right-hand side of Eq. (7.68) indicates that the defect law is logarithmic in the overlap region as well.

Investigations of pipe flow at high Reynolds numbers with increasingly accurate measurement techniques have led to some adjustments in the placement of the log law within the pipe flow and the numerical value of its constants in Eq. (7.31). To some extent this is similar to the results in channel flow where the traditional log-law location has given way to new results showing the log law starting further from the boundary. For the pipe flow it is found [6] that a log law is observed within the region $600 \leq y^+ \leq 0.12R_o^+$, as seen in Figure (7.18). According to this study $\kappa = 0.42$ and $B = 5.6$. At lower Reynolds numbers the log law is less pronounced, the same as was found in channel flow, suggesting that the log law only fully appears at sufficiently high Reynolds numbers. Measurements using particle image velocimetry (PIV) in a 111.5 m pipe at Reynolds numbers up to $R_\tau = 40{,}000$ [28] show the presence of a very similar log law in a similar y^+ range.

Another result from pipe flow experiments is that in the region beyond the viscous sublayer up to approximately $y^+ = 300$ it has been found that a power law distribution occurs in the form

$$\overline{U}^+ = 8.48(y^+)^{0.142}. \tag{7.70}$$

A similar result with exponent 0.145 is found in [28]. This power law is not to be confused with the power law scaling of the pipe flow that has been offered as a more general substitute for the log law and that is considered in the next section. It should be noted that the existence of Eq. (7.70) is not anticipated in classical discussions of the pipe flow scaling so that it represents somewhat of a departure from the traditional belief that there is a shift from Eq. (7.64) directly to a log law without a discernible scaling law in between.

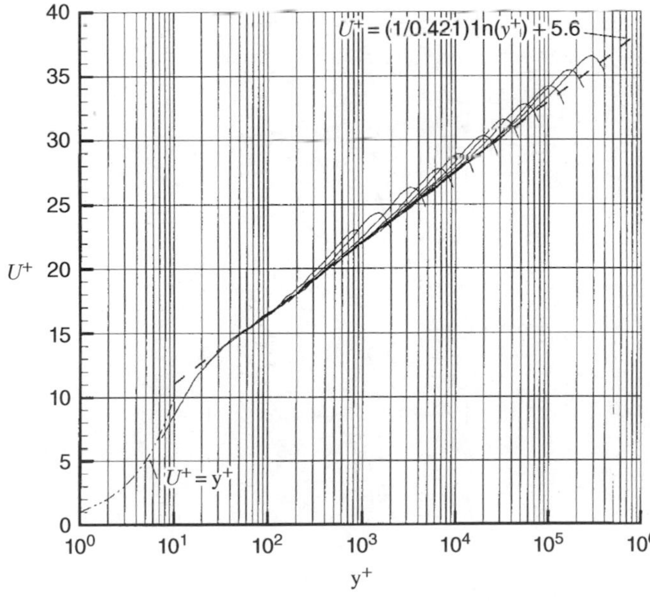

Figure 7.18 Mean velocity profiles in pipe flow [6] showing the collective approach to a log law. The curves are for Reynolds numbers between $R_e = 31 \times 10^3$ and $R_e = 18 \times 10^6$. Reprinted with permission of Cambridge University Press.

7.2.2 Power Law

Early studies of the mean velocity in pipe flows found some evidence that power laws can do a credible job of representing the observed flow behavior over the entire pipe cross-section. For example, it has been noted [25, 29, 30] that a power law of the form

$$\frac{\overline{U}}{U_{cl}} = \left(\frac{y}{R_o}\right)^{1/n}, \tag{7.71}$$

where n generally increases as the range of Reynolds number increases, can match the data, as shown in Figure 7.19 taken from [25]. In this figure, plots of the normalized pipe mean velocity field raised to the power $1/n$ fit the data with $n = 7$ for $R_e = 110{,}000$ and $n = 10$ for $R_e = 2{,}000{,}000$. Linearity of the plotted function implies power law behavior, which is seen to apply to a significant range of Reynolds numbers and covers almost the entire radius of the pipe.

Some attention has been paid in more recent work [31] to placing the power law representation of the mean velocity in pipe flow on a more formal basis that accounts for the Reynolds number dependence of the exponent. One aspect of this is to predict and accommodate the systematic way in which the $\overline{U}$ curves in Figure (7.18) rise above the log law at progressively further positions as the Reynolds number increases. To include a Reynolds number dependence at the outset is to assume incomplete similarity so that Eq. (7.71) is generalized to

$$\frac{d\overline{U}}{dy} = f(y, \tau_w, \nu, \rho, R_o), \tag{7.72}$$

where now dependence of the intermediate layer on ν and R_o is not a priori excluded. Note that Eq. (7.23) may be interpreted as anticipating this result in the sense that a Reynolds number dependence is seen to be evident outside the viscous sublayer.

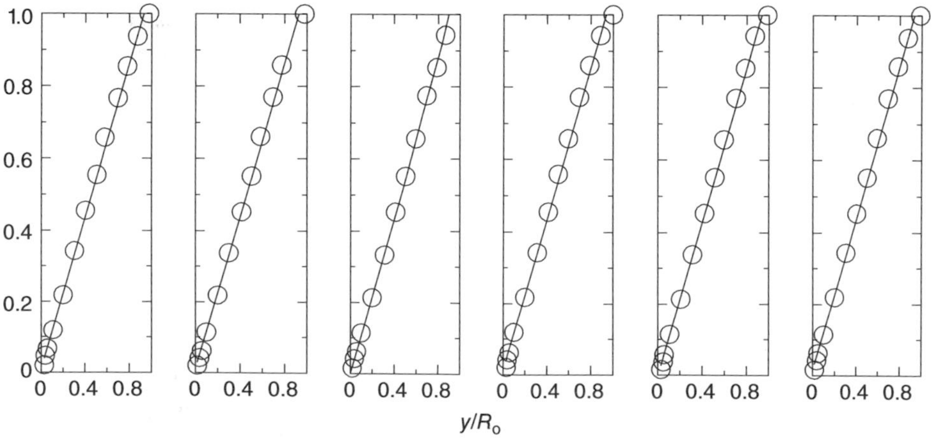

Figure 7.19 Plots of $(\overline{U}/\overline{U}_{max})^{1/n}$ in pipe flow for empirically fitted exponents, n. From left to right $1/n = 6.0, 6.6, 7.0, 8.8, 10.0$, and 10.0, and the Reynolds numbers are $4 \times 10^3, 2.3 \times 10^4, 1.1 \times 10^5, 1.1 \times 10^6, 2 \times 10^6$, and 3.2×10^6. From [25], p. 563.

By similar dimensional arguments as before, Eq. (7.72) leads to the result

$$\frac{d\overline{U}}{dy} = \frac{U_\tau}{y} f(y^+, R_\tau). \tag{7.73}$$

Since, as shown previously, R_τ is functionally related to R_e, one is at liberty to replace (7.73) with

$$\frac{d\overline{U}}{dy} = \frac{U_\tau}{y} f(y^+, R_e). \tag{7.74}$$

When f is assumed to be constant a log law is implied. Alternatively, if f obeys a power law as in

$$f(y^+, R_e) = \beta^*(R_e)(y^+)^{\alpha(R_e)} \tag{7.75}$$

for both large y^+ and R_e, then so too will $\overline{U}^+$ after integration of Eq. (7.74) using Eq. (7.75). Applying the boundary condition at the wall, a calculation gives

$$\overline{U}^+(y^+) = \beta(R_e)(y^+)^{\alpha(R_e)}, \tag{7.76}$$

where the parameter β is defined from β^* and α after the integration.

To determine an acceptable form of $\alpha(R_e)$ consider the behavior of Eq. (7.76) in the limit as $\nu \to 0$. In particular, consider pipe flow for a fixed $\partial\overline{P}/\partial x$. According to Eq. (7.57), τ_w remains constant as $\nu \to 0$, and so too does U_τ. Since $\overline{U}$ is bounded, $\overline{U}^+$ is bounded, and so the left-hand side of Eq. (7.76) is bounded as $\nu \to 0$. Consequently, the right-hand side must also be bounded as $y^+ \to \infty$ and $R_e \to \infty$. Noting the identity

$$(y^+)^{\alpha(R_e)} = e^{\alpha(R_e)\log y^+}, \tag{7.77}$$

the choice

$$\alpha(R_e) = \frac{\alpha_1}{\log R_e}, \tag{7.78}$$

where α_1 is a constant, has the advantage that the product $\alpha(R_e)\log y^+$ is given the flexibility of reaching a non-zero limit for high Reynolds numbers, as against a scaling that forces it to converge to zero. This choice also appears to lead to predictions that are most in agreement with experiments.

If it is postulated that $\beta(R_e)$ enjoys the same dependence on R_e as does α so that

$$\beta(R_e) = \beta_0 + \frac{\beta_1}{\log R_e}, \tag{7.79}$$

where β_0 and β_1 are also constants, then it is derived that

$$\overline{U}^+(y^+) = (\beta_0 \log R_e + \beta_1)(y^+)^{\frac{\alpha_1}{\log R_e}}. \tag{7.80}$$

It may be noted that the appearance of R_e in the form of its logarithm in Eq. (7.80) means that if R_e is replaced by γR_e because of a change in velocity or length scale, then $\log \gamma R_e = \log \gamma + \log R_e$, which converges to $\log R_e$ as $R_e \to \infty$. Consequently, the formula in Eq. (7.80) is unaffected at large R_e and the constants α_1 and β_1 should have a universal form.

Just as the constants in the log law in Eq. (7.31) are determined by empirical fit, so too is it necessary to find values for α_1, β_0, and β_1 by comparing Eq. (7.80) with experimental data. Such an effort [31] based on tabulated date of Nikuradze [29] over the Reynolds number range $4 \times 10^3 \to 3.24 \times 10^6$ gives the approximate values

$$\alpha_1 = 1.5 \qquad \beta_0 = 0.578 \qquad \beta_1 = 2.5. \tag{7.81}$$

Within the specified set of the Nikuradze data, Eq. (7.80) provides a succinct way of describing the measured mean velocity at different Reynolds numbers. Figure 7.20 contains a plot of the number pairs $(\log y^+, \psi)$, where

$$\psi \equiv \frac{\log R_e}{\alpha_1} \log \left(\frac{\overline{U}^+}{\beta_0 \log R_e + \beta_1} \right) \tag{7.82}$$

from many experiments at different Reynolds numbers. Since the equation $\psi = \log y^+$ is equivalent to Eq. (7.80), the latter will be satisfied to the extent that the data appears on a single line of unit slope in the $(\log y^+, \psi)$ plane. Figure 7.20 shows that this condition is generally well met apart from the region closest to the wall where the power law is not expected to apply [31].

Since the power law is meant to cover a larger region of the pipe than does the log law, it can be used to explain the systematic way in which the mean velocity field for any fixed Reynolds number departs from log-law behavior for sufficiently large y^+. Moreover, for fixed values of its constants, Eq. (7.80) gives a one-parameter family of curves depending on R_e. The individual power law curves when plotted together have an envelope that has some similarity to the log law with similar values of the constants κ and B. In this way the log-law and power-law approaches to similarity may be reconciled.

7.2.3 Streamwise Normal Reynolds Stress

One of the more interesting results to emerge from the availability of pipe flow data at high Reynolds numbers concerns the appearance of a second peak in the streamwise velocity variance $\overline{u^2}$. The existence of a peak near the wall in the vicinity of $y^+ = 15$,

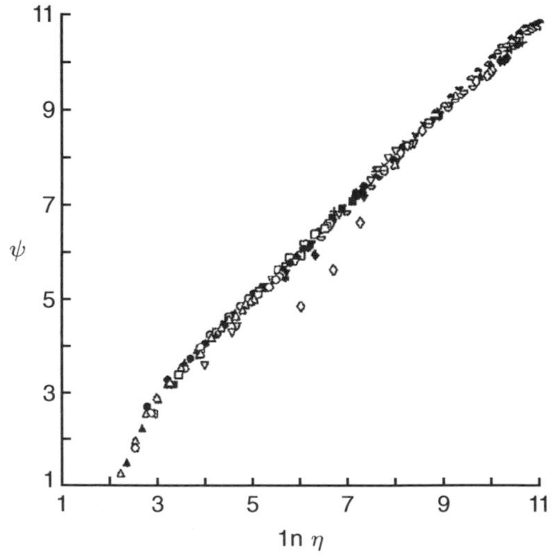

Figure 7.20 ψ vs. $\log y^+$ where $\eta \equiv y^+$ in this figure. Data are taken from 16 different Reynolds numbers from 4×10^3 to 3.24×10^6 measured in [29]. From [31]. Reprinted with permission from ASME International.

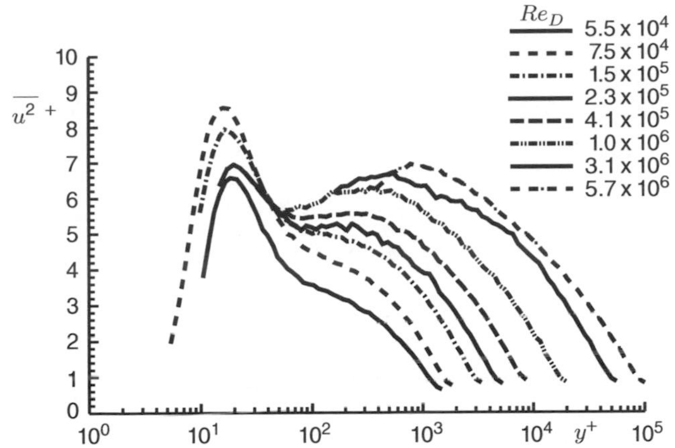

Figure 7.21 Streamwise velocity variance at high Reynolds numbers in pipe flow [32]. Reprinted with permission of Cambridge University Press.

as shown in Figure 7.8 for channel flow, is common to pipes and boundary layers as well. It also may be observed in Figure 7.7 that there is a significant increase in $\overline{u^2}$ with Reynolds number in the region beyond its peak. Measurements in pipe flow at much higher Reynolds numbers than are yet obtainable in channel flow DNS appear to show that the trend in rising $\overline{u^2}$ continues to eventually lead to the appearance of a second peak, as shown in Figure 7.21. The Reynolds numbers based on pipe diameter and mean cross-sectional velocity in Figure 7.21 go as high as 5.7M. In terms of the Kármán number defined as $R_\tau = U_\tau R_o/\nu$, the highest value is 101,000. Similar results to these, with peaks in $\overline{u^2}$ both near and away from wall that grow with Reynolds number, have been measured in pipe flow using PIV for R_τ up to 40,000 [28]. The second peak in $\overline{u^2}$ has also been observed in recent boundary layer measurements [33], though it should be cautioned that investigations of the accuracy of probe measurements [34] suggest that in the case of boundary layer flow for Reynolds numbers up to $R_\tau = 18{,}830$ (where in this case $R_\tau = U_\tau \delta/\nu$ and δ is the boundary layer thickness introduced in Chapter 8) the second peak in streamwise Reynolds stress can be attributed to measurement error caused by attenuation of small scales as they affect the u signal in the near-wall region. It can be expected that further clarity on this phenomenon, including discerning the physical mechanism behind it, will be forthcoming in future experiments and analysis.

References

1 Monty, J.P., Hutchins, N., Ng, H., Marusic, I., and Chong, M.S. (2009) A comparison of turbulent pipe, channel and boundary layer flows. *J. Fluid Mech.*, 632, 431–442.

2 Smits, A.J. (2015) Canonical wall-bounded flows: how do they differ? *J. Fluid Mech.*, 774, 1–4.

3 Smits, A.J., McKeon, B.J., and Marusic, I. (2011) High-Reynolds number wall turbulence. *Ann. Rev. Fluid Mech.*, 43, 353–375.

4 Vinuesa, R., Noorani, A., Lozano-Duran, A., Khoury, G.K.E., Schlatter, P., Fischer, P.F., and Nagib, H.M. (2014) Aspect ratio effects in turbulent duct flows studied through direct numerical simulation. *J. Turbulence*, 15, 677–706.
5 Vinuesa, R., Bartrons, E., Chiu, D., Dressler, K.M., Ruedi, J.D., Suzuki, Y., and Nagib, H.M. (2014) New insight into flow development and two dimensionality of turbulent channel flows. *Exp. Fluids*, 55, 1–14.
6 McKeon, B.J., Li, J., Jiang, W., Morrison, J.F., and Smits, A.J. (2004) Further observations on the mean velocity distribution in fully developed pipe flow. *J. Fluid Mech.*, 501, 135–147.
7 Bernard, P.S. (2015) *Fluid Dynamics*, Cambridge University Press, Cambridge.
8 Kao, T.W. and Park, C. (1970) Experimental investigations of the stability of channel flows. Part 1. Flow of a single liquid in a rectangular channel. *J. Fluid Mech.*, 43, 145–164.
9 Nishioka, M., Ida, S., and Ichikawa, Y. (1975) An experimental investigation of the stability of plane Poiseuille flow. *J. Fluid Mech.*, 72, 732–751.
10 Lee, M. and Moser, R.D. (2015) Direct numerical simulation of channel flow up to $Re_\tau \approx 5200$. *J. Fluid Mech.*, 774, 395–415.
11 Kim, J., Moin, P., and Moser, R. (1987) Turbulence statistics in fully developed channel flow at low Reynolds number. *J. Fluid Mech.*, 177, 133–166.
12 Lee, M., Ulerich, R., Malaya, N., and Moser, R.D. (2014) Experiences from leadership computing in simulations of turbulent fluid flows. *Comput. Sci. Engng.*, 16, 24–31.
13 Graham, J., Kanov, K., Yang, X., Lee, M., Malaya, N., Lalescu, C., Burns, R., Eyink, G., Szalay, A., Moser., R., and Meneveau, C. (2016) A web services-accessible database of turbulent channel flow and its use for testing a new integral wall model for LES. *J. Turbulence*, 17, 181–215.
14 Alfredsson, P., Johansson, A., Haritonidis, J., and Eckelmann, H. (1988) The fluctuating wall-shear stress and the velocity field in the viscous sublayer. *Phys. Fluids*, 31, 1026–1033.
15 von Kármán, T. (1930) *Mechanische Ähnlichkeit und turbulenz*, in *Proc. 3rd Intl. Congr. Appl. Mech.*, vol. 1 (eds C.W. Oseen and W. Weibull), AB Sveriges Litografiska Tryckenier, vol. 1, pp. 85–93.
16 Prandtl, L. (1932) Zur turbulenten ströhren und längs platten. *Ergebn. Aerodyn. Versuchanstalt*, 4, 18–29.
17 Jiménez, J. and Moser, R.D. (2007) What are we learning from simulating wall turbulence? *Phil. Trans. Roy. Soc. London A*, 365, 715–732.
18 Mizuno, Y. and Jiménez, J. (2011) Mean velocity and length-scales in the overlap region of wall-bounded turbulent flows. *Phys. Fluids*, 23, 085 112.
19 Townsend, A.A. (1976) *The Structure of Turbulent Shear Flows*, Cambridge University Press, Cambridge.
20 Moser, R.D., Kim, J., and Mansour, N.N. (1999) DNS of turbulent channel flow up to $R_\tau = 590$. *Phys. Fluids*, 11, 943–945.
21 Mansour, N.N., Kim, J., and Moin, P. (1988) Reynolds-stress and dissipation-rate budgets in a turbulent channel flow. *J. Fluid Mech.*, 194, 15–44.
22 Wu, X., Moin, P., Adrian, R.J., and Baltzer, J.R. (2015) Osborne Reynolds pipe flow: Direct simulation from laminar through gradual transition to fully developed turbulence. *Proc. Nat. Acad. Sci.*, 112, 7920–7924.
23 Moody, L.F. (1944) Friction factors for pipe flow. *ASME Trans.*, 66, 671–684.

24 White, F.M. (1986) *Fluid Mechanics*, McGraw-Hill, New York, 2nd edn.
25 Schlichting, H. (1968) *Boundary Layer Theory*, McGraw-Hill Book Co., New York, 6th edn.
26 Coles, D. (1956) The law of the wake in the turbulent boundary layer. *J. Fluid Mech.*, 1, 191–226.
27 Millikan, C.B. (1938) A critical discussion of turbulent flows in channels and circular pipes, in *Proc. 5th Intl. Conf. Appl. Mech.*, Cambridge, Mass., pp. 386–392.
28 Willert, C.E., Soria, J., Stanislas, M., Klinner, J., Amili, O., Eisfelder, M., Cuvier, C., Bellani, G., Fiorini, T., and Talamelli, A. (2017) Near-wall statistics of a turbulent pipe flow at shear Reynolds numbers up to 40000. *J. Fluid Mech.*, 826, R5–1–R5–11.
29 Nikuradze, J. (1932) Gesetzmässigkeiten der turbulenten strömung in glatten rohren, *Tech. Rep. 356*, VDI Forschungheft.
30 Hinze, J.O. (1975) *Turbulence*, McGraw-Hill, New York, 2nd edn.
31 Barenblatt, G.I., Chorin, A.J., and Prostokishin, V.M. (1997) Scaling laws for fully developed turbulent flow in pipes. *Applied Mech. Rev.*, 50, 413–429.
32 Morrison, J.F., McKeon, B., Jiang, W., and Smits, A. (2004) Scaling of the streamwise velocity component in turbulent pipe flow. *J. Fluid Mech.*, 508, 99–131.
33 Vallikivi, M., Hultmark, M., and Smits, A.J. (2015) Turbulent boundary layer statistics at very high Reynolds number. *J. Fluid Mech.*, 779, 371–389.
34 Hutchins, N., Nickels, T.B., Marusic, I., and Chong, M.S. (2009) Hot-wire spatial resolution issues in wall-bounded turbulence. *J. Fluid Mech.*, 632, 431–442.

Problems

7.1 Prove that Eq. (7.19) is true in channel flow.

7.2 Derive Eq. (7.54) for a channel flow.

7.3 Calculate the ratio of U_m/U_{cl} in pipe flow, where U_{cl} is the mean velocity in the center of a pipe, assuming that the power law Eq. (7.71) holds.

7.4 Assuming that the mean velocity in pipe flow obeys a 1/7 power law, show that Eq. (7.62) can be expressed in the form

$$f = 0.28Re_{cl}^{-1/4} \tag{7.83}$$

where Re_{cl} is the Reynolds number based on the center line velocity and the pipe radius.

7.5 The indicator function $y\, d(\log \overline{U})/dy$ can be used to reveal the presence of a log law. Find an equivalent function that can reveal the presence of power law behavior in $\overline{U}$.

8

Boundary Layers

The flow adjacent to a flat solid wall will form a boundary layer when the Reynolds number is sufficiently high. In essence, the boundary layer is a thin region over which the streamwise fluid velocity transitions from a zero value at the wall to that of the free stream traveling parallel to the wall, as shown in Figure 8.1. The boundary layer is generally filled with vorticity produced at the wall surface by the action of viscosity. In many circumstances, such as the canonical zero-pressure gradient boundary layer wherein uniform flow travels over a flat plate, the region above the boundary layer is a potential flow that is devoid of vorticity. In such cases the vorticity will be zero at a sufficient normal distance above the wall at any streamwise position.

At any given time in a boundary layer the location of the outer edge that divides potential and vortical flow varies with downstream position, as may be seen in Figure 8.2, which is a photograph of a smoke-marked turbulent flow over a flat plate in a laboratory wind tunnel. Smoke is introduced both upstream of the view in the image at the wall surface and at the interface between the freestream irrotational flow and the boundary layer. The smoke released near the wall reveals some idea of the internal chaotic structure of the boundary layer while the outer layer smoke gives a good indication of where turbulent motion ends and the outer potential flow traveling at $U_\infty(x)$ begins. The boundary layer edge is seen to take on a highly corrugated shape with turbulent flow on one side and potential flow on the other. As the flow evolves these regions convect past any given point, giving the impression that the turbulence is intermittent. In a frame of reference convecting with the interface between the turbulent and non-turbulent regions, the bulges in the boundary layer appear to be spanwise vortices with a streamwise scale on the order of the boundary layer thickness. The irrotational flow above the boundary layer moves over and around the bulges, sometimes penetrating between them deep into the boundary layer, as seen in Figure 8.2.

In the absence of extenuating circumstances, the upstream beginnings of a turbulent boundary layer, such as that illustrated in Figure 8.1, is likely to be a laminar boundary layer. At some point in its development the laminar flow become susceptible to the destabilizing influence of perturbations so that it transitions to fully turbulent flow. The laminar flow on a flat plate is accurately described by the Blasius similarity solution [2] whose streamwise extent in any flow depends on such factors as the level of free stream turbulence and the smoothness of the plate, with transition generally beginning at Reynolds number $Re_x = 400{,}000$, where $Re_x = U_\infty x/\nu$ and x is the distance from the leading edge of the plate.

Turbulent Fluid Flow, First Edition. Peter S. Bernard.

Companion website: www.wiley.com/go/Bernard/Turbulent_Fluid_Flow

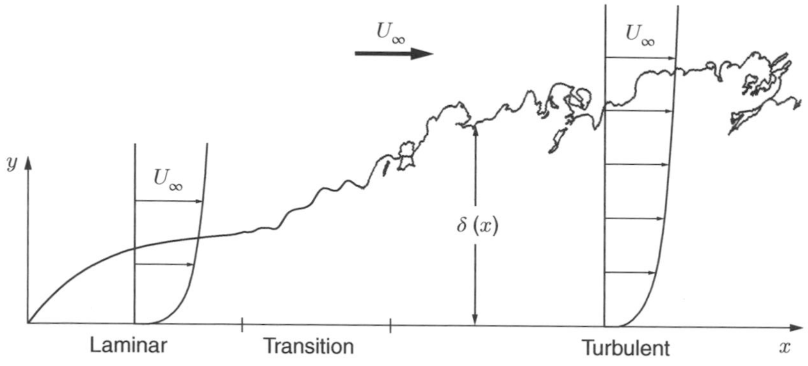

Figure 8.1 Turbulent boundary layer over a flat plate.

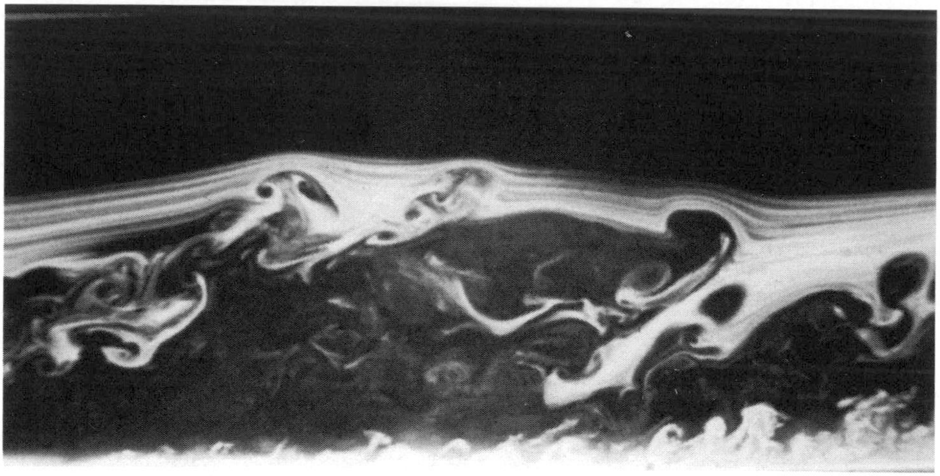

Figure 8.2 Smoke visualization of a turbulent boundary layer at $R_\theta = 3000$ [1].

The nature and extent of the transition region are influenced by the details of the perturbations that either lie within the laminar boundary layer or are external to it. For example, in the smoothest of circumstances the streamwise velocity field in the boundary layer will develop a system of 2D disturbances known as Tollmien–Schlichting (TS) waves that are of low amplitude and satisfy a linearized form of the Navier–Stokes equation [3]. As these develop in time, their amplitude grows and they acquire a 3D structure that rapidly breaks down to turbulence through a non-linear process. If the initial perturbations are of sufficiently large amplitude, transition can proceed in such a way as to bypass the linear, small amplitude instability modes directly into a non-linear development [4].

Flow statistics such as the mean velocity field have been widely studied in boundary layers. Of particular interest are similarity forms taken by the mean velocity field that are similar to those that have been studied in channel and pipe flow considered in the previous chapter. New results from both simulations and physical experiments at increasingly high Reynolds numbers have led to some adjustments to the classical view of the similarity profiles. It can be anticipated that as future studies consider even higher Reynolds numbers some further revision of these fundamental results will become necessary.

Lurking behind the particular trends in the velocity statistics are the presence of dynamical processes involving the action of coherent vortices. Some hint of this aspect of the flow was discussed in the previous chapter in the context of events that produced the Reynolds shear stress in a channel. Similar physics is a part of boundary layer flow and in this chapter we will take a more detailed look into the nature of such structure that is an essential aspect of the flow dynamics. The idea is to develop an understanding of what kinds of structure are present and what role they play in the turbulent makeup of the boundary layer. A number of the main themes in ongoing structural research will be considered in the last section of this chapter.

8.1 General Properties

Flow within the fully turbulent boundary layer can be partitioned into a viscous sublayer, intermediate layer, and outer layer similar to the way it is done for channel and pipe flows and indicated in Figure 8.3. Moreover, many of the same arguments for rationalizing

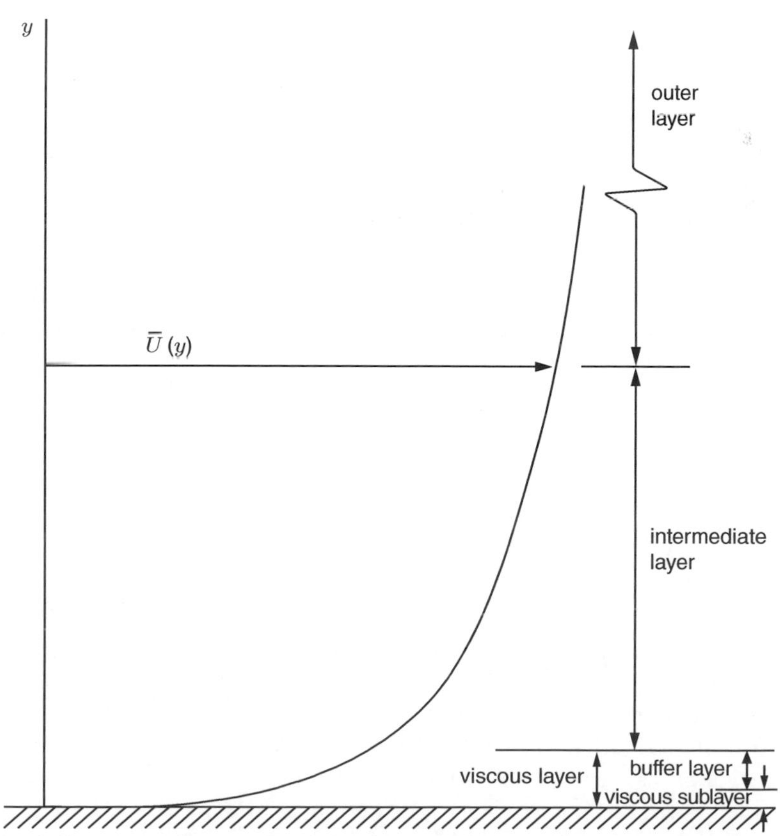

Figure 8.3 Boundary layer zones, not drawn to scale.

the existence of a log or power law for pipe and channel flow apply equally well to the boundary layer. The main difference between these flows concerns the outer flow, which is usually potential flow in the case of the boundary layer. Moreover, the thickness of the region of turbulent flow in a boundary layer is not a priori proscribed as it is for the channel and pipe. In many common situations the boundary layer thickness changes with downstream distance and it is necessary to first develop a means for predicting the thickness before the similarity properties of the boundary layer can be considered in depth.

The mean velocity field $\overline{U}(x, y)$ at a given streamwise (i.e., x) location increases from 0 at the wall surface where $y = 0$ to the local value of the potential velocity $U_\infty(x)$. One means of establishing a boundary layer thickness is to define it as the distance from the wall where the velocity is within 1% of its free stream value. Thus, denoting this as δ it follows that

$$\overline{U}(x, \delta) = 0.99\, U_\infty(x). \tag{8.1}$$

In practice, several alternative ways of characterizing the boundary layer thickness have been devised that are not dependent on the arbitrary choice of the velocity criterion in Eq. (8.1). In particular, the *displacement thickness*, $\delta_1(x)$, is defined at a given x location as the distance a flat wall would have to be shifted normal to itself so that the true flux of fluid parallel to the wall at this point will be equal to the flux of fluid in the reduced region assuming the fluid travels at the free stream value $U_\infty(x)$. In other words,

$$\int_0^\infty \overline{U}(x, y)dy = \int_{\delta_1(x)}^\infty U_\infty(x)dy. \tag{8.2}$$

Adding and subtracting the quantity $\delta_1(x)U_\infty(x)$ to the right-hand side of Eq. (8.2) and rearranging gives

$$\delta_1(x) \equiv \int_0^\infty \left(1 - \frac{\overline{U}(x, y)}{U_\infty(x)}\right) dy \tag{8.3}$$

as the displacement thickness. The integrand approaches zero for large y so that the integration is convergent.

In a similar vein, another commonly employed boundary layer measure is the *momentum thickness*, θ, defined via

$$\theta(x) \equiv \int_0^\infty \frac{\overline{U}(x, y)}{U_\infty(x)} \left(1 - \frac{\overline{U}(x, y)}{U_\infty(x)}\right) dy, \tag{8.4}$$

which has a physical interpretation as described in Problem 8.1. Similar to δ_1, the integrand in Eq. (8.4) goes to zero for large y so the scale is well defined. Reynolds numbers formed from δ, δ_1, and θ, such as $R_\theta = \theta U_\infty/\nu$, are useful for enabling comparisons of results obtained under different circumstances in different facilities and simulations.

The boundary layer measures that have been defined above describe the mean thickness of the boundary layer at a point x. In reality, the thickness of the boundary layer varies spatially and temporally from one time to the next, as is evident from Figure 8.2. To quantify how the shift between turbulent and potential flow occurs, an intermittency function, γ, can be defined as the fraction of the time when the flow at a given point is

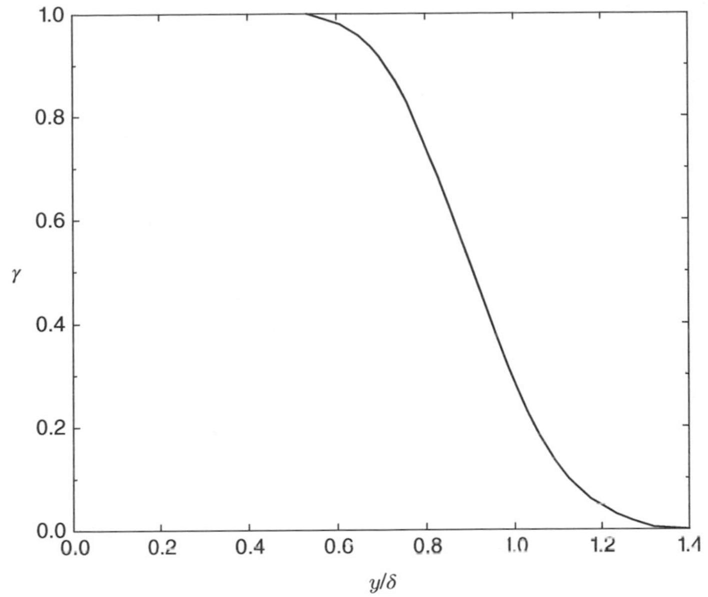

Figure 8.4 Intermittency factor in a turbulent boundary layer [5].

completely turbulent. γ is shown in Figure 8.4 for measurements in a $R_\theta \approx 1130$ boundary layer [5]. The variability in boundary layer thickness implied by Figure 8.2 ultimately is captured implicitly in the various definitions of boundary layer thickness.

An early effort at quantifying γ by Klebanoff [6] in a zero-pressure gradient (ZPG) boundary layer resulted in the expression

$$\gamma(y) = \frac{1}{2}(1 - \mathrm{erf}(5(y/\delta - 0.78))). \tag{8.5}$$

This is derived by fitting a Gaussian distribution to the instantaneous positions of the boundary layer edge, yielding a mean position 0.78δ with standard deviation 0.14δ. A subsequent formulation known as the Klebanoff intermittency function [7] has

$$\gamma_K = 1/(1 + 5.5(y/\Delta)^6), \tag{8.6}$$

which has application within some closure schemes applied to boundary layer flow. In this, Δ is a length scale that has different meanings in different models.

8.2 Boundary Layer Growth

For the classic laminar Blasius boundary layer, the availability of a similarity solution for the velocity field provides precise answers to such questions as how the boundary layer grows downstream. For example, the momentum thickness grows with $\sqrt{x}$ according to the relation [2]

$$\frac{\theta}{x} = \frac{0.664}{Re_x}, \tag{8.7}$$

while the viscous force on a flat wall of length L can be computed to be

$$F_x = 0.664\frac{\rho U_\infty^2 L}{Re_L} \tag{8.8}$$

per unit width. Finding relations equivalent to Eqs. (8.7) and (8.8) for a turbulent boundary layer depends on having knowledge gained from physical experiments and possibly direct numerical simulations of turbulent flow.

An approximate means of acquiring expressions for the boundary layer growth and surface forces makes use of an integral momentum balance applied to the mean flow in the boundary region. In this classical argument [8] boundary layer scaling equivalent to that done for laminar flow [2] is used to drop the streamwise diffusion term as well as Reynolds stress terms that are differentiated in the x direction. The result is to reduce the 2D mean momentum equations for boundary layer flow in the x and y plane to

$$\frac{\partial \overline{U}^2}{\partial x} + \frac{\partial \overline{U}\,\overline{V}}{\partial y} = -\frac{1}{\rho}\frac{\partial \overline{P}}{\partial x} + \nu\frac{\partial^2 \overline{U}}{\partial y^2} - \frac{\partial \overline{uv}}{\partial y} \tag{8.9}$$

and

$$\frac{\partial\left(\frac{\overline{P}}{\rho} + \overline{v^2}\right)}{\partial y} = 0. \tag{8.10}$$

Note that the left-hand side of Eq. (8.9) is written in flux form, which takes advantage of the continuity equation (3.10). For zero pressure gradient boundary layers $\partial\overline{P}/\partial x = 0$ in Eq. (8.9). More generally by including the mean pressure gradient the flow in arbitrary boundary layers with varying $U_\infty(x)$ can be accommodated.

Since turbulence and viscous effects are confined to the boundary layer, the outer flow satisfies the conditions required for the applicability of Bernoulli's law [2] to the effect that

$$\frac{U_0^2}{2} + \frac{P_0}{\rho} = \frac{U_\infty^2}{2} + \frac{\overline{P}}{\rho} \tag{8.11}$$

is constant on streamlines. Here, U_0 and P_0 signify reference values in the upstream far field. According to Eq. (8.11) the streamwise pressure gradient over the boundary layer satisfies

$$\frac{\partial \overline{P}}{\partial x} = -\rho U_\infty \frac{dU_\infty}{dx}. \tag{8.12}$$

To the same degree of approximation as used in deriving Eq. (8.9), the relation

$$\frac{\partial}{\partial y}\left(\frac{\partial \overline{P}}{\partial x}\right) = 0 \tag{8.13}$$

can be derived by taking an x derivative of Eq. (8.10). Consequently, it can be assumed that Eq. (8.12) applies throughout the boundary layer.

Substituting Eq. (8.12) into (8.9) and integrating normal to the wall to a point in the free stream, say $y = d$ where the streamwise velocity is everywhere equal to U_∞, yields

$$\frac{\partial}{\partial x}\int_0^d \overline{U}^2\, dy + U_\infty \overline{V}(x,d) = \int_0^d U_\infty \frac{dU_\infty}{dx} dy - \frac{\tau_w}{\rho}, \tag{8.14}$$

where the integral of the last term on the right-hand side of Eq. (8.9) vanishes since $\overline{uv}$ is zero at the wall surface and in the free stream flow. Integrating the continuity equation (3.10) across the boundary layer gives

$$\overline{V}(x,d) = -\int_0^d \frac{\partial \overline{U}}{\partial x} dy, \tag{8.15}$$

and substituting this into Eq. (8.14), commuting the x derivative with the integral, and rearranging yields

$$\frac{d}{dx}\int_0^d (\overline{U}^2 - \overline{U}U_\infty)dy = \frac{dU_\infty}{dx}\int_0^d (U_\infty - \overline{U})dy - \frac{\tau_w}{\rho}. \tag{8.16}$$

The integrations appearing in the expression can be taken out to infinity since the integrands approach zero for large y. The result after utilizing the definitions in Eqs. (8.3) and (8.4) is

$$\frac{dU_\infty^2\theta}{dx} + \frac{dU_\infty}{dx}U_\infty\delta_1 = \frac{\tau_w}{\rho}. \tag{8.17}$$

The value of this expression is in its use in relating the mean shear stress at the wall to the thickness of the boundary layer as given by the displacement and momentum thicknesses.

Some degree of empiricism must be applied to Eq. (8.17) if it is to be used for such goals as predicting the streamwise growth in the boundary layer thickness. One approach of this kind is to assume a functional form for the mean velocity field – in essence hypothesizing a similarity form – and substituting this into Eq. (8.17). For example, assume the validity of a power law form for the mean velocity field such as was done in Eq. (7.71) for pipe flow, in which case

$$\overline{U} = U_\infty\left(\frac{y}{\delta}\right)^{1/n} \tag{8.18}$$

means that according to Eq. (8.3)

$$\delta_1 = \frac{\delta}{1+n} \tag{8.19}$$

and a similar calculation with Eq. (8.4) yields

$$\theta = \frac{\delta n}{(n+1)(n+2)}. \tag{8.20}$$

Substituting these relations into Eq. (8.17) and taking $n = 7$ yields a differential equation for δ in the form

$$\frac{7}{72}\frac{dU_\infty^2\delta}{dx} + \frac{1}{8}\frac{dU_\infty}{dx}U_\infty\delta = \frac{\tau_w}{\rho}. \tag{8.21}$$

To determine δ a relation for the mean shear stress in terms of δ is required. For the particular case of constant U_∞, empirical relations such as that in Eq. (7.83) may be adapted to the present situation by changing R_o to δ wherever the latter quantity appears in R_e and R_τ. This yields an expression for τ_w in the form [9]

$$\frac{\tau_w}{\rho} = 0.0225\frac{\nu^2}{\delta^2}Re_\delta^{7/4}, \tag{8.22}$$

and substituting this into Eq. (8.21) and setting U_∞ constant gives

$$\frac{d\delta}{dx} = 0.0225 \frac{72}{7} Re_\delta^{-1/4}. \tag{8.23}$$

Taking into account the appearance of $\delta(x)$ in the definition of Re_δ, the solution of Eq. (8.23) is

$$\frac{\delta}{x} = 0.37 Re_x^{-1/5}. \tag{8.24}$$

According to this model, the growth in boundary thickness goes as $x^{4/5}$ which is a substantially higher rate than the laminar boundary layer. The result in Eq. (8.24) is consistent with a number of experimental studies for Reynolds numbers less than approximately 10^6. This result can be extended to higher Reynolds numbers by using $n = 8$ or higher in Eq. (8.18).

8.3 Log-Law Behavior of the Velocity Mean and Variance

As in the case of channel and pipe flows considered previously, there is much interest in ascertaining the high Reynolds number similarity properties of the mean velocity field in boundary layer flow. Experimental determinations of the velocity field have gone as high as $R_\tau = U_\tau \delta / \nu = 72{,}500$ [10], which should be in the final asymptotic high Reynolds number range. Some of these findings are shown in Figure 8.5 for an experiment at $R_\tau = 13{,}600$ [11, 12]. The figure suggests that the mean velocity profile approximates log-law behavior with constants $\kappa = 0.41$ and $B = 5.0$ in the region $30 \leq y^+ \leq 0.15\delta$. A plot of the log-law indicator function $y^+ \partial \overline{U}^+ / \partial y^+$, which is expected to be constant in the region where exact log-law behavior occurs, is also plotted. The latter shows an approximate constant trend beginning near $y^+ = 100$ in which there is a slight rise with y^+

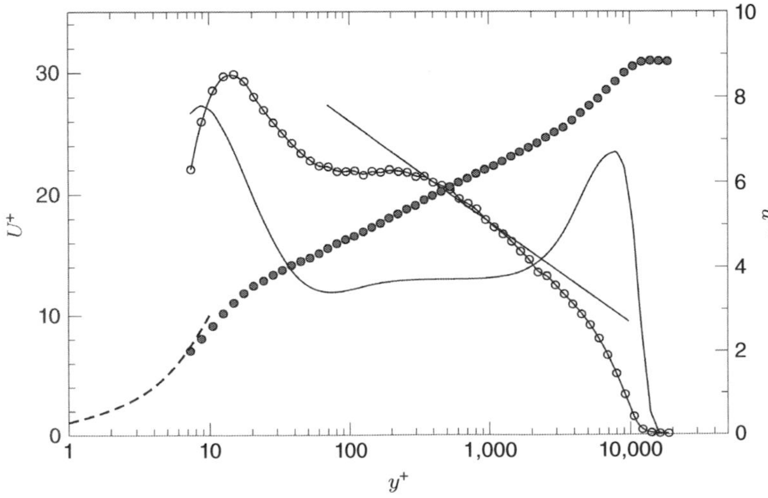

Figure 8.5 Mean and variance of the streamwise velocity in boundary layer flow at $R_\tau = 13{,}600$. •, $\overline{U}$; —, $y^+ \partial \overline{U}^+ / \partial y$; o, $\overline{u^2}^+$; straight line is a fit to Eq. (8.25). Data from [11]. Figure reproduced from [12] by permission of Annual Reviews.

suggesting that $\overline{U}^+$ departs slightly from a perfect log law. These aspects of the mean flow in boundary layers fit in with the channel flow results in Figure 7.6.

Comparing the mean velocities in channel, pipe, and boundary layer flows it may be concluded that they each contain approximate log-law regions whose position and extent within the flow are similar, though not exactly the same, in dimensionless units. Moreover, the constants κ and B differ by some small amounts between the types of flows. There are numerous factors that can be responsible for the differences between the flows, even apart from the intrinsic differences between a channel, pipe, and boundary layer. Among the former are experimental errors stemming from probe resolution, Reynolds number differences, and the effect of upstream conditions, particularly in the case of spatially growing boundary layers [13]. Consequently, it is difficult to arrive at conclusive explanations for why and in what way the log-law coefficients differ or why the log-law regions have different ranges in the different flows. When data for pipes and boundary layers at very high Reynolds numbers are considered [14], a log law in the region $3\sqrt{R_\tau} \le y^+ \le 0.15R_\tau$ holds in all cases with constants given approximately by $\kappa = 0.39$ and $B = 4.3$. This result purposefully includes information about the observed Reynolds number dependencies of the low y^+ starting point of the log law.

The scaling of the normal Reynolds stress $\overline{u^2}^+$ in boundary layers has been considered via analyses that take into account models of the turbulent structure. Among the latter is the Townsend attached eddy hypothesis [15] to the effect that the boundary layer can be viewed as formed from a collection of "attached" vortical eddies, some of whose properties will be considered in following sections. Models that apply a scaling analysis to the overlap region based on the attached eddy idea [15, 16] suggest that $\overline{u^2}^+$ also has a logarithmic form within the log-law region enjoyed by $\overline{U}^+$ and that it can be expressed as

$$\overline{u^2}^+ = B_1 - A_1 \log(y/\delta), \tag{8.25}$$

where $A_1 = 2.39$ and $B_1 = 1.03$ are empirically determined constants. The fact that the log law in Eq. (8.25) can reasonably describe the measured normal stress in the outer part of the log-law region is apparent from Figure 8.5.

8.4 Outer Layer

The occurrence of intermittency distinguishes the outer region of boundary layers from that of channel and pipe flows. Moreover, unlike channel and pipe flows, which reach a fully developed condition in which there is no streamwise dependence of the velocity statistics, boundary layers tend to contain some degree of streamwise evolution even when U_∞ is held constant. It is also the case that $\overline{V}$ tends not to vanish since in most circumstances the changes in streamwise momentum near the wall in incompressible flow must be balanced by mean convection away or toward the surface. All of these facets of boundary layer flow need to be taken into account when considering similarity solutions in the outer layer along the lines pursued in Section 8.2.

A starting point for analyzing the outer layer of boundary layers is to assume the validity of the velocity deficit scaling given by

$$U_\infty - \overline{U}(y) = F\left(y, \delta, \rho, U_\tau, \frac{dP_\infty}{dx}\right). \tag{8.26}$$

Here, x dependence is not indicated explicitly. In the outer layer viscosity has no direct role and the streamwise pressure gradient, dP_∞/dx, is included [17, 18] so as to make the analysis more general. Using dimensional analysis, it follows that

$$\frac{U_\infty - \overline{U}(y)}{U_\tau} = U_\infty^+ - \overline{U}^+(y) = F\left(\frac{y}{\delta}, \frac{\delta}{\rho U_\tau^2}\frac{dP_\infty}{dx}\right), \tag{8.27}$$

where the dimensionless group

$$\frac{\delta}{\rho U_\tau^2}\frac{dP_\infty}{dx} \tag{8.28}$$

is essentially equivalent to the so-called Clauser "equilibrium parameter," which has the same form but with δ replaced by δ_1. Experiments suggest that the mean velocity of boundary layers with different pressure gradients collapse to a single profile for equivalent values of the equilibrium parameter. Such boundary layers are said to be in turbulent equilibrium with each other.

The mean velocity shown in Figure 8.5 makes clear that $\overline{U}$ deviates from the logarithmic profile in the outer flow region $0.15 \leq y/\delta \leq 1$. However, the figure also suggests that it may be useful in determining F to require (8.27) to match the log law of the wall in the intermediate layer. This is the overlap argument seen in Section 7.2.1. In the present case this means that, in the overlap region,

$$\frac{1}{\kappa}\ln y^+ + B = U_\infty^+ - F\left(\frac{y}{\delta}\right), \tag{8.29}$$

so that $F(y/\delta)$ is the difference between the outer flow velocity U_∞^+ and the log law. For $y << \delta$,

$$F\left(\frac{y}{\delta}\right) = U_\infty^+ - \left[\frac{1}{\kappa}\ln y^+ + B\right] = \left[U_\infty^+ - \frac{1}{\kappa}\ln \delta^+ - B\right] - \frac{1}{\kappa}\ln\left(\frac{y}{\delta}\right). \tag{8.30}$$

This may be viewed as giving the functional form of F for relatively small y/δ that are in the overlap region. It then proves helpful to define a *wake function* $W(y/\delta)$ so that for all y and not just small y,

$$F\left(\frac{y}{\delta}\right) = \left[U_\infty^+ - \frac{1}{\kappa}\ln \delta^+ - B\right] - \frac{1}{\kappa}\ln\left(\frac{y}{\delta}\right) - \frac{\Pi}{\kappa}W\left(\frac{y}{\delta}\right), \tag{8.31}$$

where Π is a parameter. Note that the sum of F and $\Pi W/\kappa$ gives the difference between U_∞^+ and the log law even when extended beyond the overlap layer. Consequently, $\Pi W(y/\delta)/\kappa$ is the positive amount that $\overline{U}^+$ rises above the log law in the region beyond $y > 0.15\delta$, and according to Eq. (8.30) is zero in the region where the log law holds.

Π can be determined from the condition that $F(1) = 0$, which is implied by (8.27), so that

$$W(1)\frac{\Pi}{\kappa} = U_\infty^+ - \frac{1}{\kappa}\ln \delta^+ - B. \tag{8.32}$$

Equation (8.27) using (8.31) and (8.32) gives the Coles "law of the wake" [19]

$$\begin{aligned}\overline{U}^+ &= \frac{1}{\kappa}\ln y^+ + B + \frac{\Pi}{\kappa}W(y/\delta)\\ &= \frac{1}{\kappa}\ln y^+ + B + \frac{W(y/\delta)}{W(1)}\left(U_\infty^+ - \frac{1}{\kappa}\ln \delta^+ - B\right),\end{aligned} \tag{8.33}$$

which can be rearranged to

$$\frac{W(y/\delta)}{W(1)} = \frac{\overline{U}^+ - \frac{1}{\kappa}\ln y^+ - B}{U_\infty^+ - \frac{1}{\kappa}\ln \delta^+ - B}, \tag{8.34}$$

representing the fractional velocity deficit with respect to the log law. Commonly employed empirical forms for the wake function include [20]

$$\frac{\Pi}{\kappa} W\left(\frac{y}{\delta}\right) = \frac{2\Pi}{\kappa}\sin^2\left(\frac{\pi}{2}\frac{y}{\delta}\right) \tag{8.35}$$

and the more accurate

$$\frac{\Pi}{\kappa} W\left(\frac{y}{\delta}\right) = \frac{1}{\kappa}(1+6\Pi)\left(\frac{y}{\delta}\right)^2 - \frac{1}{\kappa}(1+4\Pi)\left(\frac{y}{\delta}\right)^3. \tag{8.36}$$

Note that for both of these choices the normalization $W(1) = 2$ is enforced. Application of either of Eqs. (8.35) or (8.36) to Eq. (8.33) provides an excellent accounting of the mean velocity in the outer region of turbulent boundary layers.

According to Eq. (8.32) Π depends on x through U_∞^+ and δ^+. In the zero-pressure gradient, smooth wall boundary layer it follows from Eq. (8.24) that

$$\delta U_\infty/\nu = 0.37 Re_x^{4/5} \tag{8.37}$$

where $Re_x \equiv xU_\infty/\nu$. Using this with Eq. (8.22) for τ_w, a calculation gives

$$U_\infty^+ = 5.89(Re_x)^{1/10}, \tag{8.38}$$

and using Eqs. (8.37) and (8.38) it follows that

$$\delta^+ = 0.0628 Re_x^{7/10}. \tag{8.39}$$

With these relations it is ascertained from Eq. (8.32) that

$$\Pi = \frac{\kappa}{2}(5.89(Re_x)^{1/10} - \frac{1}{\kappa}(-2.77 + 0.7\ln(Re_x)) - B), \tag{8.40}$$

which varies slowly with x. For example, with $\kappa = 0.4$, $B = 5.1$ and Re_x ranging from $5\times 10^6 \to 10^7$, Π varies from 0.48 to 0.63. Typically, the value $\Pi = 0.55$ is used in applications.

8.5 The Structure of Bounded Turbulent Flows

Analysis of boundary layer structure has proceeded from a number of different viewpoints with the ultimate goal of acquiring a comprehensive understanding of its role in the dynamics of the boundary layer and in explaining observed statistical trends. In addition, it is hoped that knowledge of structure, particularly in the near-wall region, can aid in deriving robust means for predicting turbulent flows.

8.5.1 Development of Vortical Structure in Transition

Structural features of turbulent boundary layers tend to first appear during transition to turbulence. While the way in which transition occurs, for example as TS waves or bypass transition of one sort or another, will affect the precise configuration of structures, the form of the local structures that eventually appear shares many of the same attributes in all transition modes. In many cases the process by which structure forms during transition can be observed in relatively clear detail. For example, the rotational regions associated with the downstream structure into which TS waves evolve is shown in Figure 8.6. The distinctive Λ shape formed by the rotational motion is displayed here via isosurfaces of a scalar marker of rotation, a technique that will be discussed in more detail in Section 8.5.3. In this kind of transition the Λ-shaped objects generally organize either into a K-type arrangement (referring to Klebanoff [21]) where they are ordered according to the wave length of the primary TS wave in the streamwise direction, or alternatively into the H-type arrangement (referring to Herbert [22]) seen in Figure 8.6. In the latter case the vortices form according to a sub-harmonic disturbance where they are in a staggered formation with double the wavelength in the streamwise direction. In some circumstances, transition occurs in the form of turbulent spots as shown in Figure 8.7, which are themselves composed of a growing collection of vortical objects that are not dissimilar from Λ vortices. The spots appear in random locations in the transitioning boundary layer, grow in size, and eventually merge to form a complete turbulent boundary layer. Measurements within the spots show statistical properties similar to the fully turbulent flow [23, 24]. In the general case of bypass transition, the appearance of vortical structure with associated Λ-like rotational regions will likely have patterns connected to the details of how the flow is disturbed.

One of the prominent characteristics of the velocity field in transition that is closely associated with the appearance of Λ vortices is the presence of elongated streamwise-oriented regions with locally reduced fluid velocity that are referred to as *low-speed streaks*. In transitional flow such structures occur with K-type or H-type arrangements of vortices, as well as in general transition modes. An example of streaks visualized in a perturbed channel flow is shown in Figure 8.8. The streaks appear as a natural response of the unstable boundary layer and are associated with motions that slowly organize the vorticity field in such a way as to ultimately lead downstream to

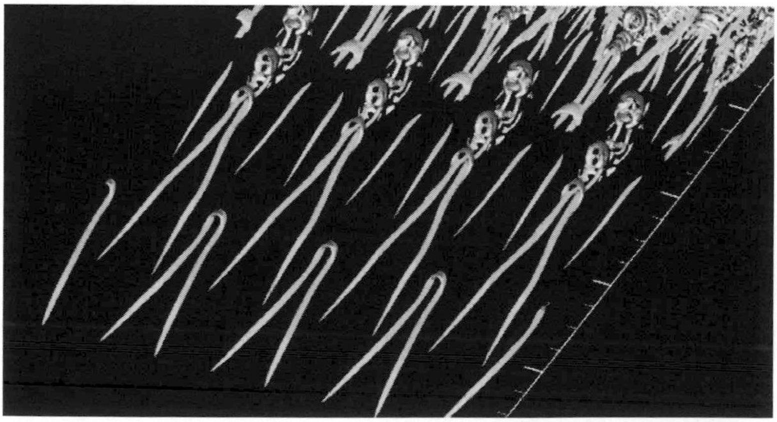

Figure 8.6 Visualization of the rotational field within Λ vortices in an H-type arrangement developing from TS waves in a boundary layer simulation [25]. Reprinted with permission of Cambridge University Press.

Figure 8.7 Turbulent spot visualized using aluminum particles in a transitioning boundary layer in water [26]. Reprinted with permission of Cambridge University Press.

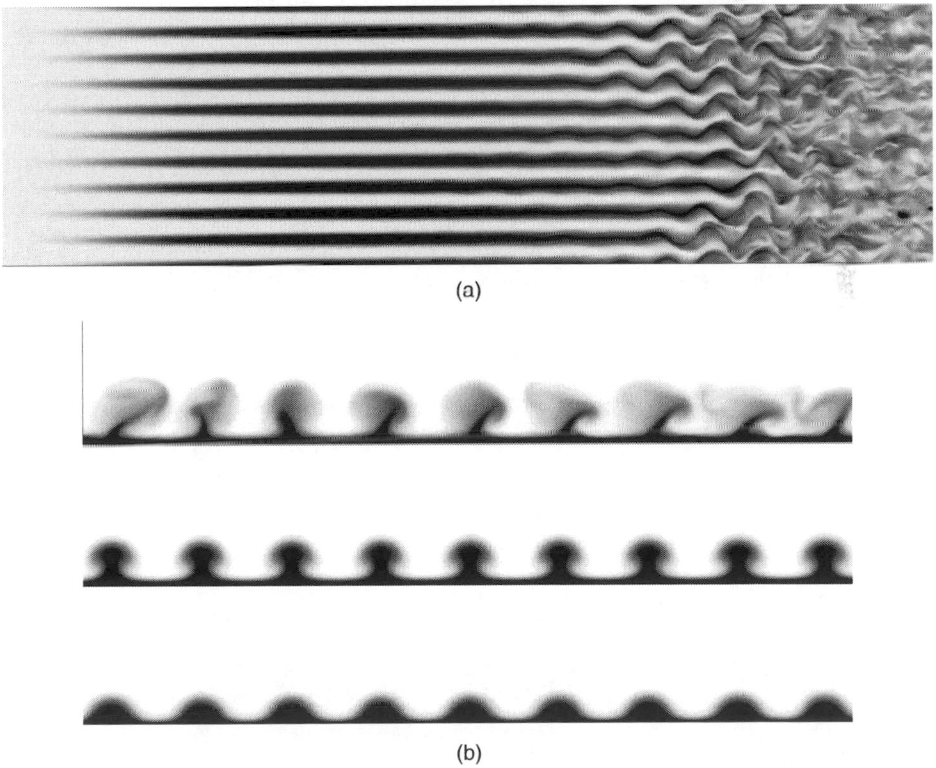

Figure 8.8 Contours of streamwise velocity in the boundary layer forming in the entrance region of a stimulated channel flow. From [27]. (a) Low-speed streaks (black) in a plane parallel to the wall $0 \leq x \leq 10$; (b) End-on view in spanwise planes at $x = 1, 4, 9$. Used with permission of Elsevier.

the appearance of vortical structure. In the particular example shown in Figure 8.8 the underlying vorticity is subject to non-linear instability that is reflected in the sinuous appearance of the streaks before they eventually break up into turbulence. Some insight into the developing vortical structure associated with the streaks is seen in Figure 8.8b by the way the velocity field is forming mushroom shaped eruptions.

8.5.2 Structure in Transition and in Turbulence

The capacity to observe the development of vortical structure in transition is much greater than it is in the turbulent flow region, where the presence of chaotic motion has the effect of obscuring the connection between the structures, the locally measured velocity field, and visualizations of the flow field via smoke or other markers. As a result, much of the research into structures has been carried out in transitional flow under the expectation that such results should also have relevance to the fully turbulent region.

Besides the observation that flow statistics in transition and the turbulent field are not dissimilar [23, 24], it is also the case that low-speed streaks show up in fully turbulent flow regions as shown in a classical result in Figure 8.9, where hydrogen bubbles are used as markers in a boundary layer flow [28]. The white areas contain dense collections of bubbles forming low-speed streaks because they are traveling slower than the surrounding fluid. Numerical simulations of turbulence also show the presence of low-speed streaks, as illustrated in Figure 8.10, taken from the plane $y^+ = 15$ in a channel flow. Streaks are observed to extend more than 1000 viscous lengths in the streamwise direction and tend to be spaced about 100 viscous lengths apart in the spanwise direction, although with a large standard deviation [29–31].

It has also been observed that besides low-speed streaks, there are other structural features that may be found both in transition and in the fully turbulent region. Among these, as will be seen below, are the mushroom-like signature of low-speed fluid ejecting outwards over the low-speed streaks seen in Figure 8.8b. Because of this similarity there

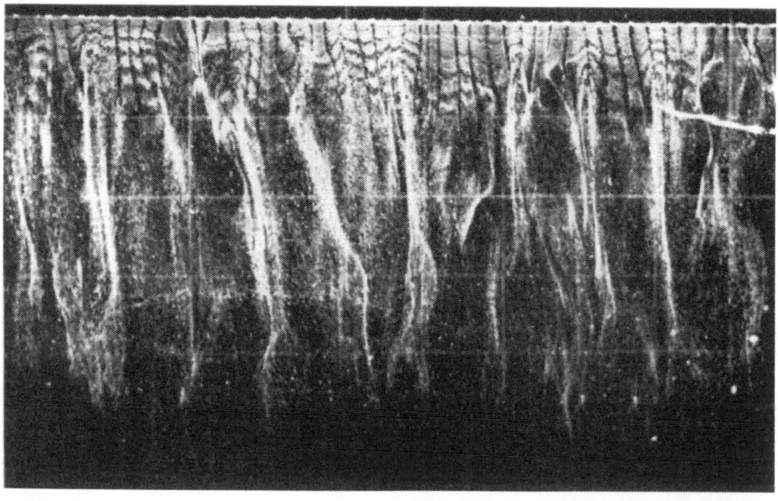

Figure 8.9 Visualization of low-speed streaks at $y^+ = 4.5$. From [28]. Reprinted with permission of Cambridge University Press.

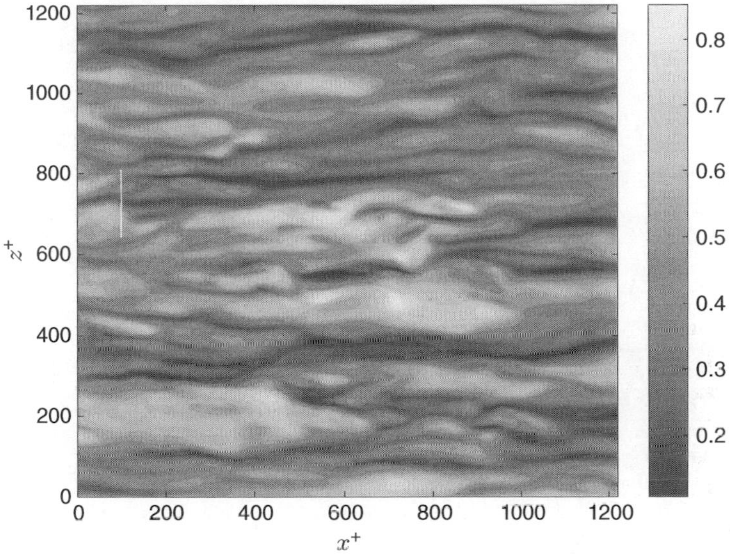

Figure 8.10 Contour plot of streamwise velocity fluctuation u in a DNS of channel flow at $R_\tau = 1000$ on the plane $y^+ = 10$ revealing the presence of low-speed streaks as thin elongated dark regions. The wider and shorter gray regions are high-speed sweeping motions. The white line is the location of the end-on views in Figures 8.25–8.27. Data taken from [32].

is justification for the practice of exploring the properties of vortical structure both in transitional and turbulent flow.

8.5.3 Vortical Structures

Of particular interest to studies of structure within fully turbulent flow has been evidence for the presence of streamwise-oriented vortices that are expected to be associated with low-speed streak formation. Among the indirect evidence for streamwise vortices is the appearance of *pockets* in smoke-marked boundary layers, as shown in Figure 8.11, in which near-wall spanwise motions scour the surface of the smoke, leaving distinctively shaped smoke-free regions. It may be imagined that streamwise vortices have the capability of causing such patterns to appear in the smoke-marked layer.

Within turbulent boundary layers streamwise vortices have also been observed indirectly in the behavior of marker particles of various kinds placed into turbulent flow. The particles are often entrained in swirling motions around a streamwise-oriented axis [34–38]. DNS allows the direct visualization of streamwise-oriented rotation through the velocity field as seen in the plot from a DNS of low Reynolds number channel flow in Figure 8.12 showing the presence of streamwise-oriented vortices near the wall. A view of all such rotational motions with a streamwise orientation at one instant in time in a channel flow simulation is shown in Figure 8.13. This figure suggests that such motion is intrinsic to the near-wall region of turbulent flows.

Some inkling of the kind of vortical structure occurring in turbulent boundary layers was observed in a classic experiment [40] in which smoke-marked images of laser light

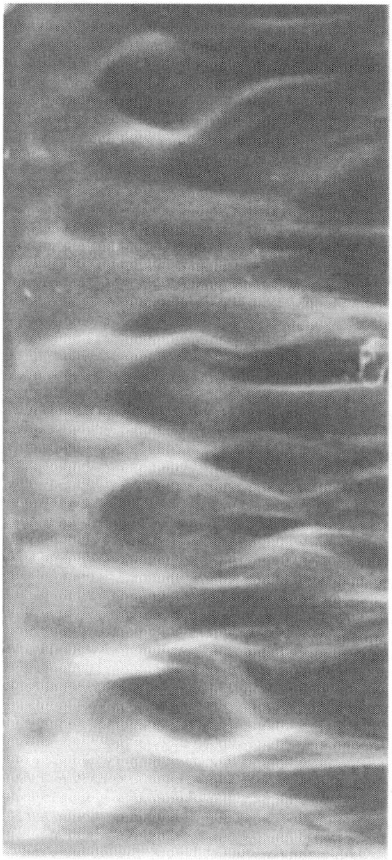

Figure 8.11 Visualization of pockets in a smoke-marked boundary layer. From [33]. Used with permission of Royal Society Publishing.

sheets are obtained on planes tilted 45° toward and away from the oncoming flow field, as shown in Figure 8.14. This reveals the presence of well-formed mushroom-like vortices tilted into the downstream direction populating the boundary layer and a coherency extending from the outer edge deep into the boundary layer. It has also been shown that there is considerable similarity between visualizations of horseshoe-shaped vortices shedding off a surface-mounted hemisphere and images of a similar shape present in turbulent flow near walls [41]. This helps justify the belief that such vortical elements appear in turbulent flow. To acquire more precise and detailed information about structure than has been described thus far, it is necessary to devise and implement schemes for locating and describing structures independent of whatever form they might take. This generally requires use of 3D velocity field data obtained from DNS or, in limited circumstances, from experiments [42].

At the outset, the development of schemes capable of identifying vortical objects in the flow is complicated by the intrinsic difficulty of arriving at a precise definition of a vortex. In fact, vortices are more easily observed than described mathematically [43] so that some degree of subjectivity tends to be inevitable in developing criteria for locating vortices in the complex flow of boundary layers. This is unlike such cases as a tornado in the atmosphere or a whirlpool visible in the surface of the ocean, objects that are readily

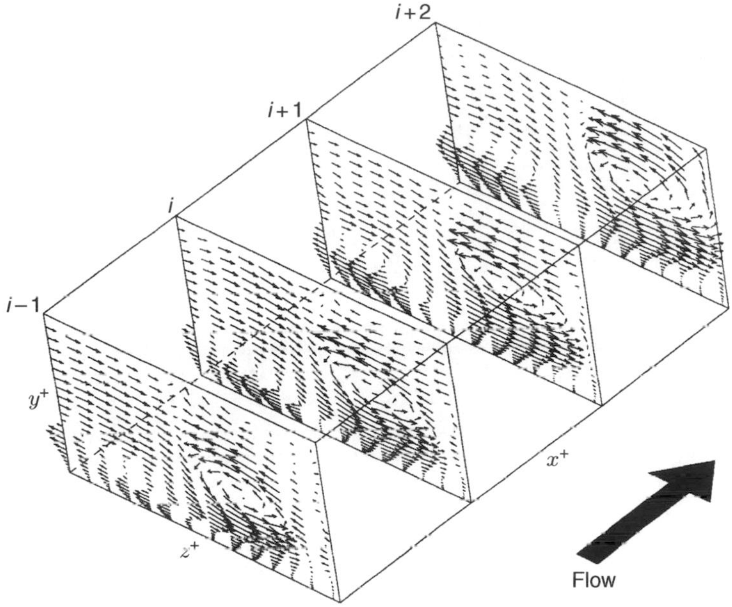

Figure 8.12 End-on velocity vector plots showing vortices in a channel flow. From [39]. Reprinted with the permission of Cambridge University Press.

Figure 8.13 Instantaneous view of quasi-streamwise vortices in channel flow with the sense of rotation indicated by light or dark shading. From [39]. Reprinted with permission of Cambridge University Press.

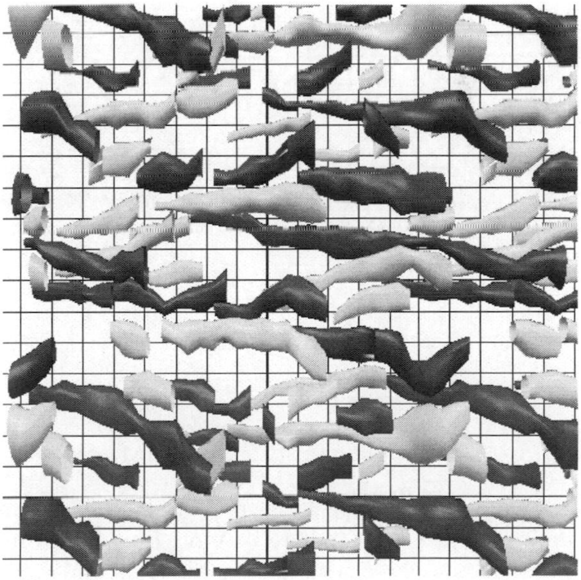

identified because of their isolation from other structures. In a similar vein, as mentioned previously, it is easier to discern structure in transition than in the fully turbulent boundary layer. For this reason much of the ensuing discussion is based on transitional structures, though it is expected that structures in the fully turbulent region share many of the same attributes.

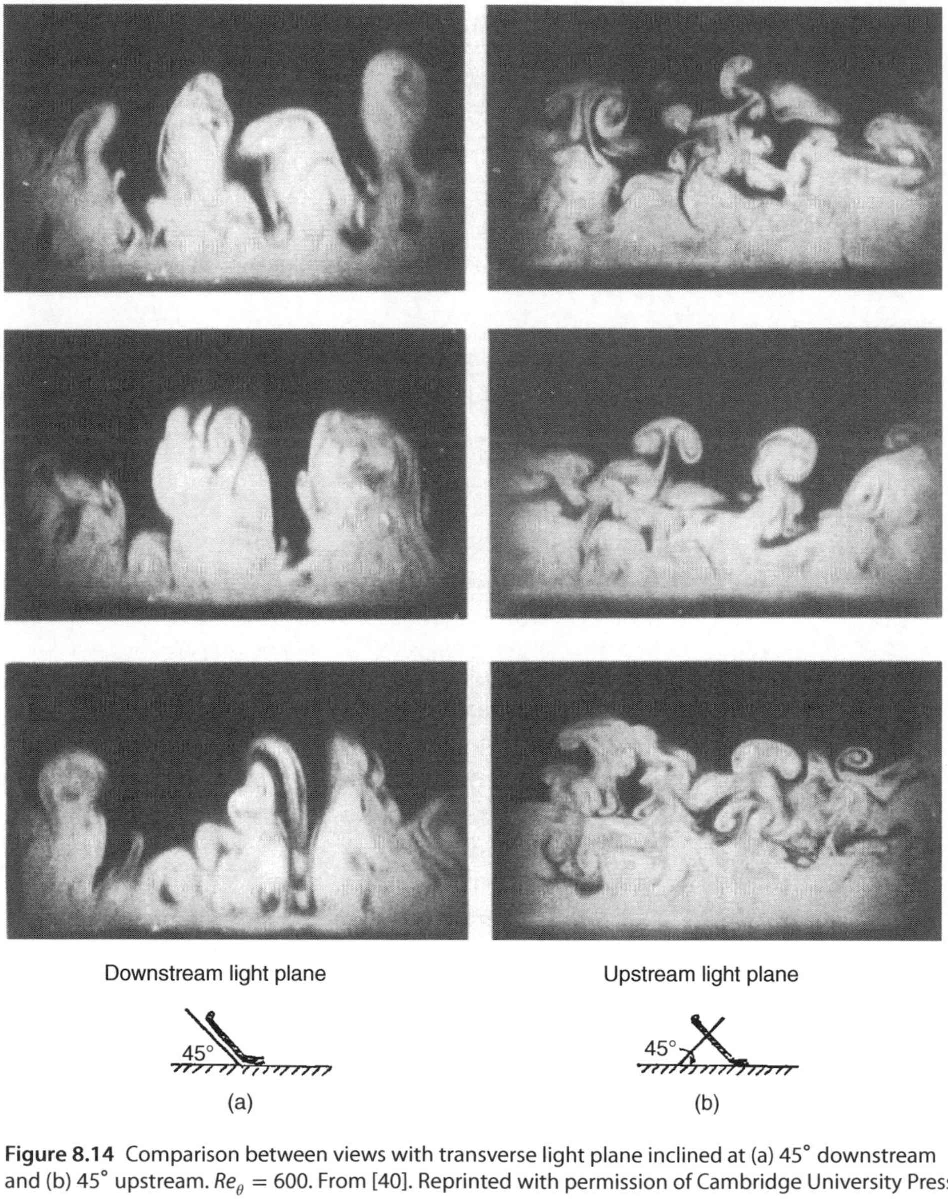

Figure 8.14 Comparison between views with transverse light plane inclined at (a) 45° downstream and (b) 45° upstream. $Re_\theta = 600$. From [40]. Reprinted with permission of Cambridge University Press.

Since a central aspect of vortical structure is the presence of rotational motion, methodologies for locating rotation within a moving fluid are instrumental for gaining some idea of where structures are located and what form they may have. In fact, it is common practice to assume that rotational regions are synonymous with structures, but this viewpoint is not necessarily supported by observation. In fact, some aspects of the structures may not involve rotational motion and so will be overlooked if rotational motion is the only criterion used in describing their form. An additional difficulty with identifying rotational regions as structures is that describing the time history of a rotational region is not rooted in the history of a physical quantity such as vorticity that obeys a conservation law. In particular, there may be more than one interpretation for explaining the origin and development of a rotational field. It will also be noted below that rotational regions have an upstream origin that may not include measurable rotation, thus complicating the dynamical analysis of the structures.

A variety of different means have been found for locating rotational motion in 3D flow fields [44, 45]. Despite their differences, the separate methodologies tend to reveal the presence of similar entities in the same locations. The success of these methods depends on their ability to locate rotation independent of the perspective of the observer, which in normal circumstances would have an effect on what is observed. For example, a camera traveling at different velocities while photographing a particle-marked channel flow will see evidence of structure in different locations with respect to the wall (see [9], p. 524).

General methods for finding swirl take advantage of properties associated with the local velocity gradient tensor. Thus, consider the linearized velocity field in the neighborhood of an observer traveling with the fluid, namely,

$$\mathbf{U}(\mathbf{x}+\mathbf{r},\mathbf{t}) = \mathbf{U}(\mathbf{x},\mathrm{t}) + (\nabla\mathbf{U})\mathbf{r} + \dots, \tag{8.41}$$

where $\nabla\mathbf{U}$ is the velocity gradient tensor with components $\partial U_i/\partial x_j$ and $\mathbf{r} = (r_1, r_2, r_3)$ forms a local coordinate system around the point $\mathbf{x}$. The eigenvalues of $\nabla\mathbf{U}$ satisfy

$$\det(\nabla\mathbf{U} - \lambda I) = 0, \tag{8.42}$$

which may be expressed as a cubic equation for λ in the form (see [46])

$$\lambda^3 + P\lambda^2 + Q\lambda + R = 0, \tag{8.43}$$

where $P = -tr(\nabla\mathbf{U}) = -\nabla\cdot\mathbf{U}$, $Q = \frac{1}{2}(tr(\nabla\mathbf{U})^2 - tr((\nabla\mathbf{U})^2))$ and $R = -\det(\nabla\mathbf{U})$ are the invariants of the tensor $\nabla\mathbf{U}$. Invariants remain the same under arbitrary rotations and translations of the coordinate system. In incompressible flow $P = 0$ and $Q = -\frac{1}{2}tr((\nabla\mathbf{U})^2)$, and in this case the discriminant of the cubic equation Eq. (8.43) is

$$D = \left(\frac{R}{2}\right)^2 + \left(\frac{Q}{3}\right)^3. \tag{8.44}$$

When D is positive it can be shown [47] that the local streamlines consist of swirling motion about the point $\mathbf{x}$. In this case, Eq. (8.43) has one real root and a pair of complex roots. Corresponding to these, $\nabla\mathbf{U}$ has one real and two complex eigenvectors, with the real eigenvector denoting the axial direction around which swirl takes place.

There are several ways of taking advantage of the properties of Eq. (8.44) in visualizing structures. One choice is to plot iso-contours of D equal to some (small) positive number. To achieve a more refined view, this can be done in conjunction with selecting a threshold in the magnitude of vorticity [47]. Another possibility is to plot contours of Q in regions where $D > 0$ [48]. Alternatively, the magnitude of the imaginary part of the complex eigenvalue, which can be seen to be within the region where $D > 0$, can be used as a marker. This is a particularly effective choice since it is directly related to the amplitude of the swirling motion.

An alternative approach to locating regions of swirl that is widely used [49] considers the symmetric tensor

$$B \equiv S^2 + W^2, \tag{8.45}$$

where $S = (\nabla \mathbf{U} + \nabla \mathbf{U}^t)/2$ is the rate-of-strain tensor and $W = (\nabla \mathbf{U} - \nabla \mathbf{U}^t)/2$ is the rotation tensor. Because it is real and symmetric, B has three real eigenvalues which can be listed as $\lambda_1 > \lambda_2 > \lambda_3$. It can be shown that regions for which $\lambda_2 < 0$ contain swirling motion, so plotting surfaces of constant $\lambda_2 < 0$ gives an idea of the shape and location of such regions.

Markers such as D and λ_2 are of great benefit in allowing the investigation of structure to be pursued with a clear view of where and in what form rotational regions occur in the flow. Numerical simulations of the turbulent flow near solid walls, whether in a channel, pipe or boundary layer, reveal the presence of many 3D volumes of swirling flow. Often these are in the form of tubes and many of these have a significant streamwise orientation, either singly or in pairs. In transition, the rotational regions generally have greater regularity than they do in the turbulent flow and so are somewhat easier to interrogate as to what is transpiring in the flow field. A common view of structure revealed in transition through λ_2 is shown in Figure 8.15. The time sequence shown here displays the evolution of a Λ vortex on the left into a hairpin-like vortex that seems to spawn a second hairpin and a third one as it convects and grows downstream. At the right of the image the structures are in late transition where there are some persistent hairpin-like shapes among a more complex region of vorticity whose structures have become less clearly identifiable. While it is tempting to refer to the clearly articulated structures in Figure 8.15 as being vortices of the same shape as the rotational regions being viewed, it must be remembered that these structures are conjoined with parts of the vorticity field that are not rotating. Thus, the actual structures involved may be configured somewhat differently from those in the figure in both shape and properties.

8.5.4 Origin of Structures

At the beginning of a transitioning boundary layer the perturbations are often infinitesimal. Whether described in terms of velocity, as they often are, or vorticity, they grow in magnitude with downstream distance to eventually produce low-speed streaks and the vortical structure associated with them. The success of this process in producing large-scale structural features out of small perturbations has to do with the fact that this process is self-reinforcing. In other words, whatever is the original change to the flow field, it enhances such changes, leading to further changes and so forth. The process by which vortical structures grow ends at the point at which the reinforcing cycle is disrupted for one reason or another. For example, the vortex may interact with neighboring vortices, causing chaotic motion that interferes with the mechanism producing the structures.

Since vortical structures are composed of vorticity, it is natural to consider the process that leads to their creation in terms of the vorticity field. Vorticity originates in the Blasius boundary layer as spanwise vorticity so that the earliest stage of the 3D process leading to structure involves reorientation of spanwise vorticity into the streamwise and wall-normal directions in order to spark the creation of objects that can be considered to be vortices. Assuming the presence of spanwise variations in the streamwise velocity, as in a TS wave or other disturbance, the regions of slow velocity will have a tendency to move away from the wall consistent with mass conservation. Any outward bulge of the spanwise vorticity creates a $\pm$ pair of wall-normal vorticity as well as a $\pm$ pair of streamwise vorticity, as shown in Figure 8.16. In the figure, a vortex line that is nominally

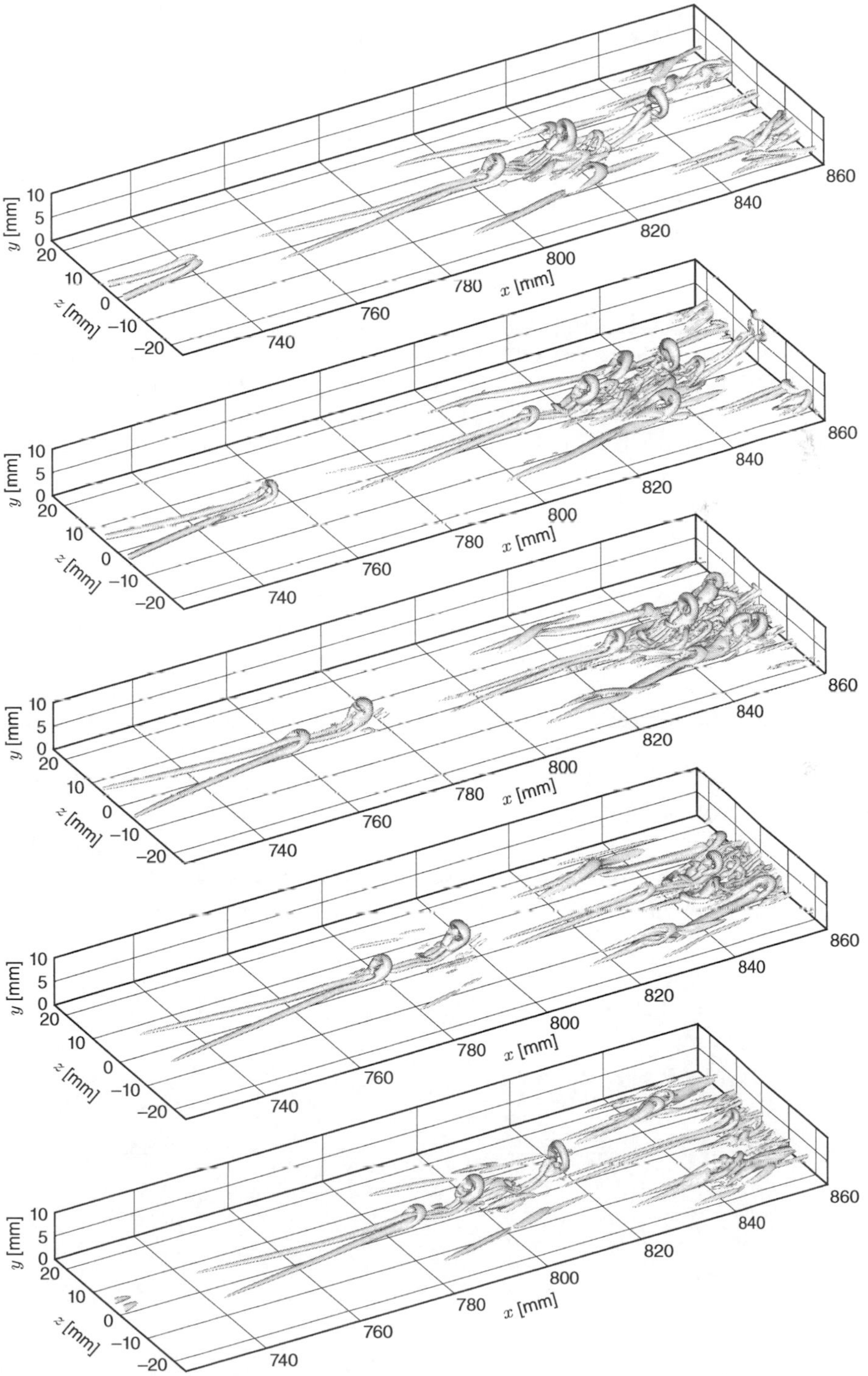

Figure 8.15 Visualization of vortices in transition using $\lambda_2 = -150$. From [50]. Reprinted with permission of Cambridge University Press.

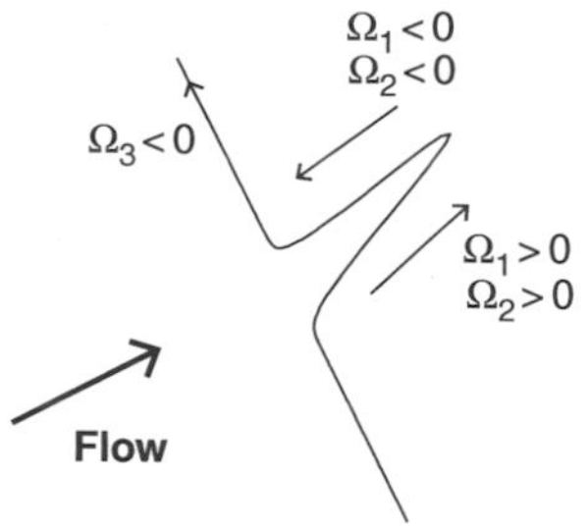

Figure 8.16 A lifted and forward sheared vortex line oriented in the $-z$ direction illustrates the creation of $\pm$ pairs of streamwise and wall-normal vorticity.

oriented in the negative z direction due to flow in the x direction over a surface $y = 0$ is lifted up by outward moving flow over a low-speed streak and then sheared forward by the faster moving fluid above the wall. Once vorticity begins to develop as shown in the figure, it contributes to the low velocity under the raised vorticity that forms a streak. This in turn contributes to the ejection of more spanwise vorticity that fuels the creation of more wall-normal and streamwise vorticity. In this way the process can be imagined to be self-sustaining. Note that this process initially occurs with the participation of the raised and sheared Ω_3 vorticity and not in the form of well-articulated structures such as modeled in Figure 8.15.

Vortex filament schemes [51, 52] in which the flow field is represented by large numbers of grid-free vortex tubes strung together forming filaments have the capacity to visualize structure within the vorticity field itself. When considered together with indicators of local rotation as used in Figure 8.15, a holistic view of structure becomes possible in which the context of how the rotational motion fits in with the entire local vorticity field is revealed. Some insights along these lines are shown in Figure 8.17, which shows an overhead view of the vortex filaments in a simulation of a transitioning boundary layer flow. Corresponding to this view are the low-speed streaks and hairpin-like regions marked by isosurfaces of λ_2 and shown in Figure 8.18. The hairpin legs bracket the low-speed regions between them. The transition in this case is somewhat rapid, reflecting the resolution parameters used in the filament calculation. Nonetheless, the

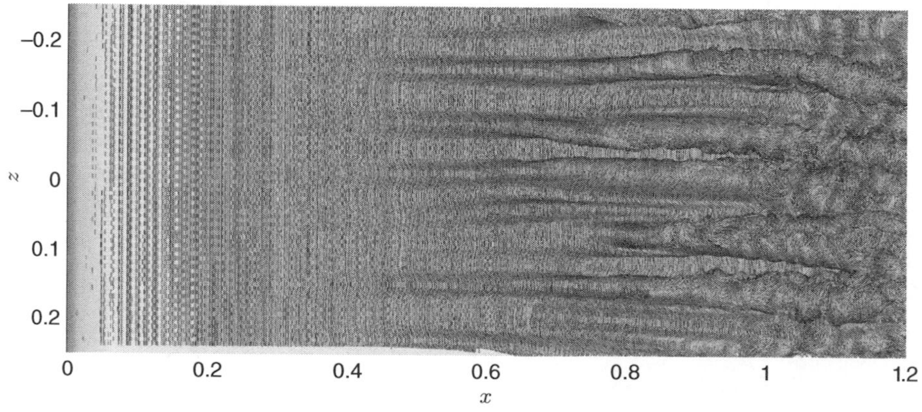

Figure 8.17 Overhead view of vortex filaments in a transitioning boundary-layer simulation using a vortex filament scheme. From [52]. Reproduced with the permission of AIAA.

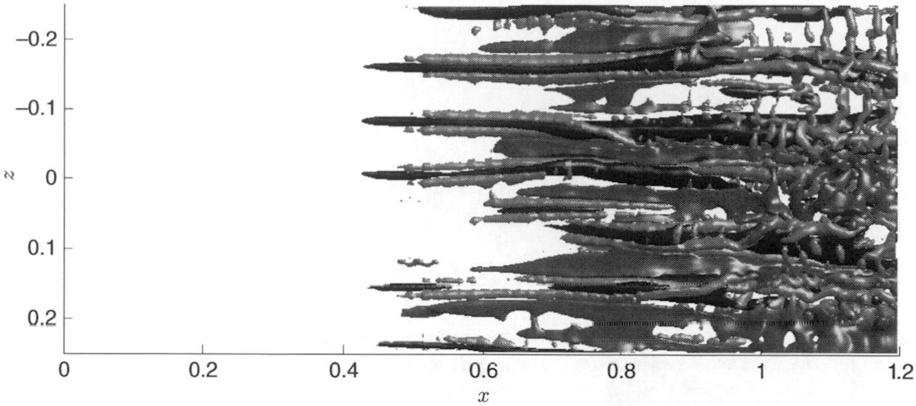

Figure 8.18 During transition, isosurfaces marking low-speed streaks (dark shading) are accompanied on one or both sides by isosurfaces of λ_2 (light shading), indicating the presence of hairpin-like rotational regions. Between the hairpins, isosurfaces of high-speed fluid (gray) reveal the presence of sweeping motions. The view is for the same time and region as in Figure 8.17. From [52]. Reproduced with the permission of AIAA.

formation of multiple hairpin-like rotational regions occurs prior to the breakup into turbulent flow.

The patterns formed by the vortex filaments displayed in Figure 8.17 correspond to the rotational regions in Figure 8.18, and from this connection it is possible to get an idea of the relationship between vorticity and the rotational structures observed in the flow. The filaments are seen to be lifted up along the extent of the low-speed streak forming a furrow-like eruption out of the wall surface vorticity. Downstream, the ejecting vorticity rolls up into structures with a mushroom-like form, reminiscent of the smoke-marked structures in Figure 8.14 and quite similar to the mushroom-like velocity signature showing up in late transition in Figure 8.8. Figure 8.19 puts together the isosurfaces marking the legs of a hairpin vortex with the vortex filaments on three planes through the structure that represent the vorticity field. It is evident that the legs of the hairpin form the lobes of the mushroom-like structure that has risen from the near-wall region. This shows that the complete structures in the late transition involve more of the local vorticity than appears in the streamwise-oriented vorticity whose rotation gets selected via scalar markers. In fact, non-rotational wall-normal vorticity forms the stem of the mushrooms and the spanwise vorticity forming the top of the hairpin-like vortices forms a shear layer whose rolling up into transverse vortices can explain the top arch-like vortices that are associated with hairpins. An example of two nested hairpins is shown in Figure 8.20, as isosurfaces of λ_2. Unseen in this visualization of the rotation field are essential contributions to the two hairpin structures from surrounding vorticity that is not associated with rotation.

A common occurrence in transitional boundary layers is the appearance of single hairpin-legs (often referred to as *canes*) as viewed through isosurfaces depicting rotational motion. While there has been some speculation as to why one- or two-legged hairpins appear, consideration of the full vorticity field responsible for hairpins as shown in Figure 8.21 provides a relatively straightforward explanation. In fact, canes can be seen to occur whenever the mushroom-like structures are tilted to one side or the other so

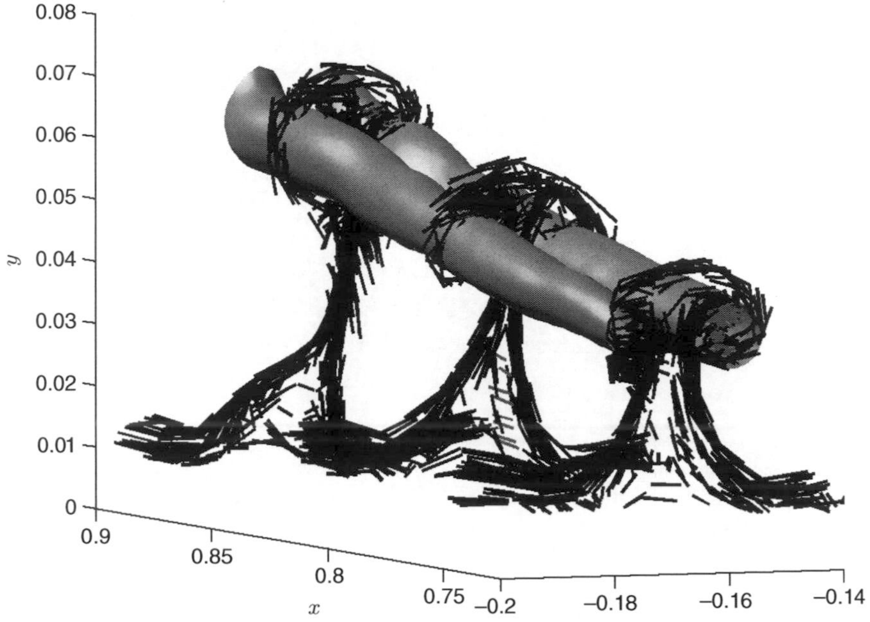

Figure 8.19 The vorticity in a lifted-up furrow adopts a mushroom-like shape whose lobes contain the streamwise-oriented rotational motion that is associated with hairpin legs. From [52]. Reproduced with the permission of AIAA.

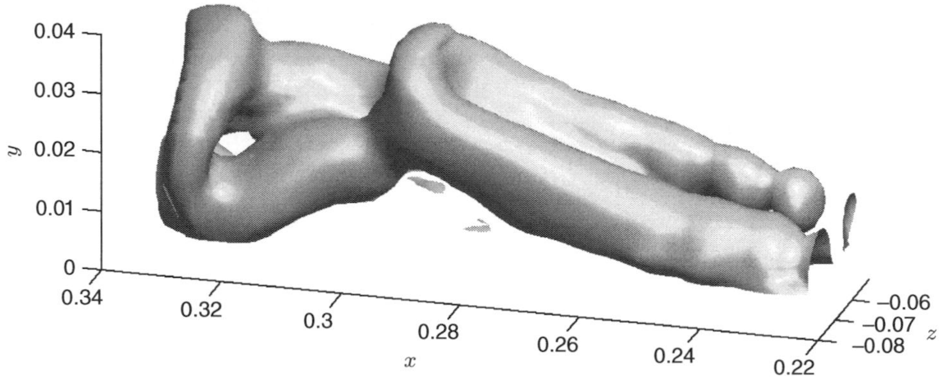

Figure 8.20 Isosurfaces of λ_2 forming what appears to be two nested hairpins that develop from vorticity raised over a low speed streak. From [52]. Reproduced with the permission of AIAA.

that the rotational field does not develop on the side closest to the wall. In noisy flows, the hairpins are more likely to be single-legged and this is likely due to the instability of the mushroom-like form, which has a tendency to tilt when perturbed.

The connection between the structures in Figures 8.17 and 8.18 and pockets is shown in Figure 8.22. In this case, a spanwise line of particles placed into the flow at regular intervals mimics the action of smoke such as is used in creating the pockets visible in Figure 8.11. The grid of lines in Figure 8.22 is made up of the streak lines consisting of

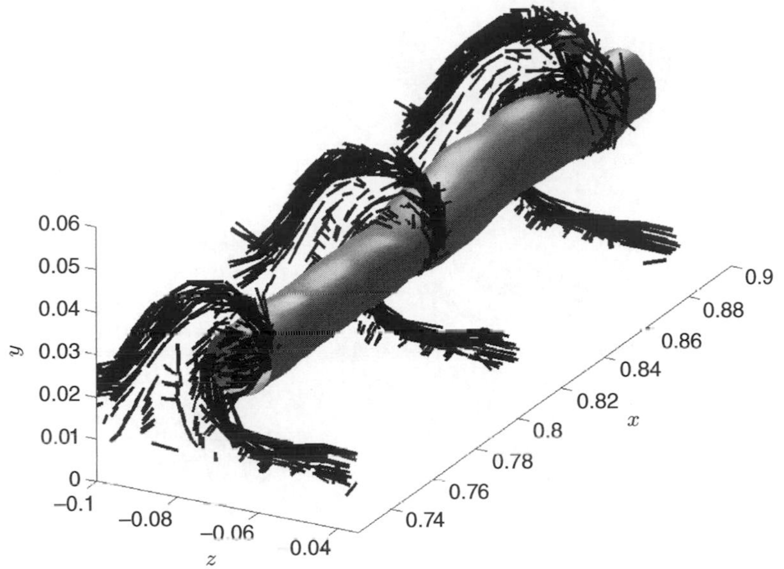

Figure 8.21 Single hairpin representing the rotational signature of a tilted furrow. From [52]. Reproduced with the permission of AIAA.

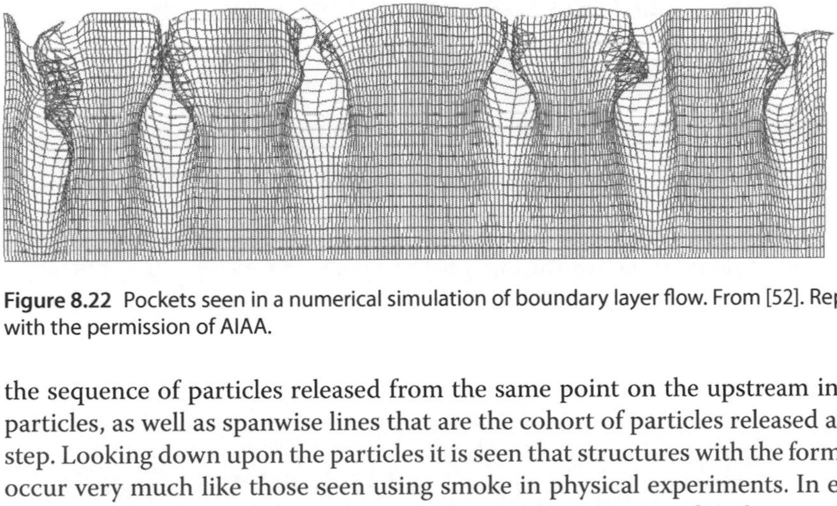

Figure 8.22 Pockets seen in a numerical simulation of boundary layer flow. From [52]. Reproduced with the permission of AIAA.

the sequence of particles released from the same point on the upstream initial line of particles, as well as spanwise lines that are the cohort of particles released at each time step. Looking down upon the particles it is seen that structures with the form of pockets occur very much like those seen using smoke in physical experiments. In essence, the scouring of smoke particles that creates the visual impression of pockets is a result of the outward lateral motion produced by the counter-rotation associated with the vortical structures responsible for the low-speed streaks. Both pairs or single hairpin legs can cause pocket-like patterns to appear.

In the end stage of the transition process the mushroom-like form of the vorticity field, which is exceedingly sensitive to slight perturbations, rapidly distorts into complex forms that come to occupy the downstream turbulent flow. Some of the various ways in which the highly unstable structures deform are shown in Figure 8.23, where it is seen that they have a tendency to tilt to one side or the other, creating a strong interaction with the wall. The process is very rapid and within a short distance what was formerly a

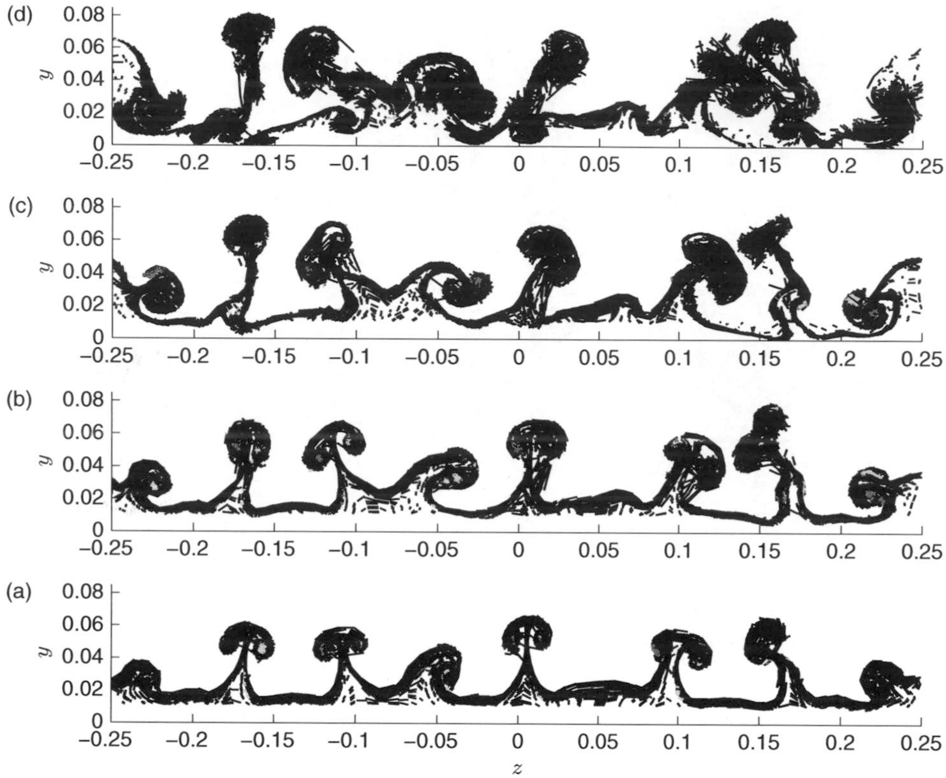

Figure 8.23 Filaments at a fixed time showing the progression to turbulent flow at the end of transition. (a)–(d) correspond to points $x = 0.8, 0.85, 0.9$, and 0.95 in the computed boundary layer. From [52]. Reproduced with the permission of AIAA.

somewhat identifiable pattern of coherent structure becomes a collection of individual vortical features lacking simple description.

8.5.5 Fully Turbulent Region

Though it technically remains an open question, it is unlikely that fully turbulent boundary layers contain an organization of vortical structure that can be given the same kind of succinct and clear-cut description as it can in transition. Some hint of a growing incoherency is visible in the downstream end of the images in Figure 8.15. A similar plot of the λ_2 isosurfaces, but for a fully turbulent boundary layer at $Re_\theta \approx 4000$, is shown in Figure 8.24. Here, it may be seen that there are numerous rotational regions, many in tubular shapes, and some of these are reminiscent of hairpins, but the overall impression is one of remarkable complexity. In fact, vortices tend to react strongly with other vortices, so that despite beginning with recognizable forms in transition, most if not all such coherency is difficult to observe in fully developed turbulence, even though there is clear evidence of the presence of rotational structures.

There has been much analysis of boundary layer structure based on such approaches as developing model calculations of vortex development in isolation from other structures [48], in observations of the details of the velocity field in 2D planes of data [54],

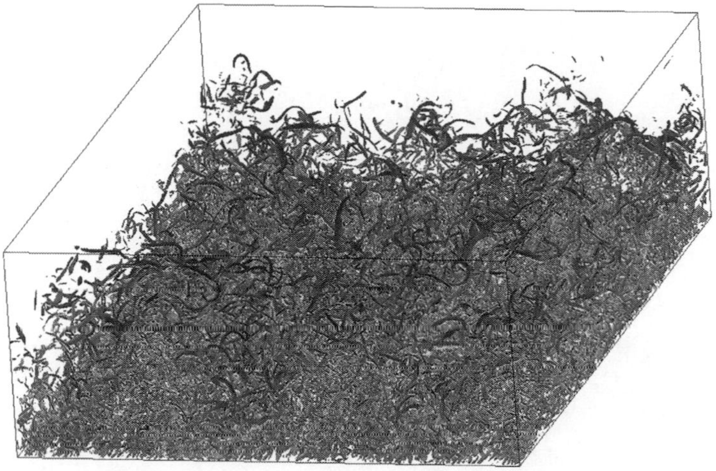

Figure 8.24 Isosurfaces of λ_2 in a fully turbulent boundary layer at $Re_\theta \approx 4000$. From [53]. Reproduced by permission of Elsevier.

and looking at conditional data sets [55]. In the latter, a subset of the full random field is averaged together based on a selection criterion such as unusually large instantaneous Reynolds shear stress. Emerging from such averages are structural forms that generally reveal hairpin-like shapes, thus helping to suggest that such objects may be present also in the instantaneous flow field. One conclusion that has been reached from these kinds of analyses is that turbulent flow contains groups of hairpin vortices, known as hairpin packets, that are created sequentially from a parent hairpin similar to the phenomenon seen in transition in Figure 8.15. However, confirmation of the presence of packets by direct examination of 3D turbulent fields via isosurfaces of rotation has not been consistently successful [56, 57]. An alternative viewpoint to that of the hairpin packet model, which derives from similar considerations, adopts the view that the coherent eddies are best described as containing clusters of intense vortices that are somewhat less organized than in the case of the packets [58].

If a clear understanding of vortex dynamics in turbulent boundary layers is not yet available, nonetheless some elements of the dynamics are amenable to analysis and can be understood. For example, as shown in Figures 8.9 and 8.10, low-speed streaks are present near solid boundaries in turbulent flow. Their existence suggests the presence of the same kind of vortical motions associated with streaks as seen in transition. An illustration of this point is shown in Figures 8.25–8.27, where velocity and vorticity contours are shown on a portion of the spanwise plane centered over a particular low-speed streak seen in Figure 8.10 in turbulent channel flow. The plane intersects the streak in Figure 8.10 at streamwise position $x^+ = 100$ and spanwise position $z^+ = 750$. The streamwise velocity field shown in Figure 8.25a has the same mushroom-like form as was seen in Figures 8.8b and 8.19 in which low-speed fluid is ejecting outward. The large ejection velocity accompanying the mushroom in Figure 8.25a is evident in Figure 8.26a, where the wall-normal velocity contours are plotted. Note also the large sweeping motions on either side of the ejection. The distance away from the wall at the top of the mushroom shape, approximately $y^+ = 70$, is comparable to the image in Figure 8.19. The spanwise velocity plot in Figure 8.27a contains a clover-leaf

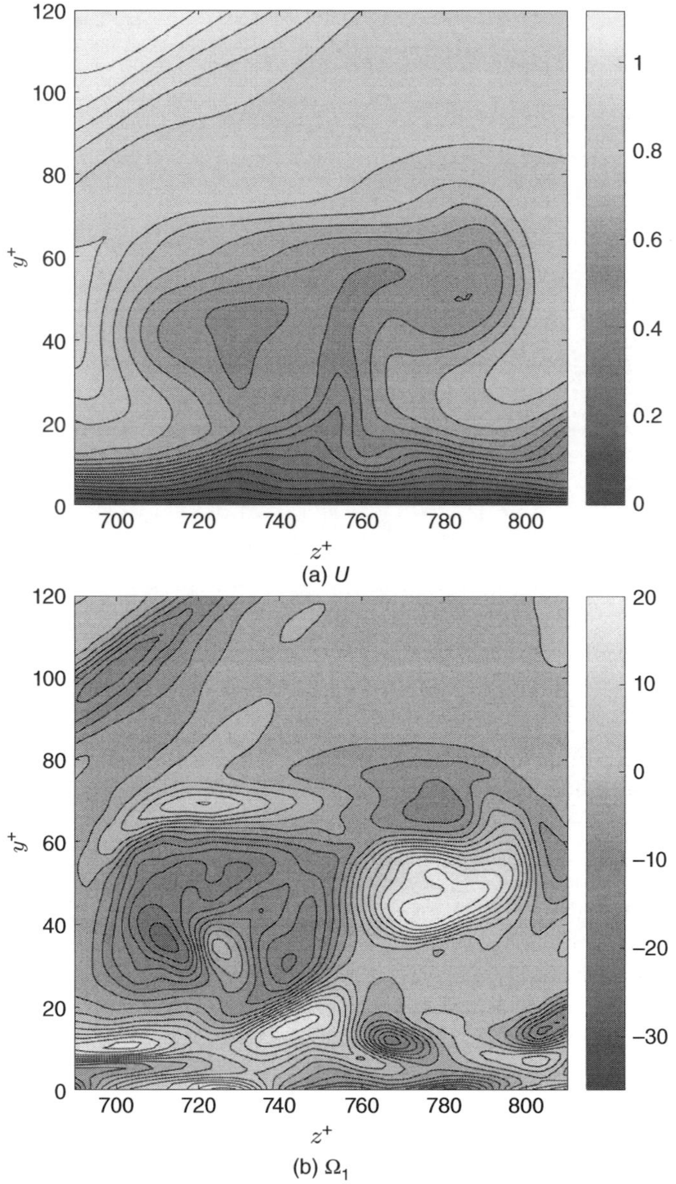

Figure 8.25 Contour plots on a plane intersecting the low-speed streak in Figure 8.10 at $x^+ = 100, z^+ = 750$ for channel flow at $R_\tau = 1000$. Data taken from [32]. (a) U; (b) Ω_1.

pattern corresponding to the presence of streamwise rotational motion. Moreover, the streamwise vorticity contour in Figure 8.25b shows the presence of strong plus and minus values that match up with the lobes of the mushroom shape in Figure 8.25a. A strong pair of plus and minus wall-normal vorticity contours appears in Figure 8.26b at the location of the "stem" of the uplifted vortex. Finally, Figure 8.27b shows that some significant spanwise vorticity lies over the top of the mushroom-like vortex so

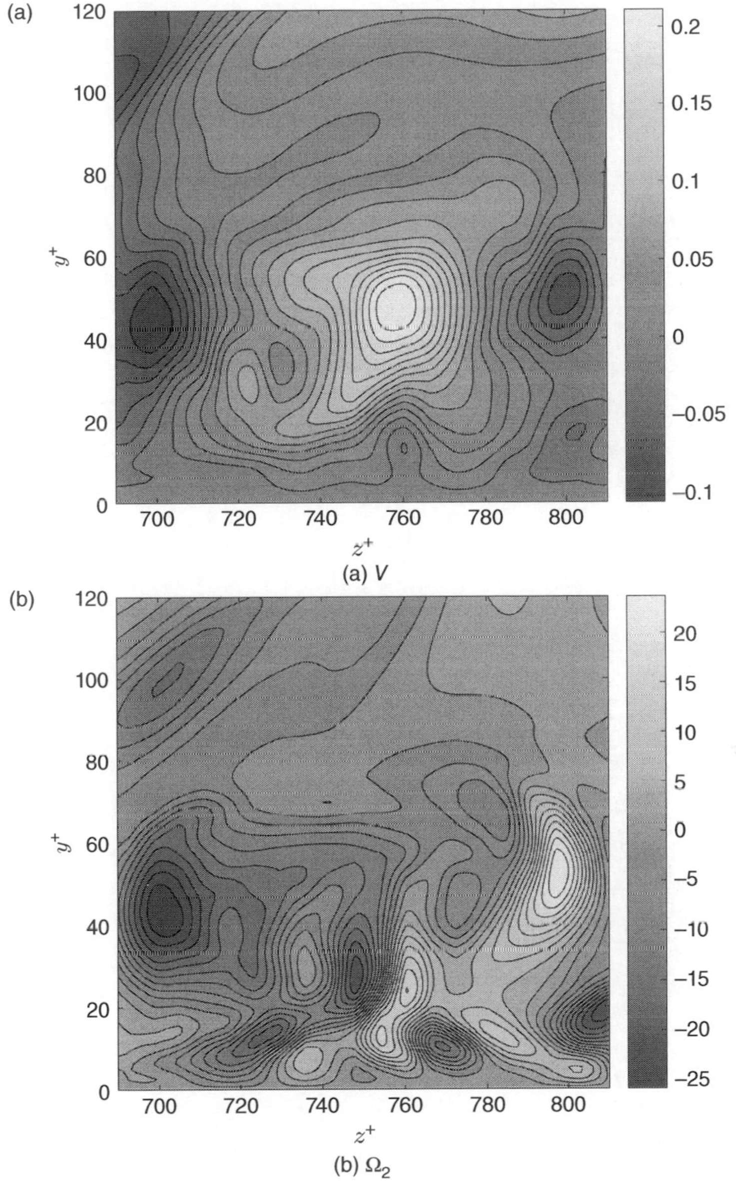

Figure 8.26 Contour plots on a plane intersecting the low-speed streak in Figure 8.10 at $x^+ = 100, z^+ = 750$ for channel flow at $R_\tau = 1000$. Data taken from [32]. (a) V; (b) Ω_2.

there is likely to be a shear layer in this position. The flow patterns in these figures are not uncommon in the channel flow data set. They are readily located by examining the local flow along the many low-speed streaks appearing in the computation. This suggests that while the chaotic motion within a fully turbulent field will distort the orderly development of structure associated with low-speed streaks, nonetheless a similar mechanism persists as in transition, thus providing a continual source of new

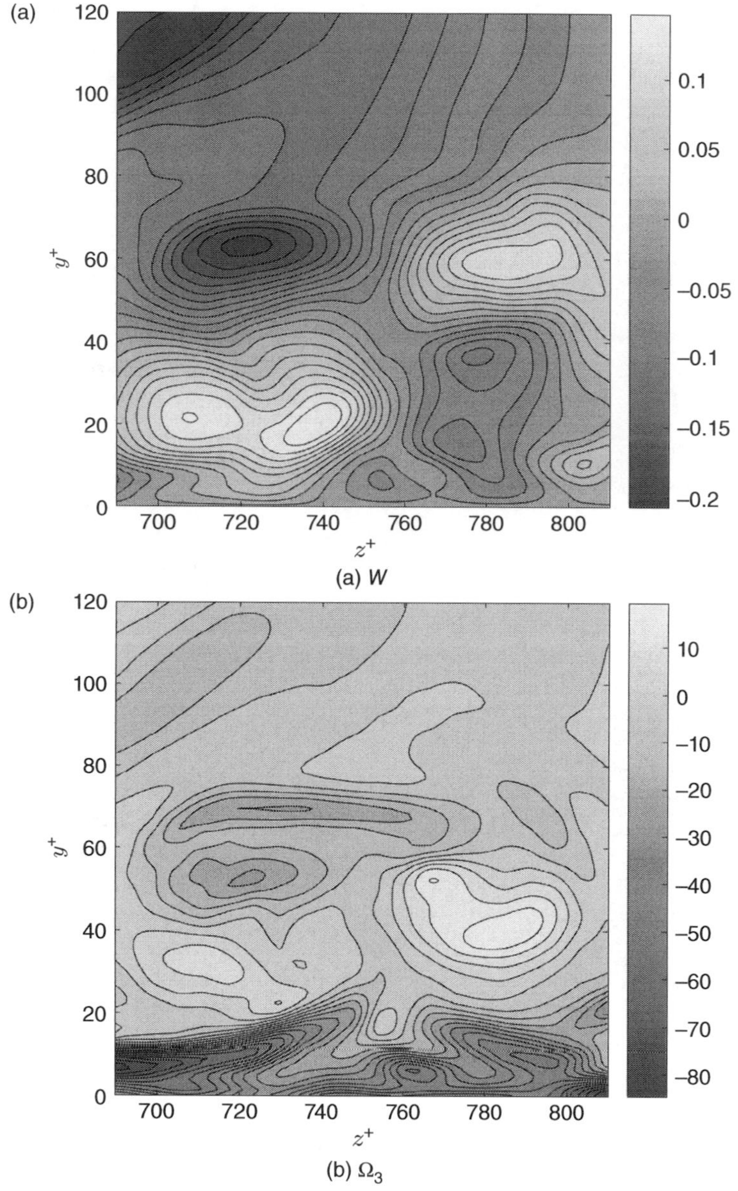

Figure 8.27 Contour plots on a plane intersecting the low-speed streak in Figure 8.10 at $x^+ = 100, z^+ = 750$ for channel flow at $R_\tau = 1000$. Data taken from [32]. (a) W; (b) Ω_3

furrow-like eruption of hairpin-like regions of rotational motion. The end result is thus a continual resupply of vortical structures to the turbulent boundary region. The presence of coherent features of the velocity field outside the near-wall region that extend for long distances has been observed in channels, pipes, and boundary layer flows. These eddy-like elements, referred to as very large-scale motions (VLSM), are generally regarded as being the end result of the further development and merger of the near-wall structures and have been the object of considerable study in recent years [59].

8.6 Near-Wall Pressure Field

The pressure force acting on solid boundaries immersed in turbulent flow is one of its most critical features. Not only is it often times the most significant force acting on bodies (e.g., in bluff body flows), but it has a direct bearing on the generation of noise and other phenomena, such as cavitation. An important aspect of the pressure near walls is the observation of localized pressure minima and maxima that convect across the solid surface. In view of the close proximity of vortical structures to the wall, a connection between the pressure signal on the wall and the structures may be likely. If so, then information about the dynamics of structure near boundaries is germane to the various phenomena surrounding pressure, such as noise generation.

The pressure fields associated with clearly defined vortices such as a tornado have distinctive minima and in early work pressure isosurfaces were used to try and identify vortices in a boundary layer [60]. While some success can be had by this approach, for the most part the complexity of the pressure surfaces near the wall prevents attributing them to well-defined structures. On the other hand, with the capability of using local rotation to mark structures it becomes possible to tie structure and pressure signals together.

One example of such a connection is shown in Figure 8.28, which is taken from a DNS of low Reynolds number channel flow [39]. Here pressure maxima moving along the surface of a turbulent channel flow are seen to be the direct effect of the passage of quasi-streamwise rotational regions, as shown previously in Figure 8.13, but now indicated by the local vorticity vector at the core of rotational structures. Figure 8.28 shows that a series of peak pressure disturbances on the surface come and go, driven by the vortex as it convects downstream in the flow. This suggests that wall pressure disturbances are primarily caused by vortices close to the wall rather than, for example, those in the outer regions of the boundary layer. It is of interest to see if a similar conclusion can be drawn for high Reynolds numbers from more recent simulations.

8.7 Chapter Summary

Turbulent flow in boundary layers has perhaps the most direct bearing on engineering work among the many ways that turbulence is manifested in applications. Between the consideration of channel and pipe flows in the previous chapter and boundary layers in the present chapter it is clear that the most complex part of bounded flows is the near-wall region, where the strong effect of viscosity gives way to fully turbulent processes involving the dynamics of vorticity. Here, the comparatively simple physics of turbulence production matching dissipation, which governs the outer flow, does not apply. Instead, a balance of many physical processes determines the magnitude and distribution of the turbulent energy.

The appropriateness of similarity laws to succinctly describe the mean velocity field in different zones of the boundary layer was considered. Evidence in support of a log law

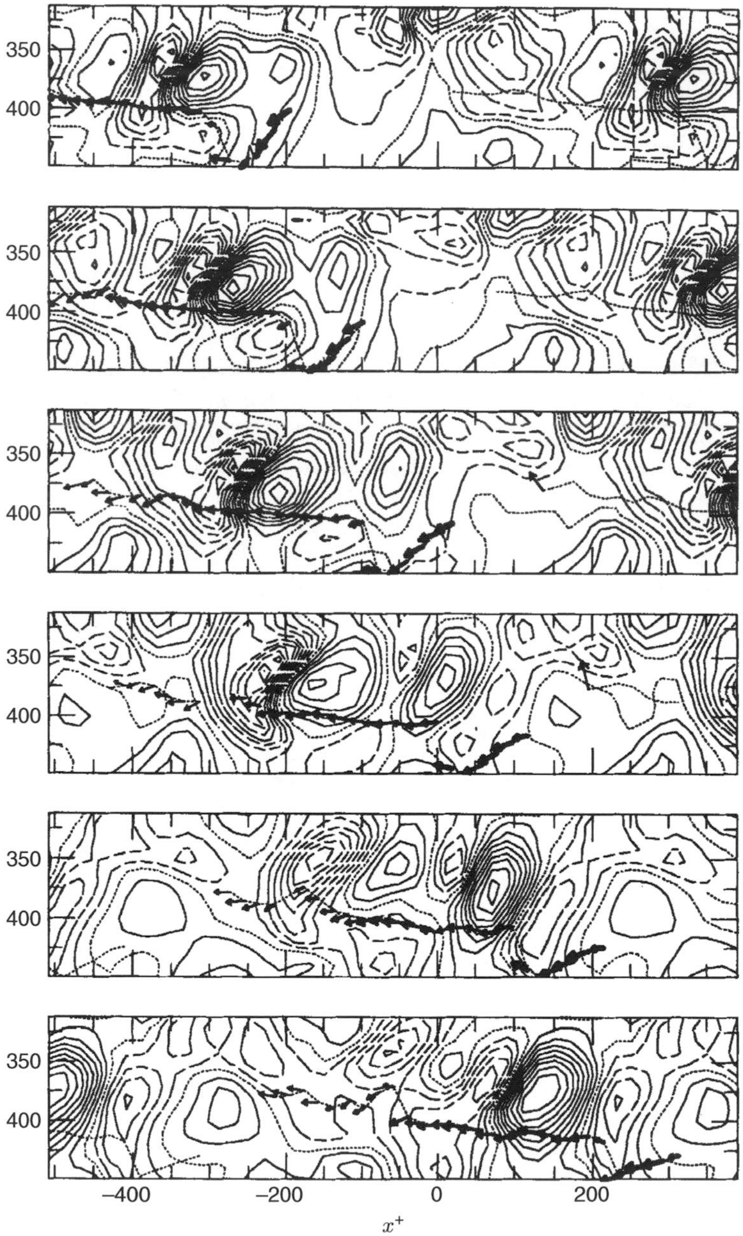

Figure 8.28 Wall pressure maxima generated by coherent vortices. From [39]. Time increases from top to bottom image. Reprinted with permission of Cambridge University Press.

or near log-law behavior has been found from high Reynolds number experiments. This fits in qualitatively with results for channels and pipes, though it is now recognized that there are some differences between these cases that reflect their individual attributes.

The complex behavior of near-wall flow involves the dynamics of vortical structures that arise from vorticity produced at the surface. A distinction was drawn between much of the knowledge of structure that is determined from analysis of transitional flows and the much more speculative understanding of structures that appear within the fully turbulent region. In the latter case, there is evidence to suggest that similar processes leading to the appearance of hairpin-like rotational regions occur in the fully turbulent region as happens in the transition region, but they tend to be highly distorted due to surrounding vortices. Much recent investigation is concerned with accounting for the presence of very large-scale eddies in the outer flow regions. While there have been some attempts at connecting their presence to the further development of vortical structures produced in the near-wall region, more study is required before a definitive understanding can be reached.

The value of gaining a good understanding of coherent vortices in turbulent flow lies in how this information can be used to better understand the transport processes that play a major role in the prediction of turbulent flow. Knowledge of structures also plays a potentially important part in devising strategies for controlling turbulence. In effect, by hindering or otherwise changing the mechanisms responsible for producing vortical structures in the near-wall flow, turbulence may be delayed or modified in beneficial ways.

References

1 Wallace, J.M., Balint, J.L., Mariaux, J.L., and Morel, R. (1983) Observations on the nature and mechanism of the structure of turbulent boundary layers, in *Proc. of 4th Engr. Mechanics Div. Specialty Conf. on recent advances in engineering mechanics and their impact on civil engineering practice*, vol. II (eds W. Chen and A. Lewis), ASCE, New York, vol. II, pp. 1198–1201.

2 Bernard, P.S. (2015) *Fluid Dynamics*, Cambridge University Press, Cambridge.

3 Schubauer, G.B. and Skramstad, H.K. (1948) Laminar-boundary-layer oscillations and transition on a flat plate, *Technical Note 909*, NACA.

4 Zaki, T.A. (2013) From streaks to spots and on to turbulence: exploring the dynamics of boundary layer transition. *Flow, Turb. Comb.*, 91, 451–473.

5 Blackwelder, R.F. and Kovasznay, L.S.G. (1972) Time scales and correlations in a turbulent boundary layer. *Phys. Fluids*, 15, 1545–1554.

6 Klebanoff, P. (1954) Characteristics of turbulence in a boundary layer with zero pressure gradient, *Technical Note 3178*, NACA.

7 Wilcox, D.C. (2006) *Turbulence Modeling for CFD*, DCW Industries, La Cañada, CA, 3rd edn.

8 Hinze, J.O. (1975) *Turbulence*, McGraw-Hill, New York, 2nd edn.

9 Schlichting, H. (1968) *Boundary Layer Theory*, McGraw-Hill Book Co., New York, 6th edn.

10 Vallikivi, M., Hultmark, M., and Smits, A.J. (2015) Turbulent boundary layer statistics at very high Reynolds number. *J. Fluid Mech.*, 779, 371–389.

11 Hutchins, N., Nickels, T.B., Marusic, I., and Chong, M.S. (2009) Hot-wire spatial resolution issues in wall-bounded turbulence. *J. Fluid Mech.*, 632, 431–442.

12 Smits, A.J., McKeon, B.J., and Marusic, I. (2011) High-Reynolds number wall turbulence. *Ann. Rev. Fluid Mech.*, 43, 353–375.

13 Marusic, I., Chauhan, K.A., Kulandaivelu, V., and Hutchins, N. (2015) Evolution of zero-pressure-gradient boundary layer from different tripping conditions. *J. Fluid Mech.*, 783, 379–411.

14 Marusic, I., Monty, J.P., Hultmark, M., and Smits, A.J. (2013) On the logarithmic region in wall turbulence. *J. Fluid Mech.*, 716, R3–1–R3–11.

15 Townsend, A.A. (1976) *The Structure of Turbulent Shear Flows*, Cambridge University Press, Cambridge.

16 Perry, A.E. and Li, J.D. (1990) Experimental support for the attached-eddy hypothesis in zero-pressure-gradient turbulent boundary layers. *J. Fluid Mech.*, 218, 405–438.

17 Clauser, F.H. (1954) The turbulent boundary layer. *Adv. Appl. Mech.*, 4, 1–51.

18 Clauser, F.H. (1954) Turbulent boundary layers in adverse pressure gradients. *J. Aero. Sci.*, 21, 91–108.

19 Coles, D.E. (1956) The law of the wake in the turbulent boundary layer. *J. Fluid Mech.*, 1, 191–226.

20 Liakopoulos, A. (1984) Explicit representations of the complete velocity profile in a turbulent boundary layer. *AIAA J.*, 22, 844–846.

21 Klebanoff, P.S., Tidstrom, K.D., and Sargent, L. (1962) The three-dimensional nature of boundary-layer instability. *J. Fluid Mech.*, 12, 1–34.

22 Herbert, T. (1988) Secondary instability of boundary layers. *Ann. Rev. Fluid Mech.*, 20, 487–526.

23 Park, G.I., Wallace, J.M., Wu, X., and Moin, P. (2012) Boundary layer turbulence in transitional and developed states. *Phys. Fluids*, 24, 035 105.

24 Wu, X., Moin, P., Wallace, J.M., and Hickey, J.P. (2017) Transitional-turbulent spots and turbulent-turbulent spots in boundary layers. *Proc. Nat. Acad. Sci.*, 114.

25 Sayadi, T., Hamman, C.W., and Moin, P. (2013) Direct numerical simulation of complete H-type and K-type transitions with implications for the dynamics of turbulent boundary layers. *J. Fluid Mech.*, 724, 480–509.

26 Cantwell, B., Coles, D., and Dimotakis, P. (1978) Structure and entrainment in the plane of symmetry of a turbulent spot. *J. Fluid Mech.*, 87, 641–672.

27 Buffat, M., Penven, L.L., Cadiou, A., and Montagnier, J. (2014) DNS of bypass transition in entrance channel flow induced by boundary layer interaction. *Eur. J. Mech. B/Fluids*, 43, 1–13.

28 Kline, S.J., Reynolds, W.C., Schraub, F.A., and Runstadler, P.W. (1967) The structure of turbulent boundary layers. *J. Fluid Mech.*, 30, 741–773.

29 Gupta, A.K., Laufer, J., and Kaplan, R.E. (1971) Spatial structure in the viscous sublayer. *J. Fluid Mech.*, 50, 493–512.

30 Smith, C.R. and Metzler, S.P. (1983) The characteristics of low-speed streaks in the near-wall region of a turbulent boundary layer. *J. Fluid Mech.*, 129, 27–54.

31 Klewicki, J.C., Metzger, M.M., Kelner, E., and Thurlow, E.M. (1995) Viscous sublayer visualizations at $R_\theta \approx 1{,}500{,}000$. *Phys. Fluids*, 7, 857–863.

32 Graham, J., Kanov, K., Yang, X., Lee, M., Malaya, N., Lalescu, C., Burns, R., Eyink, G., Szalay, A., Moser., R., and Meneveau, C. (2016) A web services-accessible database of turbulent channel flow and its use for testing a new integral wall model for les. *J. Turbulence*, 17, 181–215.

33 Falco, R.E. (1991) A coherent structure model of the turbulent boundary layer and its ability to predict Reynolds number dependence. *Phil. Trans.: Phys. Sci. and Engrg.*, 336, 103–129.

34 Corino, E.R. and Brodkey, R.S. (1969) A visual investigation of the wall region in turbulent flow. *J. Fluid Mech.*, 37, 1–30.

35 Kastrinakis, E.G., Wallace, J.M., Willmarth, W.W., Ghorashi, B., and Brodkey, R.S. (1978) On the mechanism of bounded turbulent shear flows, in *Lecture Notes in Physics*, Springer-Verlag, 75, pp. 175–189.

36 Clark, J.A. and Markland, E. (1971) Flow visualization in turbulent boundary layers. *J. Hydr. Div. ASCE*, HY19, 1653–1664.

37 Smith, C.R. and Schwartz, S.P. (1983) Observation of streamwise rotation in the near-wall region of a turbulent boundary layer. *Phys. Fluids*, 26, 641–652.

38 Kasagi, N., Hirata, M., and Nishino, K. (1986) Streamwise pseudo-vortical structures and associated vorticity in the near-wall region of a wall-bounded turbulent boundary layer. *Exp. Fluids*, 4, 309–318.

39 Bernard, P.S., Thomas, J.M., and Handler, R.A. (1993) Vortex dynamics and the production of Reynolds stress. *J. Fluid Mech.*, 253, 385–419.

40 Head, M.R. and Bandyopadhyay, P. (1981) New aspects of turbulent boundary layer structure. *J. Fluid Mech.*, 107, 297–338.

41 Acarlar, M.S. and Smith, C.R. (1987) A study of hairpin vortices in a laminar boundary layer. part 1. hairpin vortices generated by a hemisphere protuberance. *J. Fluid Mech.*, 175, 1–41.

42 Elsinga, G.E., Adrian, R.J., van Oudheusden, B.W., and Scarano, F. (2010) Three-dimensional vortex organization in a high-Reynolds-number supersonic turbulent boundary layer. *J. Fluid Mech.*, 644, 35–60.

43 Foss, J.F. (1996) Vorticity, circulation and vortices, in *FED-Vol. 238, Fluids Engineering Conference*, 3, pp. 83–95.

44 Chakraborty, P., Balachandar, S., and Adrian, R.J. (2005) On the relationship between local vortex identification schemes. *J. Fluid Mech.*, 535, 189–214.

45 Kucala, A. and Biringen, S. (2014) Spatial simulation of channel flow instability and control. *J. Fluid Mech.*, 738, 105–123.

46 Gurtin, M.E. (1981) *An Introduction to Continuum Mechanics*, Academic Press, New York.

47 Chong, M.S., Perry, A.E., and Cantwell, B.J. (1990) A general classification of three-dimensional flow fields. *Phys. Fluids A*, 2, 765–777.

48 Zhou, J., Adrian, R.J., Balachandar, S., and Kendall, T.M. (1999) Mechanisms for generating coherent packets of hairpin vortices in channel flow. *J. Fluid Mech.*, 387, 353–396.

49 Jeong, J., Hussain, F., Schoppa, F., and Kim, J. (1997) Coherent structures near the wall in a turbulent channel flow. *J. Fluid Mech.*, 332, 185–214.

50 Bake, S., Meyer, D.G.W., and Rist, U. (2002) Turbulence mechanism in Klebanoff transition: a quantitative comparison of experiment and direct numerical simulation. *J. Fluid Mech.*, 459, 217–243.

51 Bernard, P.S. (2006) Turbulent flow properties of large scale vortex systems. *Proc. Nat. Acad. Sci.*, 103, 10 174–10 179.

52 Bernard, P.S. (2013) Vortex dynamics in transitional and turbulent boundary layers. *AIAA J.*, 51, 1828–1842.

53 Schlatter, P., Li, Q., Orlu, R., Hussain, F., and Henningson, D.S. (2014) On the near-wall vortical structures at moderate Reynolds numbers. *Eur. J. Mech. B/Fluids*, 48, 75–93.
54 Christensen, K.T. and Adrian, R.J. (2001) Statistical evidence of hairpin vortex packets in wall turbulence. *J. Fluid Mech.*, 431, 433–443.
55 Adrian, R.J. (2007) Hairpin vortex organization in wall turbulence. *Phys. Fluids*, 19, 041 301.
56 Rahgozar, S. and Maciel, Y. (2016) A visual assessment of hairpin packet structures in a DNS of a turbulent boundary layer. *Eur. J. Mech. B/Fluids*, 56, 161–171.
57 Soria, J., Kitsios, V., and Atkinson, C. (2016) On the identification of intense Reynolds stress structures in wall-bounded flows using information-limited two-dimensional planar data. *Eur. J. Mech. B/Fluids*, 55, 279–285.
58 del Álamo, J.C., Jiménez, J., Zandonade, P., and Moser, R.D. (2006) Self-similar vortex clusters in the turbulent logarithmic region. *J. Fluid Mech.*, 561, 329–358.
59 Marusic, I., McKeon, B.J., Monkewitz, P.A., Nagib, H.M., Smits, A.J., and Sreenivasan, K.R. (2010) Wall-bounded turbulent flows at high Reynolds numbers: Recent advances and key issues. *Phys. Fluids*, 22, 065 103.
60 Robinson, S. (1991) Coherent motions in the turbulent boundary layer. *Ann. Rev. Fluid Mech.*, 23, 601–639.

Problems

8.1 Show that the momentum thickness represents the depth a layer of fluid traveling with velocity U_∞ would have to be so that its momentum flux is the same as the excess above the physical flux associated with imagining that the correct mass flux is occurring but that this fluid all has velocity U_∞.

8.2 Show that Eq. (8.5) is consistent with assuming that the instantaneous random edge of a boundary layer at a given point with boundary layer thickness δ obeys a Gaussian distribution with mean 0.78δ and standard deviation $\sigma = 0.14\delta$.

8.3 Prove the validity of Eqs. (8.19) and (8.20) for δ_1 and θ, respectively.

8.4 Gain access to velocity and vorticity data for a turbulent flow simulation such as the channel flow data at http://turbulence.pha.jhu.edu/. Plot isosurfaces of U, identifying low-speed streaks. By plotting contours of velocity and vorticity search for and identify additional events similar to that plotted in Figure 8.25.

8.5 Having found an event such as described in Problem 8.4, make a local plot of λ_2 contours in order to investigate the properties of rotational regions associated with the ejection event.

9

Turbulence Modeling

Traditional turbulence modeling, known as Reynolds-averaged Navier–Stokes (RANS) modeling, is concerned with the development of mathematical relations that enable the prediction of average flow quantities such as the mean velocity and Reynolds stresses through solving the averaged flow equations. In effect, the goal is to obtain closure to the exact averaged equations that otherwise contain terms that cannot be computed without additional information. For many applications RANS equations are solved for steady flows, in which case the implied averaging can be thought of over time as against the general ensemble averaging indicated in Eq. (2.10). Most of this chapter is devoted to RANS modeling in this sense, though RANS modeling is also routinely applied to unsteady mean flows in the form of unsteady RANS modeling or URANS. An introduction to URANS is included at the end of this chapter, though it is in the next chapter, devoted to large eddy simulations (LES), where it is most appropriate to describe the details of contemporary approaches to URANS. In fact, modern URANS has a close affinity with the goals and methodology of LES calculations so it is in that context in which URANS most needs to be considered.

As described previously in this volume, the phenomena encountered in RANS modeling includes turbulent transport of momentum and energy, energy production and dissipation, energy transfer between scales by vortex stretching, as well as a number of more exotic processes such as energy redistribution by the pressure force, and the production and dissipation of dissipation itself. To the extent that such individual processes can be well approximated there is hope that the collected models create a methodology for predicting the mean flow that is useful in engineering work.

In many ways the preceding chapters have shown that deriving comprehensive theories with which to represent the basic physical processes of turbulent flow is a daunting task. As a result, the various RANS models that will be considered here do not have the force of rigorous theory, as against, for example, the stress rate-of-strain law Eq. (3.2) that underlies the Navier–Stokes equation. Predictions based on modeled equations cannot be expected to be accurate in the same way that solutions to the Navier–Stokes equations are.

The notion of accuracy in turbulence modeling reflects the limitations of the underlying RANS approach. Accuracy in this case is largely a function of the end purpose of the RANS calculation. For example, if the goal is predicting the total drag force on a body to within 10% accuracy or predicting trends in the data as various parameters are changed

Turbulent Fluid Flow, First Edition. Peter S. Bernard.

Companion website: www.wiley.com/go/Bernard/Turbulent_Fluid_Flow

then the RANS solution may very well be "accurate." On the other hand, accuracy would be much harder to come by if the goal was obtaining the pressure field on the rear surface of a bluff body or predicting the correct location of separation points on smooth bodies. To help bring RANS modeling to the point where it can achieve engineering goals it is common practice to incorporate some degree of empiricism in modifying models and parameter values. This has led to a proliferation of variants of closures since there is generally no one way to improve the performance of turbulence models through empirical means. Among the many models that have been developed there is no one particular approach that is considered to be the "best" closure scheme, though some models may be better than others for predicting certain classes of flows, if for no other reason than that more effort has been placed into having them achieve enhanced accuracy in such cases.

The focus here is on the main trends in RANS modeling, with the understanding that many options exist for modifying the particular approaches given here for a variety of reasons and applications. To gain proficiency in the use of closure models requires experience, particularly in comparing the performance of various models in a variety of applications. Finding the best model and parameter selections for a particular flow – when in fact there are literally hundreds of possible models – may seem to be a daunting task. Fortunately, in practical terms, one can readily examine the performance of a few of the main closure approaches with a good likelihood that refinement or alteration of these approaches may at best yield only a small improvement in results but not necessarily a dramatic change. Consequently, it is sufficient for the present purposes to confine the discussion to several of the most well-known models covering the different types. Some of these are incorporated in commercial CFD codes and may thus be easily compared to one another in specific applications to see which is the best basic approach to take in such cases. Ultimately, the decision as to how much investment needs to be made in developing better models for a particular problem depends on the goals of the study. It also must be recognized that some complex flows are beyond the capabilities of any model that has been developed thus far.

Also affecting the decision as to which model to use, particularly in industrial applications of CFD, is the relative cost of implementing the various solution techniques. Not all closures require the same investment in computational resources, so this becomes an additional factor in deciding the complexity of models that one would like to use. It is also not strictly the case that the additional investment necessary to implement complex closure schemes provides sufficient gains in accuracy to justify their expense. On the other hand, when computational cost and time is not a primary issue, it may be more advantageous to employ LES and related techniques, as will be discussed in the next chapter.

9.1 Types of RANS Models

The primary issue faced by RANS modeling is predicting the distribution of the Reynolds stress tensor, $R_{ij} = \overline{u_i u_j}$, whose presence in the averaged momentum equation (3.9) prevents closure. One can either attempt to model R_{ij} itself by developing a constitutive law or model the exact differential equation (3.53) that governs R_{ij}. Needless to say, the first option is generally the simplest to enact and is the methodology most often adopted in engineering practice. Both of these approaches will be considered here, with a number of the most widely used methods described in some detail.

The most common way to proceed in developing a constitutive law for the Reynolds stress tensor is to mimic the arguments used in deriving the molecular transport law in Eq. (3.3), as was discussed previously in reference to Eq. (6.6). In this case

$$R_{ij} = \frac{2}{3}K\,\delta_{ij} - \nu_t\left(\frac{\partial \overline{U}_i}{\partial x_j} + \frac{\partial \overline{U}_j}{\partial x_i}\right), \tag{9.1}$$

where, as before, $\nu_t > 0$ is termed the *eddy viscosity*. Note that a trace of Eq. (9.1) yields the identity $2K = 2K$ in incompressible flow. After substitution into Eq. (3.9), (9.1) leads to the $\overline{U}_i$ equation in the form

$$\frac{\partial \overline{U}_i}{\partial t} + \overline{U}_j\frac{\partial \overline{U}_i}{\partial x_j} = -\frac{\partial}{\partial x_i}\left[\frac{\overline{P}}{\rho} + \frac{2}{3}K\right] + \nu\nabla^2\overline{U}_i + \frac{\partial}{\partial x_j}\left[\nu_t\left(\frac{\partial \overline{U}_i}{\partial x_j} + \frac{\partial \overline{U}_j}{\partial x_i}\right)\right]. \tag{9.2}$$

The appearance of K in Eq. (9.2) is in combination with $\overline{P}$, so that numerical procedures for solving Eq. (9.2), which are equivalent to those used in solving the Navier–Stokes equation, can treat $\overline{P}/\rho + 2K/3$ as an unknown in place of the pressure field itself. Since $K = 0$ at solid boundaries, having computed $\overline{P}/\rho + 2K/3$ still gives the pressure on solid surfaces that can be used to determine the force on bodies. To get $\overline{P}$ elsewhere in the flow, however, requires a separate means for predicting K.

Closures based on Eq. (9.1) are known as *eddy viscosity models* (EVM). The many different choices of this type are distinguished according to how they determine the eddy viscosity. Some models incorporate a formula for the eddy viscosity that depends only on $\overline{U}$ together with parameters selected by the user. These are referred to as *zero-equation* models since they do not introduce a need to solve additional differential equations beyond that of Eq. (9.2) and the continuity equation (3.10). Models of this type are the Prandtl mixing length theory [1], and various extensions of this idea which are designed to treat 3D mean fields, such as the models of Cebeci and Smith [2] and Baldwin and Lomax [3]. Other choices for the eddy viscosity bring in additional equations to solve, such as the Spalart–Allmaras approach [4], which is an example of a one-equation model, and the K–ϵ [5], K–ω [6], and Menter shear stress transport (SST) [7] closures, which are examples of two-equation models. The elliptic relaxation approach [8] is an example of a three-equation model.

Regardless of how ν_t is determined, the constitutive model Eq. (9.1) has fundamental limitations, some of which were discussed in Chapter 6 and some of which will be considered below. Among the problem areas associated with Eq. (9.1) is the a priori impossibility of modeling such flow phenomena as anisotropy. For example, in a unidirectional flow Eq. (9.1) implies that $\overline{u^2} = \overline{v^2} = \overline{w^2}$, which is fundamentally at odds with the computed Reynolds stresses such as were given in Figure 7.7. Models based on Eq. (9.1) also cannot apply to rotating flows since this form does not vary with system rotation, as will be considered subsequently. If a particular model for R_{ij} a priori excludes the possibility of accounting for a physical phenomenon of importance, then one must either develop generalizations of the model that can accommodate the desired physics or else consider modeling the Reynolds stress equation itself. It is also the case, however, that the fact that a closure is not a priori prevented from representing an important physical phenomenon does not guarantee that it will capture the desired effect with acceptable accuracy.

Generalizations of the linear stress rate of strain law given in Eq. (9.1) tend to assume some degree of non-linear dependence on the derivatives of the mean velocity field. Such methods are known as non-linear eddy viscosity models or NLEVM (considered in Section 9.4). In this case Eq. (9.1) generally turns out to be the first term in a non-linear expansion of R_{ij} in terms of mean velocity gradients. Another way of obtaining similar kinds of models, known as algebraic Reynolds stress models (ARM, ARSM) or explicit algebraic Reynolds stress models (EASM), represents simplifications of models made to the complete Reynolds stress equation. The various NLEVM and ARM have a lot in common including a capability for accommodating phenomena such as anisotropy. An additional benefit of these approaches is that they can be applied to rotating flows where EVM are not appropriate.

Reynolds stress equation models (RSE, RSM), also referred to as second moment closures (SMC) since the Reynolds stresses are second moments of the velocity field, develop closure to the complete equation for R_{ij} given in Eq. (3.53). Such closures are often developed by the use of constraints that limit the possible tensor forms that the component terms can take. These might include transformation laws between coordinate systems associated with different observers or ideas such as realizability in which conditions such as $K \geq 0$ are enforced for all solutions to the closed equations. The complexity of such models, however, has led to the use of a number of ad hoc simplifications to the derived closure expressions so as to make them more practical.

The fact that closures depending on Eq. (9.1) remain the most widely used suggests that the potential benefits from attempting the more complex modeling in NLEVM, ARM or RSE are limited. Among the difficulties with NLEVM or ARM is the fact that it is the derivative of R_{ij} that appears in Eq. (3.9) so that small errors or anomalies in the modeling of the Reynolds stress tensor can have a large influence on the predicted mean velocity, usually for the worse. In contrast, Eq. (9.2) has the advantage that the modeled Reynolds stress tensor is in the form of a classic diffusion operator with non-constant viscosity. Techniques for numerically approximating such operators are widely available so that closures based on Eq. (9.1) can reliably produce solutions for the mean velocity field, even if they are not necessarily accurate.

An entirely different approach toward handling the RANS problem that was originally considered by Taylor [9] is to avoid the appearance of R_{ij} by using the identity

$$\overline{u_j \frac{\partial u_i}{\partial x_j}} = \frac{\partial K}{\partial x_i} - \epsilon_{ijk}\overline{u_j\omega_k} \tag{9.3}$$

to write the momentum equation (3.9) in the vorticity transport form

$$\frac{\partial \overline{U}_i}{\partial t} + \overline{U}_j \frac{\partial \overline{U}_i}{\partial x_j} = -\frac{\partial}{\partial x_i}(\overline{P}/\rho + K) + \nu\nabla^2\overline{U}_i + \epsilon_{ijk}\overline{u_j\omega_k}. \tag{9.4}$$

In this approach a model must be sought for the vorticity flux term $\overline{u_i\omega_j}$ instead of R_{ij}. Alternatively, in the same vein one may proceed from the mean vorticity equation (3.66) in index notation to give

$$\frac{\partial \overline{\Omega}_i}{\partial t} + \overline{U}_j \frac{\partial \overline{\Omega}_i}{\partial x_j} = \overline{\Omega}_j \frac{\partial \overline{U}_i}{\partial x_j} + \overline{\omega_j \frac{\partial u_i}{\partial x_j}} + \frac{\partial}{\partial x_j}\left(\nu \frac{\partial \overline{\Omega}_i}{\partial x_j} - \overline{u_j \omega_i}\right) \tag{9.5}$$

where both the vorticity transport correlation $\overline{u_j\omega_i}$ and the stretching correlation, namely $\overline{\omega_j \partial u_i/\partial x_j}$, have to be modeled. Alternatively, Eqn. (9.5) can be rewritten in the form

$$\frac{\partial \overline{\Omega}_i}{\partial t} + \overline{U}_j \frac{\partial \overline{\Omega}_i}{\partial x_j} = \overline{\Omega}_j \frac{\partial \overline{U}_i}{\partial x_j} + \frac{\partial \overline{u_i \omega_j}}{\partial x_j} + \frac{\partial}{\partial x_j}\left(\nu \frac{\partial \overline{\Omega}_i}{\partial x_j} - \overline{u_j \omega_i}\right) \tag{9.6}$$

where only the vorticity transport correlation appears. Though some significant understanding of the physical meaning of the vorticity transport and stretching correlations is available, as discussed in Section 6.7, closure schemes based on this approach remain largely undeveloped.

A final point to be made before considering the form of individual closure models is the observation that important distinctions can be made between modeling turbulence next to boundaries versus modeling turbulence in the flow field far from the direct influence of walls. Where the former generally involves a balance of many distinct terms in the equations of motion and is highly anisotropic, turbulence away from walls is often largely a balance between turbulence production and dissipation and is at least approximately isotropic. This means that it is more difficult to model near-wall turbulence than far-field turbulence. Models that may accommodate the latter may not perform well for the former. The particular difficulties faced in modeling near-wall turbulence means that many closure schemes include special adaptations designed to accommodate near-wall flow behavior. Since there is an unlimited number of ways to devise such near-wall modeling this adds to the proliferation of possible closure expressions that may be found in the literature. Where appropriate in what follows, we will distinguish between general RANS models and the special forms that RANS models may take in the near-wall region. The latter are often referred to as being low Reynolds number models in the sense that near the wall, Reynolds numbers based on the turbulent velocity scale $\sqrt{K}$ approach zero.

9.2 Eddy Viscosity Models

Eddy viscosity models often rely on formulas for ν_t that are derived by assuming an analogy between the turbulent momentum flux created by the random eddying motion in the presence of a mean velocity gradient and the molecular momentum flux associated with the random movement of molecules in the presence of a velocity gradient. Despite the limitations of this analogy, which were discussed in Section 6.2, the need to develop practical closure models has led to models for ν_t based on this fundamental idea. Thus, the starting point in this case is to assume that Eq. (6.13) holds, which can be written as

$$\nu_t \sim \mathcal{U}\mathcal{L} \tag{9.7}$$

where $\mathcal{U}$ and $\mathcal{L}$ are appropriately chosen turbulence velocity and length scales, respectively. We now consider in detail some of the common choices for $\mathcal{U}$ and $\mathcal{L}$, including those mentioned previously.

9.2.1 Mixing Length Theory and its Generalizations

For flows that are exactly or approximately unidirectional, as in a channel, pipe or boundary layer, the main consequence of Eq. (9.1) is the assumption that

$$\overline{uv} = -\nu_t \frac{d\overline{U}}{dy}, \tag{9.8}$$

where y is distance from the solid wall. An early idea contained in the Prandtl mixing length theory [1] is to assume that

$$\mathcal{L} = l_m \tag{9.9}$$

where l_m, known as the *mixing length*, is intended to represent a distance over which turbulent eddies "carry" momentum before mixing takes place. The velocity scale is taken to be

$$\mathcal{V} = l_m \left| \frac{d\overline{U}}{dy} \right| \tag{9.10}$$

which is motivated by the same idea brought up in Section 6.2 to the effect that the velocity of fluid particles traveling the distance l_m will diverge from the velocity mean at their destination by the amount in Eq. (9.10). Using Eqs. (9.9) and (9.10) in (9.7) the eddy viscosity in the mixing length theory is

$$\nu_t = l_m^2 \left| \frac{d\overline{U}}{dy} \right| \tag{9.11}$$

and Eq. (9.8) becomes

$$\overline{uv} = -l_m^2 \left| \frac{d\overline{U}}{dy} \right| \frac{d\overline{U}}{dy}. \tag{9.12}$$

Additional considerations are necessary to derive values for l_m.

Choosing l_m is not always straightforward. For flow adjacent to a flat surface it is logical to assume that $l_m \sim y$, since distance from the wall limits the size and movement of eddies involved in turbulent transport. On the other hand, far from the wall it may be assumed that l_m is constant since the local eddies in that region are unaffected by the wall. With these conventions, mixing length theory can be seen to be consistent with some of the basic properties of turbulence mentioned in Section 7.1.4. For example, if near the wall the choice is made that

$$l_m = \kappa y, \tag{9.13}$$

where κ is the von Kármán constant, then applying this to Eqs. (7.27) and (9.12) duplicates the argument, leading to the classical log law in Eq. (7.30).

To adapt the mixing length model to the entire flow field in a channel, pipe or boundary layer it is necessary to have l_m vary between Eq. (9.13) at the wall to a constant in the interior. One simple choice of this type is to take

$$l_m = \begin{cases} \kappa y & y \leq C\delta/\kappa \\ C\delta & y > C\delta/\kappa \end{cases} \tag{9.14}$$

so that l_m is linear in y up to the location $C\delta/\kappa$, beyond which it is constant. Here, δ is the channel half-width, pipe radius or boundary layer thickness as appropriate. More elaborate forms of l_m in boundary layers have been pursued. For example, the van Driest formula [10]

$$l_m = \kappa y(1 - e^{-y^+/A^+}), \tag{9.15}$$

where $y^+ = yU_\tau/\nu$ and parameter $A^+ = 26$, reduces to κy when y^+ is sufficiently large ($\approx 2A^+$) while it has a y^2 dependence within A^+ of the wall. The latter property according to Eq. (9.12) reduces the influence of $\overline{uv}$ near the wall to y^4 as against its actual y^3 dependence. In view of Eq. (7.14), the use of Eq. (9.15) enables $\overline{U}^+ = y^+$ to hold next to the boundary besides preserving the log law in Eq. (7.31) away from the wall. In this way the trend in Figure 7.18 is accurately captured. Numerous modifications to A^+ can be found in the literature [11] that have been developed with a view to accommodating accelerating boundary layers for which there is a non-zero mean pressure gradient. The shift in scale given by Eq. (9.15), referred to as van Driest damping [10], finds wide application in a similar context for LES models near boundaries. For flows without boundaries such as jets, wakes, and mixing layers, l_m is usually taken to be a relevant local length scale at each streamwise position. For a jet this would be the local radius and for a mixing layer it would be the local half-width.

Equation (9.12) incorporating (9.15) is the standard eddy viscosity for the mixing length model. To branch out to other flows, and particularly those for which the boundary is not a flat plate, there may not be a satisfactory way of choosing l_m. In practice, l_m is generally taken as the nearest distance to a solid surface, but one can easily imagine circumstances where the geometry is complicated and this is not a relevant scale. It is also the case that if Eq. (9.11) is to apply more generally, then a suitable replacement must be found for Eq. (9.10) that is not limited to unidirectional mean flows.

Several generalizations of the mixing length idea have been developed to accommodate a wider class of flows than can be studied using Eq. (9.11). These each reduce to Eq. (9.11) for unidirectional flow. One approach is the Smagorinsky model [12] to the effect that

$$\nu_t = l_m^2(2\overline{S}_{ij}\overline{S}_{ij})^{1/2} \tag{9.16}$$

where

$$\overline{S}_{ij} = \frac{1}{2}\left(\frac{\partial \overline{U}_j}{\partial x_i} + \frac{\partial \overline{U}_i}{\partial x_j}\right) \tag{9.17}$$

is the mean rate of strain tensor and l_m is given by Eq. (9.15). Equation (9.16) has been widely adopted for use in LES and will be considered further in the next chapter.

Another general zero-equation model used in RANS computations is the Cebeci and Smith model [2], which assumes for the region close to the boundary that

$$\nu_t = l_m^2\left(\frac{\partial \overline{U}_i}{\partial x_j}\frac{\partial \overline{U}_i}{\partial x_j}\right)^{1/2} \tag{9.18}$$

and further from the walls

$$\nu_t = C_{CS}U_\infty\delta_1\gamma_K, \tag{9.19}$$

and again l_m is determined by the van Driest formula in Eq. (9.15). C_{CS} is a constant nominally taken to be 0.0168, δ_1 is the displacement thickness defined in Eq. (8.3), and

γ_K is the Klebanoff intermittency function defined in Eq. (8.6) that depends on a scale Δ. In this context the choice is made that $\Delta = \delta$ where δ is the boundary layer thickness. The inclusion of γ_K in Eq. (9.19) is meant to bring $\nu_t \to 0$ in the region outside of the turbulent boundary layer. Finally, the changeover from the inner region to the outer region model occurs at the first location where Eq. (9.18) increases to the value of Eq. (9.19).

Another widely employed model of this sort is the Baldwin–Lomax model [3], which is similar to the Cebeci–Smith model but has the advantage of not requiring input of the boundary layer thickness δ. In this model, near the wall it is assumed that

$$\nu_t = l_m^2 (\overline{\Omega}_i \overline{\Omega}_i)^{1/2} \tag{9.20}$$

where $\overline{\Omega}_i$ is the mean vorticity vector and l_m is given by the van Driest formula Eq. (9.15). An equivalent way of expressing Eq. (9.20) is as

$$\nu_t = l_m^2 (2\overline{W}_{ij}\overline{W}_{ij})^{1/2}, \tag{9.21}$$

where

$$\overline{W}_{ij} = \frac{1}{2}\left(\frac{\partial \overline{U}_i}{\partial x_j} - \frac{\partial \overline{U}_j}{\partial x_i}\right) \tag{9.22}$$

is the mean rotation tensor. In the outer region the Baldwin–Lomax model is taken to be

$$\nu_t = C_{BL} F_W \gamma_K \tag{9.23}$$

where the constant $C_{BL} = 0.0269$ and

$$F_W = \min(y_m F_m, y_m C_W U_{dif}^2 / F_m) \tag{9.24}$$

where $U_{dif} \equiv \max\sqrt{\overline{U}_i^2} - \min\sqrt{\overline{U}_i^2}$, $C_W = 0.25$, and

$$F_m = \max\left(l_m \sqrt{2\overline{W}_{ij}\overline{W}_{ij}}/\kappa\right). \tag{9.25}$$

γ_K is once again the Klebanoff intermittency function in which now the scale $\Delta \equiv y_m/0.3$ and y_m is defined as the location where the function being maximized in Eq. (9.25) achieves its maximum. The changeover from Eq. (9.21) to (9.23) occurs where the former gets larger than the latter.

Mixing length models to a large extent are tuned to the requirements of attached boundary layer flows and can be expected to work best for flows of this type. Their use can be expected to be increasingly problematic the more a flow contains phenomena such as curvature, separation, large pressure gradients, rotation, secondary flows, transient effects, and sudden changes in shear. A fundamental difficulty in the use of mixing length models is in supplying estimates of l_m in complicated geometries where there are no obvious guidelines that can be used in deciding its value. It is also the case that mixing length models predict the mean velocity field only, so if additional information such as the turbulent kinetic energy is required, alternative procedures must be employed.

9.2.2 K–ϵ Closure

Since the Reynolds stress represents a momentum flux produced by the local fluctuating fluid velocity, it is natural to regard the velocity standard deviations, $\overline{u_i^2}^{1/2}, i = 1, 2, 3$ as velocity scales with which to characterize the transport process. This idea motivates the selection of the velocity scale in Eq. (9.7) as

$$\mathcal{U} \sim \sqrt{K}. \tag{9.26}$$

If this approach is to be pursued then a means of predicting K must be found, which means that its own governing equation (3.37) must be used for this purpose.

As seen previously, the balance of physical effects that establishes K in a turbulent flow depends strongly on the dissipation rate given by ϵ defined in Eq. (3.32). Away from boundaries, as seen in Figure 7.9, K is determined by a balance between its production from the mean velocity field and its dissipation given by ϵ. It may be concluded that the K balance must be modeled together with a mechanism for predicting ϵ. In the K–ϵ closure ϵ is obtained through a model of its own exact governing differential equation (3.42).

By computing ϵ in addition to K a means is provided for determining the turbulence length scale $\mathcal{L}$ given by

$$\mathcal{L} \sim \frac{K^{3/2}}{\epsilon}. \tag{9.27}$$

$\mathcal{L}$ in this relation can be viewed as the product of $\sqrt{K}$ and the eddy turnover time, K/ϵ, so, in fact, it represents the approximate distance that fluid particles move during the events contributing to transport. Written differently, Eq. (9.27) gives

$$\epsilon \sim \frac{K^{3/2}}{\mathcal{L}}, \tag{9.28}$$

which was previously derived by dimensional arguments in Eq. (4.46).

The K–ϵ closure adopts Eqs. (9.26) and (9.27), respectively, as the relevant velocity and length scales appropriate to modeling turbulent transport, so that Eq. (9.7) yields the eddy viscosity model

$$\nu_t = C_\mu \frac{K^2}{\epsilon} \tag{9.29}$$

where C_μ is a constant. The complete K–ϵ closure consists of solving Eq. (9.2) incorporating Eq. (9.29) together with closed equations for K and ϵ that will now be considered.

While our interest here will primarily be with the traditional K–ϵ closure, it should be mentioned that there are at least two other prominent variants of the K–ϵ closure that have received considerable attention. These include the realizable K–ϵ closure [13] and the RNG (renormalization group) form of the K–ϵ closure [14]. While the overall performance of these models does not generally differ much from that of the standard K–ϵ closure, in some instances they offer an advantage. For example, as mentioned previously in regards to the stagnation point anomaly in Section 6.6, the K–ϵ closure has a tendency to overpredict energy in the vicinity of stagnation points. Modifications to the calculation of C_μ and terms in the ϵ equation that are incorporated in the realizable and RNG variants of the K–ϵ closure have the effect of limiting energy production at stagnation points so as to curtail the effects of the stagnation point anomaly.

9.2.2.1 *K* Equation

Recalling Eq. (3.37), the exact K equation can be written in the form

$$\frac{\partial K}{\partial t} + \overline{U}_j \frac{\partial K}{\partial x_j} = \mathcal{P} - \epsilon + \nu \nabla^2 K - \frac{\partial}{\partial x_i}\left(\frac{\overline{pu_i}}{\rho} + \overline{u_i(u_j^2/2)}\right), \tag{9.30}$$

where

$$\mathcal{P} = -R_{ij}\frac{\partial \overline{U}_i}{\partial x_j} \tag{9.31}$$

is the turbulent kinetic energy production term. Assuming Eq. (9.1) holds, it follows that

$$\mathcal{P} = \nu_t \frac{\partial \overline{U}_i}{\partial x_j}\left(\frac{\partial \overline{U}_i}{\partial x_j} + \frac{\partial \overline{U}_j}{\partial x_i}\right), \tag{9.32}$$

where ν_t is given by Eq. (9.29), so that no additional modeling is needed to account for turbulence production.

The pressure work and kinetic energy flux terms in Eq. (9.30) are combined into a single flux-like term that requires modeling. For wont of a formal means for analyzing the relevant physics in the $\overline{pu_i}$ and $\overline{u_i(u_j^2/2)}$ correlations, it is traditional to assume that this term obeys a gradient transport law of the form

$$\frac{1}{\rho}\overline{pu_i} + \overline{u_i(u_j^2/2)} = -\frac{\nu_t}{\sigma_K}\frac{\partial K}{\partial x_i}, \tag{9.33}$$

where the constant, σ_K, acts in this equation in the manner of a Prandtl number for turbulent diffusion of K. Since ϵ is obtained from its own equation, the closed K equation has the form

$$\frac{\partial K}{\partial t} + \overline{U}_j \frac{\partial K}{\partial x_j} = \mathcal{P} - \epsilon + \frac{\partial}{\partial x_i}\left[\left(\nu + \frac{\nu_t}{\sigma_K}\right)\frac{\partial K}{\partial x_i}\right]. \tag{9.34}$$

In this, K is determined by the balance between convection on the left-hand side, and production, dissipation, and transport on the right-hand side.

9.2.2.2 The ϵ Equation

The ϵ equation (3.42) was previously given in the form

$$\frac{D\epsilon}{Dt} = P_\epsilon^1 + P_\epsilon^2 + P_\epsilon^3 + P_\epsilon^4 + \Pi_\epsilon + T_\epsilon + D_\epsilon - \Upsilon_\epsilon, \tag{9.35}$$

with the correlations on the right-hand side defined in Eqs. (3.43)–(3.50). With the sole exception of the viscous diffusion term D_ϵ the terms on the right-hand side of Eq. (9.35) require modeling. This task is made easier by adopting previously developed models for the vortex stretching term P_ϵ^4 and the dissipation of ϵ term Υ_ϵ, given in Eqs. (5.71) and (5.72) so that

$$P_\epsilon^4 - \Upsilon_\epsilon = C_{\epsilon_3} R_T^{\frac{1}{2}} \frac{\epsilon^2}{K} - C_{\epsilon_2}\frac{\epsilon^2}{K}, \tag{9.36}$$

where $R_T = K^2/(\nu\epsilon)$ is the turbulence Reynolds number. As mentioned previously, the traditional viewpoint assumes that $C_{\epsilon_3} = 0$, in which case vortex stretching makes no independent contribution to the dissipation rate balance.

The modeling of P_ϵ^1 and P_ϵ^2 was considered in Section 6.5 in the context of homogeneous shear flow. For the present circumstances, consider

$$P_\epsilon^1 = -\epsilon_{ij}^c \frac{\partial \overline{U}_i}{\partial x_j} \tag{9.37}$$

and

$$P_\epsilon^2 = -\epsilon_{ij} \frac{\partial \overline{U}_i}{\partial x_j}, \tag{9.38}$$

where

$$\epsilon_{ij} = 2\nu \overline{\frac{\partial u_i}{\partial x_k} \frac{\partial u_j}{\partial x_k}} \tag{9.39}$$

and

$$\epsilon_{ij}^c = 2\nu \overline{\frac{\partial u_k}{\partial x_i} \frac{\partial u_k}{\partial x_j}}. \tag{9.40}$$

Note that $\epsilon_{ii} = \epsilon_{ii}^c = 2\epsilon$, and in isotropic turbulence

$$\epsilon_{ij} = \epsilon_{ij}^c = \delta_{ij} \frac{2}{3} \epsilon. \tag{9.41}$$

The simplification of ϵ_{ij} and ϵ_{ij}^c in Eq. (9.41) caused by the isotropy assumption will render $P_\epsilon^1 = P_\epsilon^2 = 0$ in incompressible flow. This places an unphysical limitation on a potentially significant anisotropic effect. On the other hand, the anisotropy associated with c_{ij} and ϵ_{ij}^c is difficult to estimate from first principles. This has led to the practice of assuming that whatever anisotropic properties ϵ_{ij} and ϵ_{ij}^c have will match that of the Reynolds stress tensor. A measure of the latter is given by the anisotropy tensor

$$b_{ij} \equiv \frac{R_{ij} - \frac{2}{3} K \delta_{ij}}{2K} \tag{9.42}$$

which is identically zero in isotropic turbulence and non-zero otherwise. Additionally, the Schwarz inequality applied to R_{ij} means that $|\overline{u_i u_j}| \leq \sqrt{\overline{u_i^2}} \sqrt{\overline{u_j^2}}$ (no sum on i and j). Using the definition of K it follows that $|b_{ij}| \leq 1$ for all i, j.

The formal assumption is now made that the deviatoric parts of ϵ_{ij} and ϵ_{ij}^c are proportional to b_{ij}, in which case

$$\frac{\epsilon_{ij} - \frac{2}{3} \epsilon \delta_{ij}}{2\epsilon} \sim b_{ij} \tag{9.43}$$

and a similar relation for ϵ_{ij}^c. Note that Eq. (9.43) is essentially the same idea as used in Eq. (6.39). Substituting these models into Eqs. (9.37) and (9.38), combining the equations, and introducing a proportionality constant C_{ϵ_1} yields

$$P_\epsilon^1 + P_\epsilon^2 = C_{\epsilon_1} \frac{\epsilon}{K} \mathcal{P}, \tag{9.44}$$

a relation that reduces to Eq. (6.40) in homogeneous shear flow.

As for the remaining terms in Eq. (9.35), P_ϵ^3 is usually not given an explicit model. Rather, its contribution is imagined to be contained within that of the other production terms. Finally, the transport terms T_ϵ and Π_ϵ are traditionally grouped together and given a gradient law treatment of the form

$$T_\epsilon + \Pi_\epsilon = \frac{\partial}{\partial x_i}\left(\frac{\nu_t}{\sigma_\epsilon}\frac{\partial \epsilon}{\partial x_i}\right), \tag{9.45}$$

where σ_ϵ is a constant with the character of a Prandtl number for the flux of ϵ. With these assumptions the final traditional modeled form of the ϵ equation is

$$\frac{\partial \epsilon}{\partial t} + \overline{U}_j \frac{\partial \epsilon}{\partial x_j} = C_{\epsilon_1}\frac{\epsilon}{K}\mathcal{P} - C_{\epsilon_2}\frac{\epsilon^2}{K} + \frac{\partial}{\partial x_i}\left[\left(\nu + \frac{\nu_t}{\sigma_\epsilon}\right)\frac{\partial \epsilon}{\partial x_i}\right]. \tag{9.46}$$

This has a similar structure to the K equation in the sense that the convection and transient terms on the left-hand side are balanced by the combined effect of production, dissipation, and transport on the right-hand side. As noted previously, the potentially significant advantage that may accrue from assuming that $C_{\epsilon_3} \neq 0$, so vortex stretching makes a separate contribution to the ϵ balance, has remained largely unexplored in this equation.

9.2.2.3 Calibration of the K–ϵ Closure

The various parameters C_μ, C_{ϵ_1}, C_{ϵ_2}, σ_K, and σ_ϵ that show up in the $\overline{U}_i$, K, and ϵ equations must be assigned numerical values. While it is not uncommon to adjust parameter values in the interest of obtaining the best possible solution for a particular turbulent flow, a set of parameter values have been developed that attempt to render the model consistent with a number of canonical flows. These are the constant values that tend to be most often associated with the implementations of the K–ϵ closure.

Since the closure equations reduce to those considered previously in Chapters 5 and 6 for the special cases of isotropic decay and homogeneous shear flow, the values of parameters C_{ϵ_1} and C_{ϵ_2} that were derived in those cases are appropriated for general flows as well. Thus $C_{\epsilon_1} = 1.44$ and $C_{\epsilon_2} = 1.92$ are standard values for the K–ϵ closure. Of course, there is no guarantee that these values will remain optimal in the presence of a general mean shearing in the flow field. For the remaining constants, values are obtained via tests of the closure against a variety of flows for which experimental data are available [5]. As a result of such studies, the parameters are taken to be $C_\mu = 0.09$, $\sigma_K = 1$, and $\sigma_\epsilon = 1.3$.

By considering the form taken by the K–ϵ closure in the constant stress region of the boundary layer discussed in Section 7.1.4, it becomes possible to develop a constraint on the model parameters whose satisfaction will imply some degree of consistency with near-wall turbulent flow. In particular, in this region it is approximately true that

$$-\overline{uv} = \frac{\tau_w}{\rho} = U_\tau^2 \tag{9.47}$$

and the turbulence production and dissipation terms in the K equation are in balance, so that

$$\mathcal{P} = -\overline{uv}\frac{d\overline{U}}{dy} = \epsilon. \tag{9.48}$$

Replacing $\overline{uv}$ in Eq. (9.48) using (9.47) yields an equation for the mean velocity derivative. Substituting this into Eq. (9.8), replacing $\overline{uv}$ using Eq. (9.47) once again, and substituting for ν_t from Eq. (9.29) gives the approximation

$$K = \frac{U_\tau^2}{\sqrt{C_\mu}}. \tag{9.49}$$

Thus, in this modeling scenario, K is constant in the equilibrium, constant stress layer. Note that Eq. (9.49) exactly solves (9.34) in the present circumstances.

To ensure that these results also fit in with the ϵ equation, note that according to Eqs. (9.47) and (9.48)

$$\epsilon = U_\tau^2 \frac{d\overline{U}}{dy}. \tag{9.50}$$

Substituting for $d\overline{U}/dy$ using the log law Eq. (7.31), gives

$$\epsilon = \frac{U_\tau^3}{\kappa y}. \tag{9.51}$$

Thus, it is seen that ϵ is decreasing throughout the constant stress layer. Incorporating Eqs. (9.49) and (9.51) in (9.46) and omitting the viscous diffusion term yields the condition

$$\frac{\sqrt{C_\mu}\sigma_\epsilon}{\kappa^2}(C_{\epsilon_2} - C_{\epsilon_1}) = 1. \tag{9.52}$$

For the standard constants listed above the left-hand side of this relation gives 1.11, showing that this constraint is reasonably well satisfied.

9.2.2.4 Near-Wall K–ϵ Models

The form of the K–ϵ closure discussed thus far has been developed without consideration of the special properties of flow in the near-wall region. One aspect of this is ensuring that K and ϵ satisfy boundary conditions that are compatible with Eqs. (9.34) and (9.46). In fact, it proves to be necessary to make a number of adjustments to the K–ϵ equations to improve their compatibility with the near-wall flow. As mentioned previously these are referred to as the low Reynolds number form of the K–ϵ closure.

At a solid surface, say at $y = 0$, $\overline{U}(0) = K(0) = 0$ while $\epsilon(0) \neq 0$. In fact, evaluation of the K equation (9.30) at the boundary gives

$$\epsilon(0) = \nu \frac{\partial^2 K}{\partial y^2}(0), \tag{9.53}$$

a relation which also follows directly from the definitions of K and ϵ using Taylor series expansions of the velocity components. An alternative way of expressing the same relation is via

$$\epsilon(0) = 2\nu \left(\frac{\partial \sqrt{K}}{\partial y}(0) \right)^2. \tag{9.54}$$

(see Problem 9.5). Eq. (9.34) is compatible with this boundary condition. Moreover, using the fact that $dK/dy(0) = 0$ it follows from Eq. (9.53) and a Taylor series for K that

$$\lim_{y \to 0} \frac{\nu K}{\epsilon y^2} = \frac{1}{2}, \tag{9.55}$$

which may be viewed as a constraint on the computed K and ϵ solutions near the wall. In practice, if a near-wall K–ϵ closure is to be considered "asymptotically consistent" with the physics of the near-wall region, then the solutions it produces for K and ϵ should satisfy Eq. (9.55). Such solutions will have a K field that approaches the wall with the correct zero slope.

The fact that some modification of the K–ϵ closure is needed near the wall is also apparent from considering the second term on the right-hand side of Eq. (9.46), which, as it stands, will be infinite so long as the boundary condition $K(0) = 0$ is enforced at the wall surface. This and other shortcomings of Eqs. (9.34) and (9.46) must be accommodated to improve the performance of the K–ϵ model near the boundary. One such tactic is to selectively replace ϵ by a reduced dissipation, $\tilde{\epsilon}$, defined, for example, as

$$\tilde{\epsilon} \equiv \epsilon - 2\nu\left(\frac{\partial\sqrt{K}}{\partial y}\right)^2, \tag{9.56}$$

which according to Eq. (9.54) is zero at the surface. Since the second term on the right-hand side of Eq. (9.56) becomes very small away from the wall this expression only modifies ϵ in the region closest to the wall. If the term ϵ^2/K in Eq. (9.46) is replaced by $\epsilon\tilde{\epsilon}/K$, then its singularity at the wall surface is avoided.

In some implementations of the K–ϵ closure, particularly in its original form [5], the pseudo boundary condition $\epsilon = 0$ is adopted, which is equivalent to replacing ϵ by $\tilde{\epsilon}$ everywhere in the closed equation. If this is done, then the K equation needs to acquire the term $-2\nu\left(\frac{\partial\sqrt{K}}{\partial y}\right)^2$ in order to make sure that the correct value of the dissipation rate is recovered at the boundary.

The eddy viscosity model in Eq. (9.29) is derived without concern for the special properties of flow near solid boundaries and must be modified to be consistent with near-wall flow. In particular, the formal analysis of the Reynolds shear stress in Section 6.3 showed that in a simple shearing flow

$$\nu_t = \mathcal{T}_{22}\overline{v^2}, \tag{9.57}$$

where $\mathcal{T}_{22}$ is a Lagrangian integral time scale. In contrast, the eddy viscosity in the K–ϵ closure can be written as

$$\nu_t = \left(\frac{2K}{3}\right)\left(\frac{3C_\mu}{2}\frac{K}{\epsilon}\right), \tag{9.58}$$

where the first factor reduces to $\overline{v^2}$ in isotropic turbulence and the second factor is a de facto model for the Lagrangian integral scale $\mathcal{T}_{22}$. Near the wall $\overline{v^2}$ is strongly affected by the damping effect of the boundary and its behavior is entirely different than that of $2K/3$, as is evident from Figure 7.7. This observation suggests that modification of Eq. (9.58) that forces the factor $2K/3$ to behave more like $\overline{v^2}$ may be a useful step to take in improving Eq. (9.8) near boundaries.

Within the confines of the K–ϵ closure there is little opportunity for adapting ν_t to the near-wall region other than to introduce a wall function f_μ in the expression

$$\nu_t = C_\mu f_\mu \frac{K^2}{\epsilon}, \tag{9.59}$$

which can be used to force the proper behavior of the eddy viscosity near solid boundaries. An alternative approach [8] is to develop a model equation specifically for $\overline{v^2}$ that is solved in conjunction with the K–ϵ equations. This is the elliptic relaxation method in which a Poisson-like equation is solved to determine the distribution of $\overline{v^2}$.

An additional consideration that can help in setting up wall functions comes from examining the K and ϵ equations in the near-wall vicinity. It was previously shown that Eq. (9.30) reduces to Eq. (9.53) at the wall surface. Taylor series expansions show that just off the surface the balance of terms proportional to y is given by

$$0 = -\epsilon + \nu\frac{\partial^2 K}{\partial y^2} - \frac{1}{\rho}\frac{\partial \overline{pv}}{\partial y}. \tag{9.60}$$

Figure 7.9 shows that the contribution of the pressure term in Eq. (9.60) is not large near the surface. Consequently, the use of the boundary condition Eq. (9.53) means that Eq. (9.60) will be reasonably well modeled near the surface so there is not a great need to develop special near-wall models for the production, transport, and pressure diffusion terms in the K equation.

The ϵ equation (9.35) at the surface is

$$\Pi_\epsilon(0) + D_\epsilon(0) = \Upsilon_\epsilon(0), \tag{9.61}$$

where $\Pi_\epsilon(0)$ is much smaller than the other two terms, as may be seen in the ϵ equation budget for channel flow in Figure 7.10. On the other hand, Π_ϵ and the other terms in Eq. (9.35) are non-negligible off the wall surface and behave in ways that are not necessarily compatible with the modeling in Eq. (9.46). To remedy these defects wall functions f_1 and f_2, respectively, are placed in the production and dissipation terms in the ϵ equation to help move these modeled terms closer toward the known wall behavior of the exact expressions. In addition, a function F is introduced as a near-wall model for Π_ϵ. The result is

$$\frac{\partial \epsilon}{\partial t} + \overline{U}_j\frac{\partial \epsilon}{\partial x_j} = C_{\epsilon_1} f_1 \frac{\epsilon}{K}\mathcal{P} - C_{\epsilon_2} f_2 \frac{\epsilon\tilde{\epsilon}}{K} + \frac{\partial}{\partial x_i}\left[\left(\nu + \frac{\nu_t}{\sigma_\epsilon}\right)\frac{\partial \epsilon}{\partial x_i}\right] + F, \tag{9.62}$$

which may be contrasted with Eq. (9.46). Note, as well, the introduction of $\tilde{\epsilon}$ in the numerator of the dissipation term to prevent this term from becoming unbounded at the wall surface.

For the same reason that there are many different models for ν_t there is great variation among the sets of wall functions that have been devised for the K–ϵ closure [15]. Ideally one would like to create tensorially correct forms for the wall functions that apply generally to all flows and all inertial observers, though this is difficult to achieve in practice. For example, the original Jones and Launder [5] model incorporates the wall functions

$$f_\mu = e^{-2.5/(1+R_T/50)} \tag{9.63}$$

$$f_1 = 1, \qquad f_2 = 1 - 0.3e^{-R_T^2} \tag{9.64}$$

and

$$F = 2\nu\nu_t\left(\frac{d^2\overline{U}}{dy^2}\right)^2. \tag{9.65}$$

The function F in this case is coordinate system dependent and not of universal applicability. With the increasing availability of DNS data that provides accurate information about the K and ϵ equation budgets more elaborate wall functions can be devised to capture specific features of the near-wall flow. Among these is the choice [15]

$$f_\mu = (1 + 3\, R_T^{-3/4})(1 + 80\, e^{-R_\epsilon})\left(1 - e^{\left(-\frac{R_\epsilon}{43} - \frac{R_\epsilon^2}{330}\right)}\right)^2 \tag{9.66}$$

$$f_1 = f_2 = 1 \tag{9.67}$$

and

$$F = e^{-\left(\frac{R_T}{40}\right)^2}\left(-0.57\frac{\epsilon\tilde{\epsilon}}{K} + 0.5\frac{(\epsilon - 2\nu K/y^2)^2}{K} - 2.25\frac{\epsilon}{K}\mathcal{P}\right), \tag{9.68}$$

where $R_\epsilon = (\nu\epsilon)^{1/4}y/\nu$. In this particular model, in contrast to the Jones and Launder model, the physical boundary condition in Eq. (9.53) (or (9.54)) is applied.

9.2.3 K–ω Models

While the eddy viscosity in Eq. (9.29) can be thought of as a result of choosing particular functions for the velocity and length scales in Eq. (9.7), it can also be viewed as the consequence of a dimensional argument assuming dependence of ν_t on K and ϵ. This motivates developing expressions for ν_t based on alternative choices of variables such as K and ω where the latter has units of inverse time and may be regarded as a vorticity strength. In this case the assumption is made that

$$\nu_t = \frac{K}{\omega} \tag{9.69}$$

where in comparison to the K–ϵ formulation this means that

$$\omega \equiv \frac{1}{C_\mu}\frac{\epsilon}{K}. \tag{9.70}$$

Since ϵ is a rate of energy dissipation, by being proportional to ϵ/K, ω may be also interpreted physically as the fractional rate of change of energy by dissipation, what is referred to as the *specific dissipation rate*. It was previously seen in Eq. (5.27) that ϵ/K is the inverse of the eddy-turnover time so that ω also has this interpretation. K–ω closures choose to determine the eddy viscosity on the basis of Eq. (9.69) so that the solution of the $\overline{U}_i$ equation is accompanied by solutions to the K equation together with one for ω.

The K–ω approach is manifested in a number of versions that have developed from its original form [6, 16]. For concreteness, and to provide a contrast with the previously considered K–ϵ equations, a recent standard form of the K–ω model is given here in detail. In this edition of the model, Eq. (9.69) is modified to be

$$\nu_t = \frac{K}{\tilde{\omega}}, \tag{9.71}$$

where

$$\tilde{\omega} = \max(\omega, C_{lim}\sqrt{2\overline{S}_{ij}\overline{S}_{ij}/\beta^*}) \tag{9.72}$$

and $C_{lim} = 7/8$ and $\beta^* = 0.09$. The modeling of the K equation in this case is no different than in the K–ϵ closure except for some minor notational differences. Specifically

$$\frac{\partial K}{\partial t} + \overline{U}_j\frac{\partial K}{\partial x_j} = \mathcal{P} - \beta^* K\omega + \frac{\partial}{\partial x_i}\left[\left(\nu + \frac{\nu_t}{\sigma_K}\right)\frac{\partial K}{\partial x_i}\right], \tag{9.73}$$

with the value $\sigma_K = 1.667$. In view of Eq. (9.70), the second term on the right-hand side of Eq. (9.73) is equivalent to ϵ, assuming that $C_\mu = 0.09$.

The equation satisfied by ω is designed to take on the appearance and properties of the ϵ equation. It is essentially an artificial creation that includes a balance between production, dissipation, and transport terms that are dimensionally appropriate for ω. Specifically, the ω equation has the form

$$\frac{\partial \omega}{\partial t} + \overline{U}_j \frac{\partial \omega}{\partial x_j} = \alpha \frac{\omega}{K} \mathcal{P} - \beta \omega^2 + \frac{\partial}{\partial x_i} \left[\left(\nu + \frac{\nu_t}{\sigma_\omega} \right) \frac{\partial \omega}{\partial x_i} \right] + \frac{\sigma_d}{\omega} \frac{\partial K}{\partial x_j} \frac{\partial \omega}{\partial x_j}. \tag{9.74}$$

Standard values for the constants in Eqs. (9.73) and (9.74) are $\alpha = 0.52, \sigma_w = 2, \beta = \beta_0 f_\beta$, $\beta_0 = 0.0708$, $f_\beta = (1 + 85\chi_\omega)/(1 + 100\chi_\omega)$, and $\chi_\omega = |\overline{\Omega}_{ij}\overline{\Omega}_{jk}\overline{S}_{ki}/(\beta^*\omega)^3|$. Finally, $\sigma_d = 0$ unless $(\partial K/\partial x_j)(\partial \omega/\partial x_j) < 0$, in which case it is 1/8.

Next to a solid boundary at $y = 0$ Eq. (9.74) reduces to approximately

$$0 = \nu \frac{\partial^2 \omega}{\partial y^2} - \beta \omega^2, \tag{9.75}$$

which has the solution

$$\omega = \frac{6\nu}{\beta y^2}. \tag{9.76}$$

This suggests that a boundary condition for Eq. (9.74) may consist of an evaluation of Eq. (9.76) at a grid point close to the surface. In practice, however, it is deemed better to use empirically derived formulas such as

$$\omega(0) = 10 \frac{6\nu}{\beta_1 (\Delta y)^2}, \tag{9.77}$$

with $\beta_1 = 0.075$ and Δy the near-wall grid spacing, a formula that is taken from the shear stress transport model described in the next section. As for the far-field boundary conditions, some latitude exists in their choice so that they tend not to be given by a definite formula that is used in all flows.

It is widely believed that the K–ω model provides better accuracy in the near-wall region than the K–ϵ closure. Moreover, the K–ω equations can be solved to the boundary without the elaborate use of wall functions, as is necessary for the K–ϵ closure. Further from the wall, however, the K–ω model displays sensitivities to ω that make it less attractive than the K–ϵ model.

9.2.4 Menter Shear Stress Transport Closure

A two-equation model known as the Menter shear stress transport (SST) closure [7] has been designed to incorporate a blending operation that modifies the K–ω closure at points away from walls so as to take on the form of the K–ϵ closure. In this way the best properties of each separate scheme are brought together into one method. As is the case for all such popular models, there are a number of editions and other modifications to the basic approach that have been derived. Here, a principal updated form of the original version is given.

For the SST model applied to constant density flow, the governing equations for K and ω are

$$\frac{\partial K}{\partial t} + \overline{U}_j \frac{\partial K}{\partial x_j} = \mathcal{P} - \beta^* K \omega + \frac{\partial}{\partial x_i} \left[\left(\nu + \frac{\nu_t}{\sigma_K} \right) \frac{\partial K}{\partial x_i} \right] \tag{9.78}$$

and

$$\frac{\partial \omega}{\partial t} + \overline{U}_j \frac{\partial \omega}{\partial x_j} = \frac{\gamma}{\nu_t} \mathcal{P} - \beta \omega^2 + \frac{\partial}{\partial x_i}\left[\left(\nu + \frac{\nu_t}{\sigma_\omega}\right)\frac{\partial \omega}{\partial x_i}\right] + 2(1 - F_1)\frac{\sigma_{\omega_2}}{\omega}\frac{\partial K}{\partial x_j}\frac{\partial \omega}{\partial x_j}, \quad (9.79)$$

where the eddy viscosity is

$$\nu_t = \frac{a_1 K}{\max(a_1 \omega, \Omega F_2)} \quad (9.80)$$

and $\Omega = (2\overline{W}_{ij}\overline{W}_{ij})^{1/2}$. Letting ϕ denote any one of the parameters $\gamma, \sigma_k, \sigma_\omega, \beta$, and eddy viscosity ν_t, then each of these varies between a near-wall state ϕ_1 and a far-wall state ϕ_2 according to formulas of the form

$$\phi = F_1 \phi_1 + (1 - F_1)\phi_2, \quad (9.81)$$

where

$$F_1 = \tanh arg_1^4 \quad (9.82)$$

and

$$arg_1 = \min\left[\max\left(\frac{\sqrt{K}}{\beta^* \omega d}, \frac{500\nu}{d^2 \omega}\right), \frac{4\rho\sigma_{\omega_2} K}{CD_{k\omega} d^2}\right], \quad (9.83)$$

where $arg_1 \geq 0$ and $0 \leq F_1 \leq 1$. Here d is distance from the boundary so that as d increases, the two expressions in the maximum in Eq. (9.83) become smaller as well as the term that the maximum is being compared to. Thus arg_1 diminishes with d, causing F_1 to approach zero and ϕ to approach the far-field value ϕ_2. The opposite behavior occurs as the wall is approached with $arg_1 \to \infty$, $F_1 \to 1$, and $\phi \to \phi_1$.

Additional parameters appearing in the above expressions are

$$CD_{K\omega} = \max\left(2\rho\sigma_{\omega_2}\frac{1}{\omega}\frac{\partial K}{\partial x_j}\frac{\partial \omega}{\partial x_j}, 10^{-20}\right), \quad (9.84)$$

$$F_2 = \tanh(arg_2^2), \quad (9.85)$$

and

$$arg_2 = \max\left(\frac{2\sqrt{K}}{\beta^* \omega d}, \frac{500\nu}{d^2 \omega}\right). \quad (9.86)$$

The constants are

$$\gamma_1 = \frac{\beta_1}{\beta^*} - \frac{\sigma_{\omega_1}\kappa^2}{\sqrt{\beta^*}} \quad (9.87)$$

$$\gamma_2 = \frac{\beta_2}{\beta^*} - \frac{\sigma_{\omega_2}\kappa^2}{\sqrt{\beta^*}} \quad (9.88)$$

and $\sigma_{k_1} = 0.85, \sigma_{k_2} = 1.0, \sigma_{\omega 1} = 0.5, \sigma_{\omega 2} = 0.856, \beta_1 = 0.075, \beta_2 = 0.0828, \beta^* = 0.09, \kappa = 0.41$, and $a_1 = 0.31$.

The close relationship between the SST model and the $K{-}\omega$ model is evident. That the SST model conforms to the $K{-}\epsilon$ model away from walls can be seen by first noting from Eq. (9.70) that nominally

$$\epsilon = C_\mu K\omega. \tag{9.89}$$

Taking a substantial derivative of Eq. (9.89) and substituting for the K and ω derivatives using Eqs. (9.78) and (9.79) gives a differential equation for ϵ of essentially the same form as Eq. (9.62). Substituting the far-field form of the constants into this expression yields an ϵ equation that conforms to that in the $K{-}\epsilon$ closure with only small differences (see Problem 9.6).

9.2.5 Spalart–Allmaras Model

To the extent that the eddy viscosity is the controlling factor in the accuracy of solutions to Eq. (9.2), two-equation models depend on the subtleties of how both K and ϵ or K and ω come together to create ν_t. The difficulty of controlling ν_t in this way has led to the development of a one-equation model by Spalart and Allmaras (SA) [4] in which a single additional differential equation for ν_t is solved in addition to Eq. (9.2).

Since the eddy viscosity does not satisfy its own conservation equation, a closed equation for ν_t has to be artificially created with a view toward producing the kind of distributions of ν_t that fit in well with a range of empirical data. To accomplish this the SA ν_t equation is built up term by term in a series of calibrations involving flows of increasing complexity. The resulting model has gone through a number of developmental iterations beyond its original form and has been widely tested in applications. Here, so as to contrast SA with other EVM, its standard form is written out. In this, the eddy viscosity is

$$\nu_t = f_{\nu_1}\tilde{\nu}, \tag{9.90}$$

where

$$f_{\nu_1} = \chi^3/(\chi^3 + c_{\nu_1}^3) \tag{9.91}$$

is a given empirical function

$$\chi \equiv \tilde{\nu}/\nu, \tag{9.92}$$

and $\tilde{\nu}$ is determined from solving the differential equation

$$\begin{aligned}\frac{D\tilde{\nu}}{Dt} &= c_{b_1}(1 - f_{t_2})\tilde{S}\tilde{\nu} - \left[c_{w_1}f_w - \frac{c_{b_1}}{\kappa^2}f_{t_2}\right]\left(\frac{\tilde{\nu}}{d}\right)^2 \\ &+ \frac{1}{\sigma}\left[\frac{\partial}{\partial x_j}\left((\nu + \tilde{\nu})\frac{\partial \tilde{\nu}}{\partial x_j}\right) + c_{b_2}\frac{\partial \tilde{\nu}}{\partial x_i}\frac{\partial \tilde{\nu}}{\partial x_i}\right].\end{aligned} \tag{9.93}$$

The various auxiliary functions are defined as

$$\tilde{S} = \sqrt{2\overline{W}_{ij}\overline{W}_{ij}} + \tilde{\nu}f_{\nu_2}/(\kappa^2 d^2)$$

$$f_{\nu_2} = 1 - \chi/(1 + \chi f_{\nu_1})$$

$$f_w = g[(1 + c_{w_3}^6)/(g^6 + c_{w_3}^6)]^{1/6}$$

$$
\begin{aligned}
g &= r + c_{w_2}(r^6 - r) \\
r &= \min[\tilde{\nu}/(\tilde{S}\kappa^2 d^2), 10] \\
f_{t_2} &= c_{t_3} e^{-c_{t_4}\chi^2}
\end{aligned}
\tag{9.94}
$$

where $\overline{W}_{ij}$ is defined in Eq. (9.22) and d is the distance to the closest wall. The nominal values of the constants are

$$
\begin{aligned}
c_{b_1} &= 0.1355 \\
\sigma &= 2/3 \\
c_{b_2} &= 0.622 \\
\kappa &= 0.41 \\
c_{w_1} &= c_{b_1}/\kappa^2 + (1 + c_{b_2})/\sigma \\
c_{w_2} &= 0.3 \\
c_{w_3} &= 2 \\
c_{v_1} &= 7.1 \\
c_{t_3} &= 1.2 \\
c_{t_4} &= 0.5.
\end{aligned}
\tag{9.95}
$$

Accompanying this basic specification of the model are boundary conditions such as the requirement that $\tilde{\nu} = 0$ at solid surfaces as well as a number of constraints, including the necessity for $\tilde{S}$ to be strictly positive. Many modifications to the approach have been developed for improved accuracy in special situations. It is beyond the scope of this discussion to delve into the calibrations that have gone into producing each term in the model and the choice of parameter values. Revised forms of the model continue to be developed, showing that it remains widely used for engineering calculations.

9.3 Tools for Model Development

To move beyond the use of what is also referred to as the Boussinesq model in Eq. 9.1 with its attendant limitations, it is desirable to have some formal methodologies available that can be applied to make model development at least somewhat systematic as against haphazard and purely ad hoc. While there is no one way to proceed in developing new closures, there are a number of steps that can be taken to direct and gain some control of the form and assessment of the models. In this section some of the traditional ideas that are commonly employed in this regard are presented, though further refinement of the theoretical basis for developing constitutive models continues to evolve (e.g., [17, 18]).

9.3.1 Invariance Properties of the Reynolds Stress Tensor

In developing constitutive laws such as one would like to derive for the Reynolds stress tensor, it is appropriate that model expressions be tensorially consistent with the quantity being modeled. For example, the way in which R_{ij} transforms from one reference frame to another should be preserved in the mathematical laws designed to represent the tensor. In the case of Eq. (9.1), to be self-consistent R_{ij} and $\overline{S}_{ij}$ should have the same transformation properties. In fact, it will be seen presently that this condition is violated. The penalty that can arise if the transformation properties are not satisfied is that the closure model will be incompatible with some classes of flows. The issue raised here is particularly relevant to the question of generalizing Eq. (9.1), since many of the types of flows for which such a simple eddy viscosity model fails are those for which the incompatibility of transform properties is germane. In developing formulas that generalize Eq. (9.1), the tensorial constraints on its form are also useful in reducing the number of possible model expressions out of which workable constituitive laws can be constructed.

To consider the invariance properties of R_{ij} imagine a turbulent flow that is being studied by two different observers, each with their own reference frame and moving relative to each other. Suppose that each observer proposes what they consider to be the same model based on the same physical ideas for modeling the Reynolds stress tensor. A fundamental question is: what constraints must be imposed on the mathematical models they propose so that they are consistent with each other? Specifically, how can it be guaranteed that the mathematical forms of the models given by each observer will naturally transform into each other in a mapping from one reference frame to the other? If a model does not satisfy this mapping, then the proposed physical law is observer dependent and open to the criticism of not being physical.

Since $R_{ij} = \overline{u_i u_j}$, it is evident that the transformation properties of R_{ij} are the same as that for the fluctuating velocity vector, u_i. The later transforms according to rules established by its own governing equation. Thus, consider an inertial observer who measures positions according to $\mathbf{x}$ and velocities according to $\mathbf{U}(\mathbf{x}, t)$ in their own reference frame. A second observer is arbitrarily accelerating and in their own reference frame measures positions $\mathbf{x}'$ and velocities as $\mathbf{U}'(\mathbf{x}', t)$. It may be shown (see [19]) that the positions of a particular fluid particle seen by the observers and denoted, respectively, by $\mathbf{x}(t)$ and $\mathbf{x}'(t)$ are related via

$$\mathbf{x}(t) = \mathbf{r}(t) + Q(t)\mathbf{x}'(t), \tag{9.96}$$

where $\mathbf{r}(t)$ is the position vector connecting the origin of the two reference frames and $Q(t)$ is a rotation tensor that compensates for the different orientations of the observers. Q is an orthogonal tensor so that $Q^{-1} = Q^t$ represents rotation in the opposite sense as Q. Clearly,

$$QQ^t = I. \tag{9.97}$$

The vector

$$\mathbf{x}^*(t) = Q(t)\mathbf{x}'(t) \tag{9.98}$$

may be interpreted as being the position vector seen by the second observer and expressed in the orientation of the inertial observer, so that Eq. (9.96) becomes

$$\mathbf{x} = \mathbf{r}(t) + \mathbf{x}^*. \tag{9.99}$$

If there is no relative rotation between the observers, then $Q(t) = I$. If the second observer is inertial, then Eq. (9.99) reduces to

$$\mathbf{x} = \mathbf{x}^* + \mathbf{U}_c t, \tag{9.100}$$

where $\mathbf{U}_c$ is a constant, translational velocity.

Letting an over dot denote time differentiation, it may be shown from Eq. (9.97) that $(\dot{Q}Q^{-1})^t = -\dot{Q}Q^{-1}$ so that $\dot{Q}Q^{-1}$ is a skew-symmetric tensor. In this case it may be shown [19] that an axial vector $\mathbf{\Omega}_0$ exists such that

$$(\dot{Q}Q^{-1})\mathbf{v} = \mathbf{\Omega}_0 \times \mathbf{v} \tag{9.101}$$

for all vectors $\mathbf{v}$. $\mathbf{\Omega}_0$ represents the instantaneous angular velocity or rotation rate of the non-inertial coordinate system. Taking a time derivative of Eq. (9.96) and using (9.101) yields the velocity transformation

$$\mathbf{U}(t) = \dot{\mathbf{r}}(t) + \mathbf{U}^*(t) + \mathbf{\Omega}_0 \times \mathbf{x}^*, \tag{9.102}$$

where $\mathbf{U}^* = Q\mathbf{U}'$ and $\mathbf{U}'$ is the velocity seen by the second observer. A time derivative of Eq. (9.102) yields the transformation law for accelerations in the form

$$\mathbf{a} = \ddot{\mathbf{r}} + 2(\mathbf{\Omega}_0 \times \mathbf{U}^*) + (\mathbf{\Omega}_0 \times (\mathbf{\Omega}_0 \times \mathbf{x}^*)) + \dot{\mathbf{\Omega}}_0 \times \mathbf{x}^* + \mathbf{a}^*. \tag{9.103}$$

The acceleration $\mathbf{a}$ refers to that of a material fluid element, which is given in terms of $\mathbf{U}$ as $\mathbf{a} = D\mathbf{U}/Dt$ and similarly for $\mathbf{a}^* = D\mathbf{U}^*/Dt$. With this interpretation, Eq. (9.103) may be substituted into the Navier–Stokes equation producing a relation valid for arbitrary observers in the form

$$\begin{aligned}\frac{\partial \mathbf{U}^*}{\partial t} + (\nabla \mathbf{U}^*)\mathbf{U}^* &= -\frac{1}{\rho}\nabla P^* + \nu\nabla^2\mathbf{U}^* - \ddot{\mathbf{r}} - 2(\mathbf{\Omega}_0 \times \mathbf{U}^*)\\ &\quad -(\mathbf{\Omega}_0 \times (\mathbf{\Omega}_0 \times \mathbf{x}^*)) - (\dot{\mathbf{\Omega}}_0 \times \mathbf{x}^*).\end{aligned} \tag{9.104}$$

The last four terms on the right-hand side of Eq. (9.104) account for rectilinear accelerations of the reference frame, the Coriolis force, the centrifugal force, and the effect of changes in rotation rate, respectively.

Subtracting the average of Eq. (9.104) from itself yields an equation for the velocity fluctuation in the form

$$\begin{aligned}\frac{\partial \mathbf{u}^*}{\partial t} + (\nabla \mathbf{u}^*)\overline{\mathbf{U}}^* &= -(\nabla \mathbf{u}^*)\mathbf{u}^* - (\nabla \overline{U}^*)\mathbf{u}^* - \frac{1}{\rho}\nabla p^* + \nu\nabla^2\mathbf{u}^*\\ &\quad -\nabla \cdot R^* - 2(\mathbf{\Omega}_0 \times \mathbf{u}^*),\end{aligned} \tag{9.105}$$

where $R^* = \overline{\mathbf{u}^* \otimes \mathbf{u}^*}$ and $\nabla \cdot R^*$ is the tensor divergence of R^*, that is, a vector with components $\partial u_i^* u_j^*/\partial x_j$. The presence of the Coriolis term in Eq. (9.105) is all that prevents the velocity fluctuation from satisfying the identical equation for all observers. Observers who are not rotating so that $\mathbf{\Omega}_0 = 0$ and consequently experience at most a rectilinear acceleration, share the identical velocity fluctuation equation. This means that any proposed Reynolds stress model should have the same tensor form depending on $\nabla\overline{\mathbf{U}}$, for example, for all observers experiencing rectilinear accelerations. This is referred to as extended Galilean invariance and is a significant constraint on the allowable tensor forms of Reynolds stress models. Note that in contrast to extended Galilean invariance, material frame indifference (MFI) is when a tensor property has the identical form for all observers, including those who are rotating.

Applying constraints such as extended Galilean invariance to the development of Reynolds stress models has the advantage of reducing the complexity of the resulting models. For example, since R_{ij} satisfies extended Galilean invariance so too must the quantities on which it depends. The velocity field $\overline{\mathbf{U}}$ does not satisfy this condition so R_{ij} should not depend on it directly. On the other hand, the mean velocity gradient tensor in Eq. (9.105) does satisfy extended Galilean invariance, as can be ascertained by differentiating Eq. (9.102) and taking note of Eq. (9.99) to yield

$$(\nabla \mathbf{U})^*_{ij} = (\nabla \mathbf{U})_{ij} + \epsilon_{ijk}\Omega_{0_k}. \tag{9.106}$$

Thus, it is reasonable for R_{ij} to depend on $\nabla\overline{\mathbf{U}}$. Returning to Eq. (9.1), the right-hand side contains twice the mean rate of strain tensor (i.e., $2\overline{S}_{ij} = \partial \overline{U}_i/\partial x_j + \partial \overline{U}_j/\partial x_i$). It may be shown from the transformation property of $(\nabla\overline{\mathbf{U}})$ in Eq. (9.106) that $\overline{S}_{ij}$ satisfies MFI so that the left- and right-hand sides of Eq. (9.1) have different invariance properties. In the case of system rotation, the left-hand side of Eq. (9.1) responds to the change while the right-hand side is unaffected.

A similar analysis of the rotation tensor $\overline{W}_{ij}$ from its definition shows that it has the same invariance properties as $\nabla\overline{\mathbf{U}}$ given in Eq. (9.106) so that it satisfies extended Galiliean invariance and not MFI. Since

$$\nabla\overline{\mathbf{U}}^* = \overline{S}^* + \overline{W}^*, \tag{9.107}$$

both sides of this relation are sensitive to changes in the rotation tensor. By generalizing Eq. (9.1) to depend on both $\overline{S}$ and $\overline{W}$ individually, the Reynolds stress model is made to be consistent with invariance properties of R_{ij}.

For the special case of 2D turbulence it may be shown [20] that the Coriolis term in Eq. (9.105) can be written as a gradient of a potential that can be absorbed inside the pressure term. This means that **u** in this case satisfies MFI since the form of its governing equation does not change from one rotating observer to the next. Consequently, Reynolds stress models should ideally satisfy MFI in the limit of 2D turbulence [21]. Some turbulent flows approximate 2D conditions, as, for example, rapidly rotating systems for which the Taylor–Proudman theorem [22] shows that the motion becomes largely 2D. 2D turbulence is also approximated to some extent near solid boundaries where the wall-normal Reynolds stress component is much smaller than the wall-parallel components.

To develop a constitutive model in which MFI is enforced it is helpful to rewrite Eq. (9.105) in the form

$$\frac{\partial \mathbf{u}^*}{\partial t} + (\nabla \mathbf{u}^*)\overline{\mathbf{U}}^* - (\nabla\overline{\mathbf{U}}^*)\mathbf{u}^* = -(\nabla \mathbf{u}^*)\mathbf{u}^* - 2(\overline{S}^* + \overline{W}^a)\mathbf{u}^* - \frac{1}{\rho}\nabla p^* + \nu\nabla^2\mathbf{u}^* - \nabla\cdot R^*, \tag{9.108}$$

where $\overline{W}^a_{ij} \equiv \overline{W}^*_{ij} + \epsilon_{jik}\Omega_{0_k}$ is the absolute mean vorticity tensor. The left-hand side of Eq. (9.108) is the frame indifferent "Oldroyd" derivative [23] of $\mathbf{u}^*$

$$\frac{D_o \mathbf{u}^*}{Dt} \equiv \frac{\partial \mathbf{u}^*}{\partial t} + (\nabla \mathbf{u}^*)\overline{\mathbf{U}}^* - (\nabla\overline{\mathbf{U}}^*)\mathbf{u}^*, \tag{9.109}$$

whose form satisfies MFI as does $\overline{S}^*$ and $\overline{W}^a_{ij}$ as previously noted. Consequently, if Reynolds stress models are constructed using $D_o\mathbf{u}^*/Dt$, $\overline{S}^*$ and $\overline{W}^a$, then they can be constrained to be frame indifferent. This property has sometimes been exploited in deriving models.

At a practical level the advantages of satisfying MFI in the 2D limit may not materialize in the treatment of general flows which are fully 3D. In fact, flows that mimic 2D turbulence may be far removed from the general 3D state of turbulent flows, so that enforcing MFI in the 2D limit may actually be harmful by overly restricting the forms of models and thereby reducing their range of applicability.

9.3.2 Realizability

As additional complexity is brought into Reynolds stress modeling so that more physics can be accounted for than is possible with Eq. (9.1), the likelihood increases that solutions will display such unphysical properties as negative energy. This is especially a concern in the case of non-linear constitutive laws that may include multiple solutions to the governing differential equations. The idea of designing closure models that are a priori prevented from having negative energy is one aspect of what is referred to as enforcing *realizability*. More broadly, with a realizable closure one would like to prevent negative values of the normal Reynolds stress components R_{11}, R_{22}, and R_{33} from occurring, as well as guaranteeing that the of-diagonal components, R_{12}, R_{23}, and R_{13} satisfy the Schwarz inequality, to the effect that

$$|R_{\alpha\beta}| \leq \sqrt{\overline{u_\alpha^2}}\sqrt{\overline{u_\beta^2}} \tag{9.110}$$

for $\alpha \neq \beta$. The fact of satisfying realizability does not guarantee that a closure will be superior in a practical sense to closures that are not realizable. For example, insisting that transient solutions to the closure equations be realizable may have the unintended consequence of causing the steady-state solution to be inaccurate. It is even possible that a realizable steady-state solution of interest cannot be reached via a time integration that is not allowed to transient through non-realizable states. In such cases, the added requirement of realizability may be harmful and not helpful to the modeling process. Since there also may be more than one way to design realizability into models [24], there is some degree of arbitrariness in selecting one particular realizable form over another.

9.3.3 Rapid Distortion Theory

Another approach for limiting the possible form of closures is by enforcing their compatibility with rapid distortion theory (RDT) [25]. RDT describes the short-time linear response of the turbulent field to sudden changes in the external conditions that affect the mean field. In the RDT limit quadratic terms consisting of products of the velocity fluctuation vector and its derivatives are assumed to be negligible in comparison to the remaining terms so that a strictly linear analysis of $\mathbf{u}$ is possible. For example, fluid in turbulent motion in a pipe would experience a rapid change in circumstances if it were to pass through a narrowing exit nozzle. In such situations, if the time scale associated with significant changes in the mean field is much smaller than that associated with non-linear processes such as energy redistribution and decay, then the latter phenomena can be neglected. In other words, self-interactions of the turbulent field with itself are deemed to occur at a slow rate in RDT compared to the rapid effect of the mean flow on the turbulence so that their neglect can be justified.

To quantify the necessary conditions for RDT to be relevant, consider initially homogeneous turbulence which is suddenly subjected to a mean shearing $S = d\overline{U}/dy > 0$. One may pursue a RDT analysis if

$$\frac{1}{S} << \frac{K}{\epsilon}, \tag{9.111}$$

which states that the time scale of the shearing is much less than the eddy turnover time. Without the non-linear term, the velocity fluctuation equation for incompressible flow derived similar to Eq. (9.105) is

$$\rho\left(\frac{\partial \mathbf{u}}{\partial t} + (\nabla \mathbf{u})\overline{\mathbf{U}}\right) = -\rho(\nabla\overline{\mathbf{U}})\mathbf{u} - \nabla p + \mu\nabla^2\mathbf{u}. \tag{9.112}$$

This is a linear equation in **u** which can be analyzed in many fruitful ways, particularly if the mean field is of a relatively simple structure. Many examples of such analyses are available in the literature. One particular situation which is often used in model calibration is that in which an initially isotropic turbulent flow is suddenly subjected to mean shear [26]. RDT analysis in this case leads to a specific form of the pressure-strain term Π defined in Eq. (3.54) that can be used in developing models. In another example [27], it may be shown that an initially isotropic turbulent flow subjected to a sudden rotation remains isotropic. It is generally the case that the possibility of gaining consistency with RDT is reserved for second moment closures [28], since simpler models, such as EVM, contain a predetermined instantaneous response to changes in the mean field, as is evident from Eq. (9.1). Whether or not it is always wise to be concerned with the RDT limit when designing SMC is not entirely clear. As in the case of realizability, the RDT limit may be far removed from the circumstances of a flow of interest, so forcing compatibility with the RDT limit might hinder rather than help the performance of the method. For this reason, consistency with RDT predictions is not a universal goal of model development.

9.4 Non-Linear Eddy Viscosity Models

Alternatives to the EVM assumption for the most part posit the appropriateness of non-linear (usually quadratic) tensor forms in modeling R_{ij} that are consistent with the constraints described in Section 9.3. A variety of strategies have been employed for reducing the complexity of the most general tensor expressions to practical form by invoking ideas from realizability, RDT, and other approaches. The end result of such development is several closely related NLEVM schemes, some of whose particular attributes will be described here.

A useful starting point for constructing non-linear generalizations of Eq. (9.1) is to hypothesize that

$$R(\mathbf{x}, t) = F(\overline{\mathbf{U}}(\mathbf{y}, s) - \overline{\mathbf{U}}(\mathbf{x}, s), K, \epsilon), \tag{9.113}$$

where $s \leq t$ represents times up to and including the present at time t and $\mathbf{y}$ represents an arbitrary point in the flow domain. The velocity difference between two points is used here, as against the velocity at a single point, since the former satisfies extended Galilean invariance consistent with R. To generate non-linear models from Eq. (9.113) one approach [29] is to substitute a truncated Taylor series expansion for $\overline{U}(\mathbf{y}, s) - \overline{U}(\mathbf{x}, s)$, and then require form invariance under the extended Galilean group. Through second-order terms in $\overline{S}$ and $\overline{W}$ this yields

$$\begin{aligned} R = {} & \frac{2}{3}KI - 2C_\mu\frac{K^2}{\epsilon}\overline{S} + A_1\frac{K^3}{\epsilon^2}\left(\overline{S}^2 - \frac{1}{3}|\overline{S}|^2 I\right) + A_2\frac{K^3}{\epsilon^2}\left(\overline{W}^2 - \frac{1}{3}|\overline{W}|^2 I\right) \\ & + A_3\frac{K^3}{\epsilon^2}(\overline{W}\,\overline{S} - \overline{S}\,\overline{W}) + A_4\frac{K^3}{\epsilon^2}\frac{D\overline{S}}{Dt}, \end{aligned} \tag{9.114}$$

where

$$\left(\frac{D\overline{S}}{Dt}\right)_{ij} \equiv \frac{\partial \overline{S}_{ij}}{\partial t} + \overline{U}_k \frac{\partial \overline{S}_{ij}}{\partial x_k} \tag{9.115}$$

and it may be noticed that the first two terms on the right-hand side of Eq. (9.114) are equivalent to Eq. (9.1) with ν_t given by Eq. (9.29). The expressions containing K and ϵ in front of the various terms are inserted to ensure dimensional consistency. Equation (9.114) is a general form out of which a number of different models have been derived [30].

An alternative methodology for deriving NLEVM is one that is commonly employed in deriving expressions used in algebraic Reynolds stress models, as will be seen in Section 9.6. In this, the starting point is a hypothesis as to the functional dependence of R that is consistent with extended Galilean invariance. In this case it may be assumed that

$$R = F(\overline{S}, \overline{W}, K, \epsilon). \tag{9.116}$$

For any given observer it is a property of tensors that Eq. (9.116) must transform under rotations of the coordinate system according to

$$QRQ^{-1} = F(QSQ^{-1}, QWQ^{-1}, K, \epsilon), \tag{9.117}$$

where Q is an orthogonal tensor. Tensor functions of tensors that satisfy Eq. (9.117) are isotropic and are restricted to being functions of a limited set of tensors known as the integrity basis of F [31]. In this instance the integrity basis consists of ten independent tensors. Among these are the tensors on the right-hand side of Eq. (9.114) apart from the last term containing the substantial derivative. The integrity basis tensors include both cubic and quartic combinations of $\overline{S}$ and $\overline{W}$ that can be added to Eq. (9.114) to create a more general expression. The omitted cubic terms in Eq. (9.114), which are included in some models [32], have the form

$$\begin{aligned} &A_5 \frac{K^4}{\epsilon^3} \left[\overline{W}^2 \overline{S} + \overline{S}\,\overline{W}^2 + |\overline{W}|^2 \overline{S} - \frac{2}{3}\, \mathrm{tr}(\overline{W}\,\overline{S}\,\overline{W}) I \right] \\ &\quad + A_6 \frac{K^4}{\epsilon^3} (\overline{W}\,\overline{S}^2 - \overline{S}^2 \overline{W}). \end{aligned} \tag{9.118}$$

These provide potentially new ways of accounting for such effects as mean streamline curvature and swirl that are not necessarily attainable via Eq. (9.114).

While the coefficients $A_i, i = 1 \ldots 4$ in Eq. (9.114) are nominally constants, for increased generality they may be assumed to be functions of tensor invariants of $\overline{S}$ and $\overline{W}$ such as $|\overline{S}|^2$ and $|\overline{W}|^2$ whose values are the same for all observers. In fact, by this means additional non-linear effects can be readily brought into Eq. (9.114) without introducing additional tensor forms. This will be seen to be an important consideration in the development of ARM methods.

In another approach toward generalizing Eq. (9.114), the condition that MFI be enforced in the limit of 2D turbulence [33] has been applied to the hypothesis that

$$R = F\left(\overline{S}, \frac{D_o \overline{S}}{Dt}, K, \epsilon\right), \tag{9.119}$$

where the Oldroyd derivative associated with $\overline{S}$, namely,

$$\frac{D_o\overline{S}_{ij}}{Dt} = \frac{\partial \overline{S}_{ij}}{\partial t} + \overline{U}_k \frac{\partial \overline{S}_{ij}}{\partial x_k} - \frac{\partial \overline{U}_i}{\partial x_k}\overline{S}_{kj} - \frac{\partial \overline{U}_j}{\partial x_k}\overline{S}_{ki}, \tag{9.120}$$

is included in Eq. (9.119) since it contains quadratic terms in the mean velocity that are consistent with the goal of developing a model that is quadratic in the mean velocity and its derivatives. After applying the requirement that Eq. (9.119) be isotropic it is found that

$$\begin{aligned} R = {} & \frac{2}{3}KI - 2C_\mu \frac{K^2}{\epsilon}\overline{S} - 4C_D C_\mu^2 \frac{K^3}{\epsilon^2}\left(\overline{S}^2 - \frac{1}{3}|\overline{S}|^2 I\right) \\ & - 4C_E C_\mu^2 \frac{K^3}{\epsilon^2}\left(\frac{D_o\overline{S}}{Dt} - \frac{1}{3}\frac{D_o \operatorname{tr}\overline{S}}{Dt} I\right), \end{aligned} \tag{9.121}$$

in which no higher than quadratic terms have been kept. This relatively compact expression is commonly referred to as the "non-linear K–ϵ model" [33] and is among the more commonly used closures of this type. Calibrating against experimental measurements of the normal Reynolds stresses in channel flow gives $C_D = C_E = 1.68$.

In summary, Eqs. (9.114) and (9.121) contain as special cases most of the quadratic generalizations of Eq. (9.1) which have been derived. Such models offer greater flexibility in comparison to Eq. (9.1) without the need for significantly greater numerical effort in calculating mean flow fields than is involved in applying the standard K–ϵ closure. NLEVM closures based on these approaches can accommodate a greater range of physical phenomena than can be modeled with Eq. (9.1). For example, such models can accommodate situations where it is essential to include anisotropic effects, as in secondary flow in the corner of a duct flow. In addition, low Reynolds number modifications to NLEVM have been developed to account for flow near boundaries. Despite the many apparent advantages of NLEVM as a methodology, this approach is nonetheless local in nature and cannot be expected to completely capture the non-local physics of turbulent transport. This has encouraged the development of closures to the Reynolds stress equations themselves, since by their very nature these provide what is, in effect, non-local representations of the Reynolds stresses.

9.5 Reynolds Stress Equation Models

Reynolds stress equation (RSE) models, also known as second moment closures (SMC), effect a closure to the exact Reynolds stress equation

$$\frac{\partial R_{ij}}{\partial t} + \overline{U}_k \frac{\partial R_{ij}}{\partial x_k} = -R_{ik}\frac{\partial \overline{U}_j}{\partial x_k} - R_{jk}\frac{\partial \overline{U}_i}{\partial x_k} - \epsilon_{ij} - \frac{\partial \beta_{ijk}}{\partial x_k} + \Pi_{ij} + \nu\nabla^2 R_{ij}, \tag{9.122}$$

that was previously derived in Eq. (3.53). To accomplish this, model expressions are sought for the pressure-strain correlation Π_{ij} defined in Eq. (3.54), the "transport" correlation β_{ijk} defined in Eq. (3.55), and the anisotropic dissipation rate ϵ_{ij} defined in Eq. (9.39). Since $K = R_{ii}/2$ it can be expected that there will be many similarities between modeling the R_{ij} and K equations. In fact, the most noteworthy difference between the equations for K and those of the individual R_{ii} is the pressure-strain term, which has no counterpart in the K equation. Since this term is responsible for the anisotropy of the Reynolds stresses, and the desire to compute such effects is a rationale for developing RSE closures in the first place, the modeling of the pressure-strain term is a central focus of most efforts at modeling Eq. (9.122).

9.5.1 Modeling of the Pressure-Strain Correlation

Some insight into what to consider when modeling the pressure-strain correlation can be obtained by expanding out its exact expression via a substitution for the pressure deriving from the Navier–Stokes equation. Thus, taking the divergence of the Navier–Stokes equation (3.6) and subtracting from it the divergence of the mean velocity equation (9.2) yields, after some simplification, the following Poisson equation for the pressure fluctuation field:

$$\frac{1}{\rho}\nabla^2 p = -2\frac{\partial u_j}{\partial x_i}\frac{\partial \overline{U}_i}{\partial x_j} - \frac{\partial u_i}{\partial x_j}\frac{\partial u_j}{\partial x_i}. \tag{9.123}$$

For an infinite domain without boundaries the fundamental solution of Eq. (9.123) is (e.g., see [34])

$$p(\mathbf{x},t) = \frac{\rho}{4\pi}\int_{\Re^3}\frac{1}{|\mathbf{x}-\mathbf{x}'|}\left(2\frac{\partial u_j}{\partial x_i}(\mathbf{x}')\frac{\partial \overline{U}_i}{\partial x_j}(\mathbf{x}') + \frac{\partial u_i}{\partial x_j}(\mathbf{x}')\frac{\partial u_j}{\partial x_i}(\mathbf{x}')\right)d\mathbf{x}'. \tag{9.124}$$

If boundaries are present Eq. (9.124) must be modified by the addition of terms containing surface integrals. Regardless of the circumstances, however, it is usual to consider just Eq. (9.124) when analyzing the pressure-strain term with a view towards its modeling.

Substituting Eq. (9.124) into Π_{ij} gives

$$\Pi_{ij} = A_{ij} + M_{ijkl}\frac{\partial \overline{U}_k}{\partial x_l}, \tag{9.125}$$

where

$$A_{ij}(\mathbf{x}) = \frac{1}{4\pi}\int_{\Re^3}\frac{1}{|\mathbf{x}-\mathbf{x}'|}\overline{\left(\frac{\partial u_i}{\partial x_j}(\mathbf{x}) + \frac{\partial u_j}{\partial x_i}(\mathbf{x})\right)\frac{\partial u_k}{\partial x_l}(\mathbf{x}')\frac{\partial u_l}{\partial x_k}(\mathbf{x}')}d\mathbf{x}' \tag{9.126}$$

and

$$M_{ijkl}(\mathbf{x}) = \frac{1}{2\pi}\int_{\Re^3}\frac{1}{|\mathbf{x}-\mathbf{x}'|}\overline{\left(\frac{\partial u_i}{\partial x_j}(\mathbf{x}) + \frac{\partial u_j}{\partial x_i}(\mathbf{x})\right)\frac{\partial u_l}{\partial x_k}(\mathbf{x}')}d\mathbf{x}'. \tag{9.127}$$

Note that the second term in Eq. (9.125) is strictly speaking valid only in homogeneous shear flow, since the mean velocity gradient term in Eq. (9.124) has been taken outside of the integral in order to produce a more tractable expression. The first of the terms in Eq. (9.125) is referred to as the "slow" term and the second as the "fast" term. This nomenclature refers to the fact that if one were to suddenly change the mean velocity field, then it is the second of the terms in Eq. (9.125) that would immediately respond to the change, while the first one, since it does not depend explicitly on the mean velocity field, can be expected to adapt more slowly as the turbulent velocity field evolves as a whole.

One has the option of modeling Π_{ij} as one entity, as will be done here, or via separate models for the fast and slow parts. In either case the main use of Eq. (9.125) is as a motivation in setting up dimensional models, so the end result is essentially independent of whether or not A_{ij} and M_{ijkl} are treated separately. Moreover, if one models the pressure-strain correlation as a whole it is often clear by inspection which terms in the model may be attributed to the fast and slow parts of Eq. (9.125), if one wishes to do so.

The modeling of Π_{ij} depends on a hypothesis as to which mean characteristics of the turbulent flow field it should functionally depend upon. In view of its role in affecting the extent to which the flow is isotropic, it is natural to insist that it depends on the tensor b defined in Eq. (9.42). Its dependence on mean velocity gradients is also apparent from Eq. (9.125). Thus, one is led to the assumption that Π is a tensor function of the tensors $b, \overline{S}$, and $\overline{W}$, specifically

$$\frac{\Pi}{\epsilon} = F\left(b, \frac{K}{\epsilon}\overline{S}, \frac{K}{\epsilon}\overline{W}\right), \tag{9.128}$$

where the factor K/ϵ is used to non-dimensionalize the mean velocity derivative tensors. Since Eq. (9.125) is linear in the mean velocity gradient, the functional F is limited to terms which are linear in its second two arguments. Moreover, since the trace of the left-hand side of Eq. (9.128) is zero, so too must that of the right-hand side, and F must be chosen accordingly.

As in Eq. (9.116), the model in Eq. (9.128) should be isotropic so that for an arbitrary orthogonal tensor Q,

$$\frac{Q\Pi Q^{-1}}{\epsilon} = F\left(QbQ^{-1}, \frac{K}{\epsilon}Q\overline{S}Q^{-1}, \frac{K}{\epsilon}Q\overline{W}Q^{-1}\right). \tag{9.129}$$

The most general form of Π, non-linear in b, and linear in $\overline{S}, \overline{W}$ that satisfies Eq. (9.129) can be constructed from the integrity bases for this case and is given by [35]

$$\begin{aligned}\Pi = {} & A_0\epsilon b + A_1\epsilon\left(b^2 - \frac{1}{3}|b|^2 I\right) + A_2 K\overline{S} + (A_3\,\mathrm{tr}(b\overline{S}) + A_4\,\mathrm{tr}(b^2\overline{S}))Kb \\ & + (A_5\,\mathrm{tr}(b\overline{S}) + A_6\,\mathrm{tr}(b^2\overline{S}))K\left(b^2 - \frac{1}{3}|b^2|I\right) + A_7 K(b\overline{S} + \overline{S}b \\ & - \frac{2}{3}\,\mathrm{tr}(b\overline{S})I) + A_8 K\left(b^2\overline{S} + \overline{S}b^2 - \frac{2}{3}\,\mathrm{tr}(b^2\overline{S})I\right) + A_9 K(\overline{W}b - b\overline{W}) \\ & + A_{10}K(\overline{W}b^2 - b^2\overline{W}).\end{aligned} \tag{9.130}$$

The coefficients $A_0 \dots A_{10}$ are most generally functions of the invariants of the real symmetric tensor b given by $I \equiv \mathrm{tr}(b) = 0$, $II \equiv \mathrm{tr}(b^2)$, and $III \equiv \mathrm{tr}(b^3)$. As it is, Eq. (9.130) is a formidable expression whose use in closure schemes is not practical nor necessary [36]. In practice, for any particular class of flows, such as those for which $\overline{S} \neq 0$ and $\overline{W} \neq 0$, it can be shown that a subset of the integrity basis (five elements in the case of 3D flow) can be used as independent basis functions to represent the Π tensor. In addition, the coefficients of the terms in Eq. (9.130) can be made to depend on the invariants of b, which aids in accommodating complex flow physics such as streamline curvature and system rotation.

9.5.2 LRR Model

Among the strategies that have been devised for simplifying and calibrating Eq. (9.130) is one due to Launder, Reece, and Rodi (LRR) [37] in which linearity in b is assumed so that $A_1 = A_3 = A_4 = A_6 = A_8 = A_{10} = 0$. Such an assumption is based on the belief that the magnitude of b is sufficiently small so that quadratic and higher terms can be neglected. Besides the fact that the components of b are smaller than unity, as noted previously in regard to Eq. (9.42), a stronger bound on the components of b can be found by considering the magnitude of its eigenvectors. In particular, since b is real and symmetric it has an orthogonal basis of real eigenvectors $\mathbf{b}_i, i = 1, 2, 3$ with eigenvalues $b_i, i = 1, 2, 3$. These are bounded according to

$$-\frac{1}{3} \leq b_i \leq \frac{2}{3}, \quad i = 1, 2, 3, \tag{9.131}$$

with the lower limit deriving from the fact that the eigenvalues of the positive semi-definite, real, symmetric matrix R must be non-negative (see Problem 9.8). The upper limit follows from an application of the Schwarz inequality to R_{ij}. Furthermore, experiments show that the largest of the three eigenvalues has magnitude less than 0.25. In the coordinate system corresponding to the eigenvectors, b is diagonal with b_1, b_2, b_3 residing along the main diagonal. If Eq. (9.130) is expressed in the coordinate system associated with $\mathbf{b}_i, i = 1, 2, 3$, then the terms containing b introduce powers of $b_i, i = 1, 2, 3$. In view of Eq. (9.131) it can be argued that the assumption of linearity in Eq. (9.130) may have some justification since quadratic and higher terms contain small numerical factors.

With these considerations the LRR model has the basic form

$$\Pi = -C_1 \epsilon b + C_2 K \overline{S} + C_3 K \left(b\overline{S} + \overline{S} b - \frac{2}{3} \operatorname{tr}(b\overline{S}) I \right) + C_4 K (\overline{W} b - b \overline{W}), \tag{9.132}$$

where the constants have been renamed for convenience. It may be noticed that the only term in Eq. (9.132) that is non-zero when $b = 0$ is that with coefficient C_2. Turbulence with vanishing b is homogeneous and isotropic, and in this case $\overline{S} = 0$ as well, so Eq. (9.132) is compatible with this limiting flow. For a situation in which the flow is initially homogeneous and isotropic, and suddenly a non-zero value of shearing as given by $\overline{S}$ is applied, then the flow evolves at first within the RDT regime toward an anisotropic state with $b \neq 0$. Evidently, it is only the term with C_2 in Eq. (9.132) that can accommodate this kind of flow development in the LRR model.

The situation that just been described – in which a non-zero mean shearing is suddenly applied to homogeneous isotropic turbulence – is one for which Π can be determined. In particular [26], with the help of the incompressible flow identities

$$M_{iikl} = M_{ijkk} = 0 \tag{9.133}$$

and the homogeneous shear flow relations

$$M_{ikkj} = M_{jkki} = 2R_{ij}, \tag{9.134}$$

(see Problem 9.13) it may be shown that

$$\Pi_{fast} \sim \frac{4}{5} K \overline{S} \tag{9.135}$$

in these circumstances. Thus Eq. (9.132) is compatible with (9.135) so long as $C_2 = 4/5$. Eq. (9.135) is referred to as the *Crow constraint*.

The opposite situation to that in which Eq. (9.135) holds occurs when the shearing that produces an initially homogeneous, anisotropic turbulence is suddenly removed, resulting in a slow return to isotropy. This is a phenomenon that cannot be modeled with a NLEVM since the instant that shear is removed such models predict that the flow reverts instantaneously to an isotropic state. To model this flow in the context of the LRR model, note that Eq. (9.132) reduces in this case to

$$\Pi_{slow} = -C_1 \epsilon b, \tag{9.136}$$

which is a linear model for the return to isotropy. Such a linear model was originally introduced by Rotta [38], and in the current notation C_1 is known as the Rotta constant. Comparisons with experiment suggest that within the limitations of this model $C_1 \approx 3$ is an optimal value. The LRR model has been calibrated for this flow with values of C_1 varying from 2.8 to 3.6. If Eq. (9.136) is used in (9.122) it can be shown that the computed solution for each separate component of b decays according to a power law at the same rate when the mean shear is removed. To accommodate the likely possibility that different components of b decay at different rates models for Π must include the non-linear part of the slow term in Eq. (9.130), specifically the term with coefficient A_1.

The determination of the values of C_3 and C_4 in the LRR model reflect the way in which the general form of the model in Eq. (9.132) is developed. In particular, a linear tensorial relation is postulated between M_{ijkl} and R_{ij}, with Eq. (9.134) as motivation. By using Eq. (9.133) the model can be simplified and when substituted into Eq. (9.125) yields terms identical to those containing C_3 and C_4 in Eq. (9.132), but with the simplification that

$$C_3 = \frac{18C + 12}{11} \tag{9.137}$$

and

$$C_4 = \frac{20 - 14C}{11}, \tag{9.138}$$

where C is the sole constant to be determined.

Another aspect of model calibration, which can be applied generally to all Reynolds stress equation models and specifically to determining the constants C and C_1 in the LRR model, is to select parameters that allow an optimal prediction of the equilibrium state of homogeneous shear flow. For this purpose, Eq. (9.122) and its trace, which is an equation for K, can be used to develop a dynamical equation for b from its definition in Eq. (9.42) by substituting into the relation

$$\frac{Db_{ij}}{Dt} = \frac{1}{2}\left(\frac{K\frac{DR_{ij}}{Dt} - \frac{2}{3}K\frac{DK}{Dt}\delta_{ij} - (R_{ij} - \frac{2}{3}K\delta_{ij})\frac{DK}{Dt}}{K^2}\right). \tag{9.139}$$

The asymptotic limit of b at large times consists of an equilibrium state with constant components of b_{ij}, which occurs for both scenarios discussed previously in Section 6.5, that is, the equilibrium states of b are the same whether or not K and the components of R_{ij} grow exponentially at large times or converge to constant values.

Calibration of RSE models using the b_{ij} equation is done by dropping the convection term on the left-hand side of Eq. (9.122) and the transport and viscous diffusion terms on the right-hand side, and similarly for the K equation. In addition, the viscous dissipation term in the R_{ij} equation is given the simple isotropic model in Eq. (9.41) and the pressure-strain term is replaced by whatever model is being considered. In the case of the LRR model Eq. (9.132) is used together with Eqs. (9.137) and (9.138).

The equilibrium state of the system is achieved when the transient term on the left-hand side of Eq. (9.139) is zero. For a homogeneous shear flow $\overline{U}(y)$, the only non-zero components of $\overline{S}$ are $\overline{S}_{12} = \overline{S}_{21} = S/2$ where $S \equiv d\overline{U}/dy$ is constant everywhere and the only non-zero components of $\overline{W}$ are $\overline{W}_{12} = -\overline{W}_{21} = S/2$. Applying these relations to the equations for b_{11}, b_{22}, b_{33} in the case of LRR yields simple algebraic equations whose solution is

$$\begin{aligned}
(\overline{u^2} - 2K/3)/(2K) &= (8 + 12C)/(33C_1)\\
(\overline{v^2} - 2K/3)/(2K) &= (2 - 30C)/(33C_1)\\
(\overline{w^2} - 2K/3)/(2K) &= (-10 + 18C)/(33C_1).
\end{aligned} \tag{9.140}$$

Reasonable agreement with data [39] can be had with $C = 0.4$ and $C_1 = 1.5$. In this case, according to Eqs. (9.137) and (9.138), $C_3 = 1.75$ and, $C_4 = 1.31$. Evidently, optimizing C_1 for the return to isotropy flow and for the anisotropic equilibrium state of homogeneous shear flow leads to somewhat different values of C_1.

If just the two linear terms on the right-hand side of Eq. (9.132) are kept, then

$$\Pi = -C_1 \epsilon b + \frac{4}{5} K\overline{S} \tag{9.141}$$

and the resulting closure is known as the isotropization of production (IP) model [40, 41]. This simplified form of the LRR model with $C_3 = C_4 = 0$ is desirable for its lack of complexity owing to the absence of non-linear terms.

9.5.3 SSG Model

Speziale, Sarkar, and Gatski [42] approached the problem of developing a practical pressure-strain model out of Eq. (9.130) by including non-linear terms that facilitate the accommodation of a wider range of physics than is possible with a linear model such as LRR. To enhance the practicality of the modeling, the terms depending on A_8 and A_{10} in Eq. (9.130) are omitted based on the observation that in the equilibrium state of homogeneous shear flow it can be shown that the cubic tensors associated with these terms are proportional to the quadratic tensors appearing in the terms with coefficients A_7 and A_9. This state of affairs fits in with the previous observation that not all ten terms in Eq. (9.130) are necessary for representing the pressure-strain correlation as a general combination of basis tensors.

The specific form taken by the SSG model is

$$\begin{aligned}
\Pi = &-(C_1\epsilon + C_1^* \mathcal{P})b + C_2\epsilon\left(b^2 - \frac{1}{3}|b|^2 I\right)\\
&+(C_3 - C_3^*|b|)K\overline{S}\\
&+C_4 K\left(b\overline{S} + \overline{S}b - \frac{2}{3}\,\mathrm{tr}(b\overline{S})I\right)\\
&+C_5 K(\overline{W}b - b\overline{W}),
\end{aligned} \tag{9.142}$$

where the terms depending on C_1^* and C_3^* are designed both to bring their respective terms up to a quadratic b dependence matching that of the other terms, but also to allow for a better accounting of transient effects in homogeneous shear flows. Note that the term containing C_1^* is quadratic in b since $\mathcal{P}$ contains a linear term in b. This term thus has the same b dependence as the quadratic term with coefficient C_2. The rationale for including the term with C_3^* is to add a linear b dependence matching the other part of the fast model as represented by the terms with C_4 and C_5. These modifications to Eq. (9.130) are consistent with the fact that coefficients are allowed to be functions of the invariants $II = \text{tr}(b^2)$ and $III = \text{tr}(b^3)$.

The SSG model can be made consistent with the Crow constraint by selecting $C_3 = 4/5$. Unlike the LRR model, Eq. (9.142) includes a non-linear part for the slow term in the expression containing C_2. C_1 and C_2 are found by applying SSG to the return to isotropy problem where all other terms in Eq. (9.142) are identically zero. In this circumstance, Eq. (9.122) containing Eq. (9.142) can be manipulated [43] to create coupled dynamical equations for the invariants II and III, in the form

$$\frac{dII}{d\tau} = -2[(C_1 - 2)II - C_2 III] \tag{9.143}$$

$$\frac{dIII}{d\tau} = -3\left[(C_1 - 2)III - C_2\frac{II^2}{6}\right], \tag{9.144}$$

where τ is the same time coordinate as appeared in Eq. (5.32). Examination of Eqs. (9.143) and (9.144) makes clear that the condition $C_1 > 2$ is necessary to ensure that solutions of the system of equations do not become unbounded. In addition, the long-time solutions of these equations, no matter what the initial state, should be isotropic turbulence for which $b = 0$ and so too are II and III. It may be shown that for this to happen $C_2 \leq 3(C_1 - 2)$. It is also interesting to note that this same condition enforces realizability to the effect that the normal Reynolds stresses remain positive during the decay. Comparing predictions of Eqs. (9.143) and (9.144) with experimental data suggests that the best agreement for any given value of C_1 occurs when C_2 has its maximum allowable value of $3(C_1 - 2)$. The choice of $C_1 = 3.4$ appears to be the best overall, in which case $C_2 = 4.2$.

The success of this part of the calibration of the SSG model is illustrated in Figure 9.1 wherein the normalized invariants $II^{1/2}$ and $III^{1/3}$ are plotted versus each other during the return to isotropy process. A curved trajectory matching the data given in [44] is achieved with the SSG model. This is attributable to keeping a quadratic form for the slow term and hence different decay rates for different components of b. This result is to be contrasted with LRR and other linear models, which, as shown in the figure, predict an unphysical straight line return to isotropy.

Despite the advantages of the quadratic model evident in Figure 9.1, in practice the added terms often make only a small contribution to the pressure-strain term and can be dropped without significant penalty. In fact, the advantage of including non-linear terms is often not sufficiently significant to warrant the extra expense of including them. For example, when the prediction of the Reynolds stress components themselves are considered, as shown in Figure 9.2, both the LRR and SSG models have the same degree of success in modeling the return to isotropy, though the latter model is slightly more accurate.

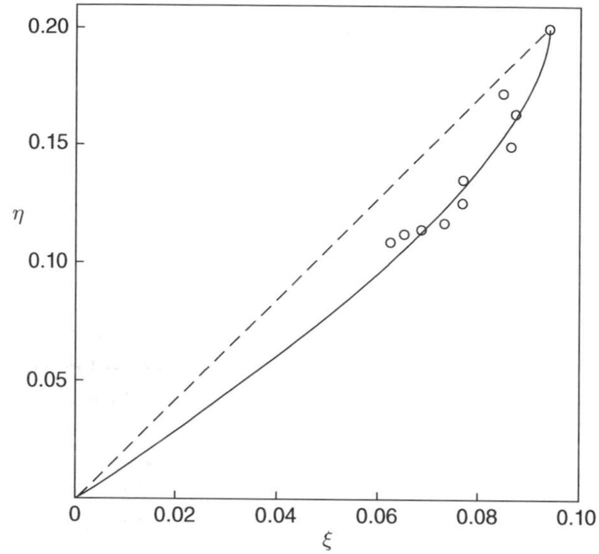

Figure 9.1 Prediction of the return to isotropy in plane strain. —, SSG model; – –, LRR model; o, experiments [44]. Here, $\xi = III^{1/3}$ and $\eta = II^{1/2}$. From [42]. Reprinted with the permission of Cambridge University Press.

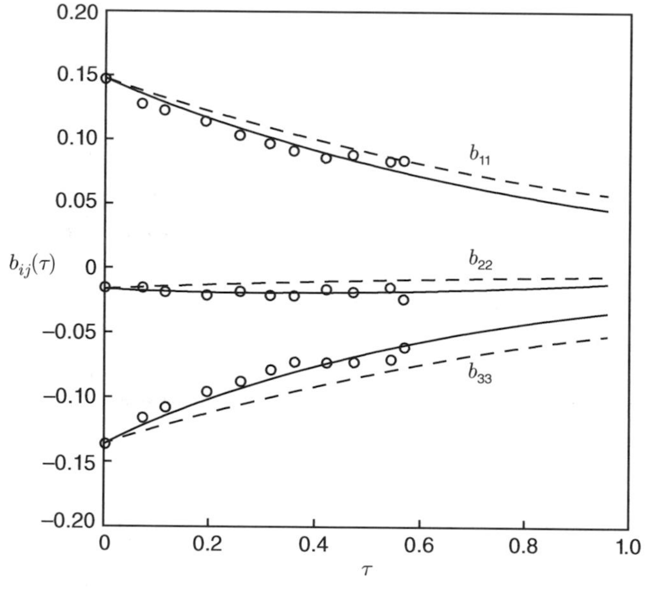

Figure 9.2 Normal components of *b* during return to isotropy. —, SSG model; – –, LRR model; o, experiments [44]. From [42]. Reprinted with the permission of Cambridge University Press.

The remaining constants in the SSG model are determined via a somewhat more elaborate consideration of homogeneous shear flow than was done in the case of LRR. Previously, LRR compared the model predictions to equilibrium states in a simple shear flow with $d\overline{U}/dy = S$, a constant. For SSG, these flows are widened to include the family of 2D mean flows $\overline{U}(y)$, $\overline{V}(x)$ such that

$$\overline{S}_{12} = \overline{S}_{21} = \frac{1}{2}\left(\frac{d\overline{U}}{dy} + \frac{d\overline{V}}{dx}\right) = S \tag{9.145}$$

and

$$\overline{W}_{12} = -\overline{W}_{21} = \frac{1}{2}\left(\frac{d\overline{U}}{dy} - \frac{d\overline{V}}{dx}\right) = \Omega. \tag{9.146}$$

Since

$$\frac{d\overline{U}}{dy} = S + \Omega \tag{9.147}$$

$$\frac{d\overline{V}}{dx} = S - \Omega \tag{9.148}$$

for $S = \Omega$ this recovers the simple shear flow considered in LRR. For $S = 0$ the flow is a rotation around the origin at angular velocity Ω oriented in the plane perpendicular to the mean field. A calculation of the streamlines associated with this mean velocity field shows that they are ellipses for $\Omega > S$ and hyperbolic paths otherwise.

The constants C_1^*, C_3^*, C_4, and C_5 in the SSG model are calibrated using data from physical experiments of homogeneous shear flows [45] together with the RDT prediction of the behavior of rotating shear flows [46]. As in the calibration of LRR it is not possible to select constant values that enable exact agreement with the measured or theoretical results, so that some optimization is necessary. This has yielded the values $C_1^* = 1.8$, $C_3^* = 1.3$, $C_4 = 1.25$, and $C_5 = 0.4$.

Further improvements to models such as SSG can be obtained by applying a more refined analysis of equilibrium states to the determination of coefficients. For example, within the context of homogeneous shear flows [36] it is possible to formally calculate the exact equilibrium properties of the closed dynamical set of equations for the elements of b. Such analysis reveals the way in which fundamental properties of the equilibrium solution depend on the relative magnitudes of shearing and rotation. The parameter

$$\eta_1 \equiv \frac{\overline{S}_{ij}\overline{S}_{ij}}{\overline{S}_{ij}\overline{S}_{ij} + \overline{W}_{ij}\overline{W}_{ij}} \tag{9.149}$$

is useful for distinguishing the flow regimes since for $\eta_1 < 0.5$ rotation dominates shear and the mean streamlines are ellipses, while for $\eta_1 > 0.5$ shear dominates and the streamlines are open-ended. Numerical simulations and theoretical analyses show that for all $\eta_1 > 0$, no matter how small the shearing, there will eventually be exponential growth in K at long times, even if rotation acts to dampen the growth for smaller times. Relaminarization only occurs when $\eta_1 = 0$. Improved performance of models such as SSG can be achieved by altering them to be consistent with the full range of properties of the equilibrium state.

In the case of the SSG model with its standard coefficient values it is found that there is a bifurcation in the predicted long time equilibrium solution at $\eta_1 \approx 0.35$. The bifurcation is such that for $\eta_1 < 0.35$ the solution relaminarizes at long times, while for $\eta_1 > 0.35$ there is an eventual exponential growth in energy. This means that the SSG model with its standard values is incompatible with the physics of strong rotation. To some extent this reflects the fact that SSG is calibrated exclusively using data for $\eta_1 > 0.5$ where rotational effects are not fully engaged in controlling the equilibrium flow.

The SSG model can be made to agree with the correct equilibrium behavior by moving the bifurcation point separating energy growth and decay to $\eta_1 = 0$ through a change in the model parameter C_5. Specifically, it can be shown that the proper bifurcation occurs when $C_5 = 2$. Thus, an empirical model for C_5 that accommodates this condition and also keeps the model unchanged for non-elliptical flows with $\eta_1 > 0.5$ is via the formula

$$C_5 = \begin{cases} 0.4 & \eta_1 \geq 0.5 \\ 2 - 1.6\left(\dfrac{\eta_1}{1-\eta_1}\right)^{3/4} & \eta_1 < 0.5 \end{cases}. \tag{9.150}$$

In this, for strain-dominated flows $C_5 = 0.4$, the value traditionally used in the SSG model. A continuous change of C_5 to the value of 2 occurs as η_1 decreases to zero.

The calibration process can also be expanded to better accommodate the various extreme states to which the models must apply. For example, one would like models such as SSG to respond well to imposed shear in RDT according to the Crow constraint at far from equilibrium conditions when b_{ij} is small, yet also be applicable to flows near equilibrium. In the case of the former it was seen above that it is necessary that $C_3 = 4/5$. To also allow accommodation of the equilibrium state, SSG incorporates the coefficient $C_3 - C_3^*|b|$ in Eq. (9.142). With $C_3^* = 0.36$ this has been calibrated based on predictions in the equilibrium range. A more careful means of negotiating the RDT limit at one extreme and the equilibrium condition at the other [36] is to make use of a parameter such as

$$\omega_1 \equiv \frac{\epsilon}{\sqrt{\eta_1}K}, \tag{9.151}$$

which is a generalized form of the ratio of time scales previously considered in Eq. (9.111). The discussion surrounding Eq. (9.111) indicates that $\omega_1 \to 0$ in the RDT limit while for equilibrium conditions it converges to a value, say, ω_∞. Then to improve the modeling of the extreme states and the transition between them, the factor $C_3 - C_3^*|b|$ is replaced by the expression

$$C_3 = \begin{cases} 0.36 & \dfrac{\omega_1}{\omega_\infty} \geq 1 \\ \dfrac{4}{5} - 0.44\left(\dfrac{\omega_1}{\omega_\infty}\right)^{1/4} & \dfrac{\omega_1}{\omega_\infty} < 1. \end{cases} \tag{9.152}$$

At equilibrium, $C_3 = 0.36$ while in the RDT limit $C_3 = 0.8$.

The benefit of Eqs. (9.150) and (9.152) is illustrated in Figure 9.3 comparing the prediction of K in a homogeneous shear flow using the SSG model with the traditional versus the updated coefficients. For this particular flow, $\eta_1 = 0.26$, which is in the elliptic range where the traditional SSG model, as confirmed by the figure, is known to fail in estimating the proper growth of kinetic energy. On the other hand, with revised coefficients the calculation captures the trend in K as predicted via DNS [47].

9.5.4 Transport Correlation

The turbulent transport correlation, β_{ijk}, defined in Eq. (3.55) is the sum of the Reynolds stress transport term $\overline{u_i u_j u_k}$ and the pressure-velocity correlation $\overline{pu_i}\delta_{jk}$. The former is symmetric in the sense that its value is unchanged by any reordering of the indices. The pressure-velocity correlation, on the other hand, is not symmetric. This means that β_{ijk}

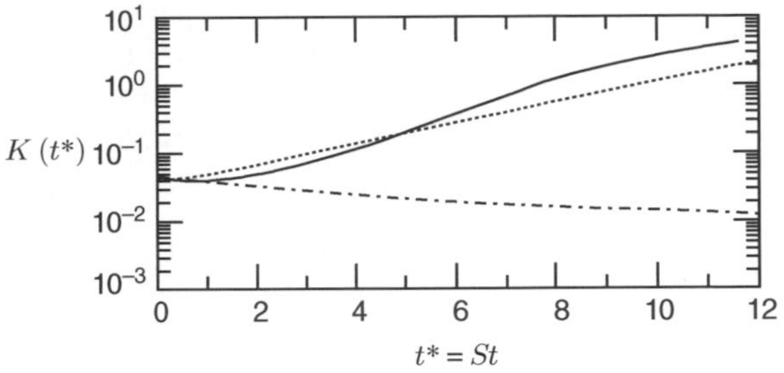

Figure 9.3 Prediction of K in homogeneous shear flow with $\eta_1 = 0.26$: —, DNS [47];– –, SSG model with Eqs. (9.150) and (9.152); – · –, SSG model with traditional coefficients. From [36]. Reprinted with permission of Cambridge University Press.

can only be thought of as symmetric if the neglect of $\overline{pu_i}$ can be justified. DNS measurements in a channel flow [48], for example, show that there is some justification for taking β_{ijk} to be symmetric away from boundaries, even if the pressure term is not exactly zero. This observation helps justify the common practice of treating β_{ijk} as if it were exactly symmetric and thus modeling it in a similar fashion. Among the popular approaches adopting this point of view is that due to Hanjalic and Launder [49] which assumes that:

$$\beta_{ijk} = -C_s \frac{K}{\epsilon}\left(R_{im}\frac{\partial R_{jk}}{\partial x_m} + R_{jm}\frac{\partial R_{ik}}{\partial x_m} + R_{km}\frac{\partial R_{ij}}{\partial x_m}\right). \tag{9.153}$$

This model is often used with the LRR and SSG closures. An alternative expression is

$$\beta_{ijk} = -C_s \frac{K^2}{\epsilon}\left(\frac{\partial R_{jk}}{\partial x_i} + \frac{\partial R_{ik}}{\partial x_j} + \frac{\partial R_{ij}}{\partial x_k}\right), \tag{9.154}$$

which follows from (9.153) after invoking an isotropy assumption. In both these models it is usually assumed that $C_s = 0.11$. Another approach, often used with the IP model, is the non-symmetric model due to Daly and Harlow [50], namely,

$$\beta_{ijk} = -C_s \frac{K}{\epsilon} R_{km}\frac{\partial R_{ij}}{\partial x_m}, \tag{9.155}$$

where $C_s = 0.22$.

In view of the intractability of the correlations in β_{ijk}, there has been little progress to date in developing alternative forms of the transport correlation that better agree with DNS or experimental results. However, it is also the case that it is not generally perceived that developing better transport models should be a high priority compared to more pressing issues, such as better accommodating the pressure-strain term.

9.5.5 Complete Second Moment Closure

To summarize what is entailed in a second moment closure we gather together the complete set of equations that are solved using such methods. This includes the mean velocity equation

$$\frac{\partial \overline{U}_i}{\partial t} + \overline{U}_j\frac{\partial \overline{U}_i}{\partial x_j} = -\frac{\partial \overline{P}/\rho}{\partial x_i} + \frac{\partial}{\partial x_j}\left(\nu\frac{\partial \overline{U}_i}{\partial x_j} - R_{ij}\right), \tag{9.156}$$

the continuity equation,

$$\frac{\partial \overline{U}_i}{\partial x_i} = 0, \tag{9.157}$$

and the Reynolds stress equation. To give concrete expression to the latter, we incorporate the modeling out of which both the LRR and SSG models are postulated. In particular, after substituting Eqs. (9.41), (9.142), and (9.153) into (9.122) it is found that

$$\begin{aligned}\frac{\partial R_{ij}}{\partial t} + \overline{U}_k \frac{\partial R_{ij}}{\partial x_k} &= -R_{ik}\frac{\partial \overline{U}_j}{\partial x_k} - R_{jk}\frac{\partial \overline{U}_i}{\partial x_k} - \frac{2}{3}\epsilon\delta_{ij} \\ &+ \frac{\partial}{\partial x_k}\left[C_s \frac{K}{\epsilon}\left(R_{im}\frac{\partial R_{jk}}{\partial x_m} + R_{jm}\frac{\partial R_{ik}}{\partial x_m} + R_{km}\frac{\partial R_{ij}}{\partial x_m}\right)\right] \\ &- (C_1\epsilon + C_1^* \mathcal{P})b_{ij} + C_2\epsilon(b_{ik}b_{kj} - \frac{1}{3}II\delta_{ij}) \\ &+ \left(\frac{4}{5} - C_3^*(II)^{1/2}\right)K\overline{S}_{ij} \\ &+ C_4K(b_{ik}\overline{S}_{jk} + b_{jk}\overline{S}_{ik} - \frac{2}{3}b_{kl}\overline{S}_{lk}\delta_{ij}) \\ &+ C_5K(b_{ik}\overline{W}_{jk} + b_{jk}\overline{W}_{ik}) + \nu\nabla^2 R_{ij}. \end{aligned} \tag{9.158}$$

A trace of equation Eq. (9.158) gives an implied equation for K. Also included in the RSE model is an equation for the dissipation rate that usually takes its standard form as it appears in the K–ϵ closure, namely,

$$\frac{\partial \epsilon}{\partial t} + \overline{U}_j \frac{\partial \epsilon}{\partial x_j} = C_{\epsilon_1}\frac{\epsilon}{K}\mathcal{P} - C_{\epsilon_2}\frac{\epsilon^2}{K} + \frac{\partial}{\partial x_i}\left[\left(\nu + \frac{\nu_t}{\sigma_\epsilon}\right)\frac{\partial \epsilon}{\partial x_i}\right]. \tag{9.159}$$

The complete set of constants includes $C_1, C_1^*, C_2, C_3^*, C_4, C_5, C_{\epsilon_1}, C_{\epsilon_2}$, and σ_ϵ though alternative choices are also possible as, for example, as contained in Eqs. (9.150) and (9.152).

The system of equations (9.156)–(9.159) represents, in the most general 3D flow, 11 equations for the three components of $\overline{\mathbf{U}}$, the mean pressure $\overline{P}$, six components of R, and the dissipation ϵ. Such systems can be difficult to solve numerically due to the complex non-linear coupling between the variables that is exacerbated by the fact that the Reynolds stresses appear in differentiated form in the mean velocity equation (9.156). Small discrepancies in predicting the Reynolds stresses can be magnified into numerical instability or strongly unphysical behavior. The algebraic Reynolds stress models discussed in Section 9.6 attempt to make use of much of the physics that has gone into the derivation of Eq. (9.158) but without the complexities arising from the transient behavior of the equation.

9.5.6 Near-Wall Reynolds Stress Equation Models

As in the case of two-equation models, second moment closures generally need some form of modification to be applicable in the near-wall region. Without such special treatment predictions of canonical wall-bounded flows tend to be significantly degraded. It is also the case that the anisotropy of the near-wall flow is pronounced and fundamental to the physics of turbulent flow and thus should have some direct bearing on the near-wall form adopted by RSE models.

One general principle that can guide model development near the wall is to have the modeling reflect some of the basic balances between terms that occur in the Reynolds stress equation in this region. Thus, from Taylor series expansions it follows that through terms of $O(y)$ near the wall under steady-state conditions, Eq. (9.122) reduces to

$$\epsilon_{ij} = D^p_{ij} + \Pi_{ij} + \nu\nabla^2 R_{ij} \tag{9.160}$$

where

$$D^p_{ij} \equiv -\frac{1}{\rho}\frac{\partial \overline{pu_i}}{\partial x_k}\delta_{jk} - \frac{1}{\rho}\frac{\partial \overline{pu_j}}{\partial x_k}\delta_{ik} \tag{9.161}$$

is the pressure diffusion term. Formerly, this term was included as part of

$$-\frac{\partial \beta_{ijk}}{\partial x_k} = -\frac{\partial \overline{u_i u_j u_k}}{\partial x_k} + D^p_{ij}. \tag{9.162}$$

Near the wall, it is only this part of β_{ijk} that is significant.

It may be shown that the $(1,1), (3,3)$, and $(1,3)$ components in (9.160) are $O(1)$ near a surface at $y = 0$, the $(2,2)$ component is $O(y^2)$, and the remainder are $O(y)$. These are trends that one hopes to capture in turbulence models, though it often proves difficult to achieve them in practice. Another set of conditions is given by the asymptotic boundary relations as $y \to 0$ to the effect that

$$\frac{\epsilon}{K} = \frac{\epsilon_{11}}{R_{11}} = \frac{\epsilon_{33}}{R_{33}} = \frac{\epsilon_{13}}{R_{13}} = \frac{1}{2}\frac{\epsilon_{12}}{R_{12}} = \frac{1}{2}\frac{\epsilon_{23}}{R_{23}} = \frac{1}{4}\frac{\epsilon_{22}}{R_{22}}. \tag{9.163}$$

These relations express the anisotropy just adjacent to the wall surface, and are readily derived from Taylor series expansions of the numerator and denominators [51] of each expression. Ideally, closure models for ϵ_{ij} should be able to satisfy these conditions.

The availability of DNS simulation data for canonical flows such as the channel means that closure development can aim for a much more comprehensive matching of near-wall conditions than follows from reproducing the asymptotic conditions at the wall itself [52]. The idea in this case is to design models for the individual terms in Eq. (9.160) that match DNS studies and then re-examine their performance when the complete model is applied to the flows used in calibration. Despite the possibility of matching DNS results in a priori testing, a posteriori tests tend not to be as promising due to inevitable errors that arise in computing such quantities as ϵ that affect all aspects of the solution. It must be expected that a posteriori tests of new models will generally reveal areas of concern that require additional model development.

Of particular importance in near-wall modeling is capturing the trends in Π_{ij} and the dissipation tensor ϵ_{ij}. The latter can be expressed as the sum of isotropic and deviatoric parts according to

$$\epsilon_{ij} = \frac{2}{3}\epsilon\delta_{ij} + \left(\epsilon_{ij} - \frac{2}{3}\epsilon\delta_{ij}\right). \tag{9.164}$$

Near-wall models are often sought for the combined terms [53–55]

$$\Pi_{ij} - \left(\epsilon_{ij} - \frac{2}{3}\epsilon\delta_{ij}\right), \tag{9.165}$$

with ϵ determined from its own equation. Alternatively, DNS data can be used to help construct models for the separate terms [56, 57] in Eq. (9.165). While this can potentially bring in more physics than an analysis of the combined effects, the models resulting from such attempts tend to be complicated and involve extensive formulas that limit their usefulness and generality [52].

Regarding special treatments for the remaining terms in the Reynolds stress equation, the first term on the right-hand side of Eq. 9.162 that depends on the triple velocity correlation tends to be much smaller than the other terms and is not specially modeled. The term D^p_{ij} is somewhat prominent near the wall, as mentioned previously, and is thus a more likely target for near-wall modeling. Many, but not all, models consider D^p_{ij} independently of Π_{ij} [52, 55, 58, 59]. For those that do, the feasibility of this kind of effort is improved by the availability of DNS data, though the complexity of the required models can be daunting and an obstacle to computation.

9.6 Algebraic Reynolds Stress Models

The equilibrium properties of b, the isotropy tensor, in homogeneous shear flow were previously shown in Sections 9.5.2 and 9.5.3 to play a role in the calibration of SMC such as the LRR and SSG models. The analysis in these cases was limited to considering b and its properties for a constant mean shearing and rotation of the flowfield. This same equilibrium idea can be generalized beyond the restriction of homogeneous turbulence to include non-homogeneity in the mean velocity gradients by assuming that

$$\frac{Db}{Dt} = 0, \tag{9.166}$$

with the interpretation that equilibrium refers to a small local flow region convecting with the mean velocity. The solution for b depends on the local mean velocity gradient and so may vary in non-homogeneous turbulence. As before, to simplify the analysis it is assumed that the transport and diffusion terms in the R_{ij} equation can be neglected, but not the production term. The result from Eq. (9.139) is that

$$K\left(-R_{ik}\frac{\partial \overline{U}_j}{\partial x_k} - R_{jk}\frac{\partial \overline{U}_i}{\partial x_k} - \epsilon_{ij} + \Pi_{ij}\right) - R_{ij}(\mathcal{P} - \epsilon) = 0. \tag{9.167}$$

After specification of Π_{ij} according to one of the SMC models, and replacing R_{ij} with $R_{ij} = 2Kb_{ij} + 2/3K\delta_{ij}$, which comes from Eq. (9.42), Eq. (9.167) becomes an algebraic equation in the variable b_{ij} that can be solved for given values of the mean velocity gradient tensor. For example, for the particular choice of Eq. (9.142) with the quadratic term depending on C_2 omitted, Eq. (9.167) becomes

$$\begin{aligned} b\left(C_1 - 2 + (C_1^* + 2)\frac{\mathcal{P}}{\epsilon}\right) &= \left(C_3 - \frac{4}{3} - C_3^*|b|\right)\frac{K}{\epsilon}\overline{S} \\ &\quad + (C_4 - 2)\frac{K}{\epsilon}\left(b\overline{S} + \overline{S}b - \frac{2}{3}(b : \overline{S})I\right) \\ &\quad + (C_5 - 2)\frac{K}{\epsilon}(\overline{W}b - b\overline{W}). \end{aligned} \tag{9.168}$$

This is an algebraic relation for b whose solution constitutes an ARM. In this way each ARM corresponds to a particular choice of SMC.

The factor

$$\mathcal{P} \sim b\overline{S} \tag{9.169}$$

appearing on the left-hand side of Eq. (9.168) depends on b so that the term containing it is non-linear in b while the terms on the right-hand side are linear in b. In some analyses it is assumed that the ratio $\mathcal{P}/\epsilon$ is equal to its equilibrium value for homogeneous shear flows, so that in this case Eq. (9.168) is linear. Otherwise, it is a non-linear equation.

As ARM methods were originally developed [60], solutions for b were found by numerically solving relations such as Eq. (9.168) at all points in the flow field. Besides the potentially difficult numerical problems that may occur in finding a solution – especially for the non-linear case – the computational expense involved is considerable so that the practicality of the method is limited.

Subsequent analysis has shown [61] that it is also possible to find the exact solution to Eq. (9.168) in both the linear and non-linear cases. This is accomplished via representation theory in which the fact that b is determined as the solution to an equation depending on $\overline{S}$ and $\overline{W}$ in Eq. (9.168) means that it can be written as a linear combination of the integrity basis tensors formed from $\overline{S}$ and $\overline{W}$. For a general 3D mean velocity field where there are ten integrity basis tensors an explicit solution for b can be obtained for the linear formulation of Eq. (9.168) [35], though the resulting equation is not practical owing to its length and complexity. In contrast, if the mean field is taken to be 2D, since there are only three integrity basis tensors in this case, b can be expressed in the more practical formula

$$b_{ij} = \frac{G_1 K}{\epsilon}\overline{S}_{ij} + \frac{G_2 K^2}{\epsilon^2}(\overline{S}_{ik}\overline{W}_{kj} - \overline{W}_{ik}\overline{S}_{kj}) + \frac{G_3 K^2}{\epsilon^2}\left(\overline{S}_{ik}\overline{S}_{kj} - \frac{1}{3}\overline{S}_{mn}\overline{S}_{mn}\delta_{ij}\right). \qquad (9.170)$$

Substituting Eq. (9.170) into (9.168) and gathering terms corresponding to each of the three basis tensors results in three coupled equations for G_1, G_2, and G_3. If Eq. (9.168) is modeled as linear, then the equations for G_1, G_2, and G_3 are linear. Otherwise, a solvable cubic equation appears for G_1 whose solutions can be used to obtain G_2 and G_3. The exact functional forms of G_1, G_2, and G_3 are of manageable complexity.

The solution for the linear approximation contained in Eq. (9.168) is vulnerable to a singularity when the flow is far from equilibrium, though this may be prevented via the use of a regularization process. On the other hand, the exact solutions for G_1 in the non-linear case has multiple roots and some care must be taken in making sure that the physical root is extracted. Though the formulas derived in these cases assume 2D mean flow, they are typically applied to arbitrary 3D mean flows as well.

9.7 URANS

The flow in one complete cycle in the movement of a piston and valves in an internal combustion engine cylinder may be regarded as one realization of a turbulent flow, so that averaging in this case may be thought of as over the ensemble of engine cycles. Alternatively, it is possible that a time average over a small time interval within the period of the piston motion can be attempted. Either way, the mean velocity calculated from solving the traditional RANS equations in such situations must of necessity be non-steady and so may be regarded as a URANS solution. For a typical RANS model the calculation will be fully deterministic so that random large-scale structures having to do with the evolution of the turbulent field will be absent from the solution. For the engine cylinder computation the transient structures which occur in the URANS solution are driven by the external boundary conditions and so are not in the same category as the random structures that cannot be accommodated by the URANS modeling.

Sometimes flows contain a natural process by which some aspects of the flow field are cyclical in nature, for example vortex shedding behind a cylinder. In the turbulent case the shedding may be at a more or less fixed frequency which is determined via an interaction between 3D random turbulent structures of small scale and the large shed vortices. RANS calculations of the cylinder flow using traditional models are very likely to give a symmetric steady flow in which there is no vortex shedding. Even if the RANS model does behave as a URANS in this situation by producing shed vortices, the solution will not necessarily be physical. For example, it is likely to predict the wrong Strouhal number because the dynamical effect of random motions on causing large-scale shedding is excluded. Unlike the engine calculation, in the absence of transient boundary conditions to control the non-steady behavior, the neglect of small 3D structures in the RANS model will likely have a strong effect on altering the physics of the computed solution, even if it is non-steady.

In recent years variations of the URANS models have been developed that allow for the inclusion of transient, random large-scale structures in a similar fashion as occurs in LES. Such schemes generally are based on lowering the local eddy viscosity of the method to the point that diffusion no longer suppresses instabilities so that structures appear via the natural evolution of disturbances within the flow field. In the case of URANS this step towards LES simulation is accomplished without employing parameters associated with local meshing, so that in this way URANS can be considered to be a separate approach from that used in LES [62]. An example of this approach will be seen in the next chapter.

9.8 Chapter Summary

This chapter has tried to put into perspective the major themes in turbulence closure development. A few of the most commonly encountered RANS models, including the K–ϵ, K–ω, Menter SST, and Spalart–Allmaras approaches, have been described in detail. Also considered, and more likely to be used in niche areas, are the SMC, ARM, and NLEVM methods, which aim to accommodate physical phenomena that may be a priori excluded from representation within EVM. Such higher order methods tend to come with significant extra computational cost and complexity. Moreover, their vulnerability to unanticipated non-physical solutions resulting from model non-linearity requires that such schemes be used cautiously. In fact, the slow rate of development in improving RANS models, the difficulty of working with complicated models, and the increasing availability of large-scale computing have fueled interest in the LES methods discussed in the next chapter. It also will be seen there that in a recent trend some of the EVM RANS models that have been discussed here are now employed in the near-wall region of LES calculations. These are the hybrid LES/RANS models described in the next chapter.

References

1 Prandtl, L. (1925) Über die ausgebildete turbulenz. *ZAMM*, 5, 136–139.
2 Cebeci, T. and Smith, A.M.O. (1974) *Analysis of Turbulent Boundary Layers*, Academic Press, New York.
3 Baldwin, B.S. and Lomax, H. (1978) Thin-layer approximation and algebraic model for separated turbulent flows. *AIAA Paper No. 78-257*.
4 Spalart, P.R. and Allmaras, S.R. (1992) A one-equation turbulence model for aerodynamic flows. *AIAA Paper No. 92-439*.
5 Jones, W.P. and Launder, B.E. (1972) The prediction of laminarization with a two-equation model of turbulence. *Intl. J. Heat and Mass Transfer*, 15, 301–314.
6 Wilcox, D.C. (2006) *Turbulence Modeling for CFD*, DCW Industries, La Canada, CA, 3rd edn.
7 Menter, F.R. (1994) Two-equation eddy-viscosity turbulence models for engineering applications. *AIAA J.*, 8, 1598–1605.
8 Durbin, P.A. (1991) Near wall turbulence models without damping functions. *Theor. Comp. Fluid Dyn.*, 3, 1–13.
9 Taylor, G.I. (1932) The transport of vorticity and heat through fluids in turbulent motion. *Proc. Roy. Soc.*, 135A, 685–705.
10 van Driest, E.R. (1956) On turbulent flow near a wall. *J. Aero. Sci.*, 23, 1007–1011.
11 Granville, P.S. (1989) A modified van Driest formula for the mixing length of turbulent boundary layers in pressure gradients. *ASME J. Fluids Engrg.*, 111, 94–97.
12 Smagorinsky, J. (1963) General circulation experiments with the primitive equations. *Mon. Weather Rev.*, 91, 99–165.
13 Shih, T.H., Liou, W.W., Shabbir, A., Yang, Z., and Zhu, J. (1995) A new $K-\epsilon$ eddy viscosity model for high Reynolds number turbulent flows. *Computers and Fluids*, 24, 227–238.
14 Yakhot, V. and Orszag, S.A. (1986) Renormalization group analysis of turbulence. I. Basic theory. *J. Sci. Comput.*, 1, 3–51.
15 Sarkar, A. and So, R.M.C. (1997) A critical evaluation of near-wall two-equation models against direct numerical simulation data. *Intl. J. Heat Fluid Flow*, 18, 197–208.
16 Wilcox, D.C. (2008) Formulation of the K–ω turbulence model revisited. *AIAA J.*, 46, 2823–2838.
17 Hamba, F. (2006) Euclidean invariance and weak-equilibrium condition for the algebraic Reynolds stress model. *J. Fluid Mech.*, 569, 399–408.
18 Luca, I. and Sadiki, A. (2008) New insight into the functional dependence rules in turbulence modelling. *Int. J. Engrg. Sci.*, 46, 1053–1062.
19 Bernard, P.S. (2015) *Fluid Dynamics*, Cambridge University Press, Cambridge.
20 Speziale, C.G. (1981) Some interesting properties of two-dimensional turbulence. *Phys. Fluids*, 24, 1425–1427.
21 Speziale, C.G. (1985) Modeling the pressure gradient-velocity correlation of turbulence. *Phys. Fluids*, 28, 69–71.
22 Pedlosky, J. (1979) *Geophysical Fluid Dynamics*, Springer Verlag, New York.

23 Speziale, C.G. (1989) Turbulence modeling in non-inertial frames of reference. *Theor. Comp. Fluid Dyn.*, 1, 3–19.

24 Durbin, P.A. and Speziale, C.G. (1994) Realizability of second-moment closure via stochastic analysis. *J. Fluid Mech.*, 280, 395–407.

25 Hunt, J.C.R. and Carruthers, D.J. (1990) Rapid distortion theory and the 'problems' of turbulence. *J. Fluid Mech.*, 212, 497–532.

26 Crow, S.C. (1968) Viscoelastic properties of fine-grained incompressible turbulence. *J. Fluid Mech.*, 33, 1–20.

27 Reynolds, W.C. (1989) Effects of rotation on homogeneous turbulence. *Proc. Australasian Conf. Fluid Mech.*

28 Mishra, A.A. and Girimaji, S.S. (2010) Pressure-strain correlation modeling: Towards achieving consistency with rapid distortion theory. *Flow Turbulence Combust.*, 85, 593–619.

29 Speziale, C.G. (1991) Analytical methods for the development of Reynolds-stress closures in turbulence. *Ann. Rev. Fluid Mech*, 23, 107–157.

30 Yoshizawa, A. (1984) Statistical analysis of the deviation of the Reynolds stress from its eddy-viscosity representation. *Phys. Fluids*, 27, 1377–1387.

31 Smith, G.F. (1971) On isotropic functions of symmetric tensors, skew-symmetric tensors and vectors. *Intl. J. Eng. Sci.*, 9, 899–916.

32 Apsley, A.D. and Leschziner, M.A. (1997) A new low-Reynolds-number non-linear two-equation turbulence model for complex flows. *Intl. J. Heat Fluid Flow*, 19, 209–222.

33 Speziale, C.G. (1987) On nonlinear K–l and K–ϵ models of turbulence. *J. Fluid Mech.*, 178, 459–475.

34 Riley, K.F., Hobson, M.P., and Bence, S.J. (2006) *Mathematical Methods for Physics and Engineering*, Cambridge University Press, New York, 3rd edn.

35 Gatski, T.B. and Speziale, C.G. (1993) On explicit algebraic stress models for complex turbulent flows. *J. Fluid Mech.*, 254, 59–78.

36 Girimaji, S.S. (2000) Pressure-strain correlation modeling of complex turbulent flows. *J. Fluid Mech.*, 422, 91–123.

37 Launder, B.E., Reece, G.J., and Rodi, W. (1975) Progress in the development of a Reynolds stress turbulence closure. *J. Fluid Mech.*, 68, 537–566.

38 Rotta, J.C. (1951) Statistische theorie nichthomogener turbulenz. *Z. Phys.*, 129, 547–572.

39 Champagne, F.H., Harris, V.G., and Corrsin, S. (1970) Experiments on nearly homogeneous turbulent shear flow. *J. Fluid Mech.*, 41, 81–139.

40 Gibson, M.M. and Launder, B.E. (1978) Ground effects on pressure fluctuations in the atmospheric boundary layer. *J. Fluid Mech.*, 86, 491–511.

41 Naot, D., Shavit, A., and Wolfshtein, M. (1973) Two-point correlation model and the redistribution of Reynolds stress. *Phys. Fluids*, 16, 738–740.

42 Speziale, C.G., Sarkar, S., and Gatski, T.B. (1991) Modeling the pressure-strain correlation of turbulence: An invariant dynamical systems approach. *J. Fluid Mech.*, 227, 245–272.

43 Sarkar, S. and Speziale, C.G. (1990) A simple nonlinear model for the return to isotropy in turbulence. *Phys. Fluids A*, 2, 84–93.

44 Choi, K.S. and Lumley, J.L. (1984) Return to isotropy of homogeneous turbulence revisited, in *Turbulence and Chaotic Phenomena in Fluids* (ed. T. Tatsumi), North Holland, New York, pp. 267–272.

45 Tavoularis, S. and Karnik, U. (1989) Further experiments on the evolution of turbulent stresses and scales in uniformly sheared turbulence. *J. Fluid Mech.*, 204, 457–478.

46 Bertoglio, J.P. (1982) Homogeneous turbulent field within a rotating frame. *AIAA J.*, 20, 1175–1181.

47 Blaisdell, G.A. and Shariff, K. (1996) Simulation and modeling of elliptic streamline flow. *Center for Turbulence Research, Proc. Summer Program.*

48 Mansour, N.N., Kim, J., and Moin, P. (1988) Reynolds-stress and dissipation-rate budgets in a turbulent channel flow. *J. Fluid Mech.*, 194, 15–44.

49 Hanjalic, K. and Launder, B.E. (1976) Contribution towards a Reynolds-stress closure for low-Reynolds-number turbulence. *J. Fluid. Mech.*, 74, 593–610.

50 Daly, J. and Harlow, F.H. (1970) Transport equations in turbulence. *Phys. Fluids*, 13, 2634–2649.

51 Launder, B.E. and Reynolds, W.C. (1983) Asymptotic near-wall stress dissipation rates in a turbulent flow. *Phys. Fluids*, 26, 1157–1158.

52 Gerolymos, G.A., Lo, C., Vallet, I., and Younis, B.A. (2012) Term-by-term analysis of near-wall second-moment closures. *AIAA J.*, 50, 2848–2864.

53 Gerolymos, G.A. and Vallet, I. (2001) Wall-normal-free near-wall Reynolds-stress closure for 3D compressible separated flows. *AIAA J.*, 39, 1833–1842.

54 Suga, K. (2004) Modeling the rapid part of the pressure-diffusion process in the Reynolds stress transport equation. *J. Fluids Engrg.*, 126, 633–641.

55 So, R.M.C., Aksoy, H., Sommer, T.P., and Yuan, S.P. (1994) Development of a near-wall Reynolds-stress closure based on the SSG model for the pressure strain. *NASA Contractor Report 4618.*

56 Jakirlić, S. and Hanjalić, K. (2002) A new approach to modeling near-wall turbulence energy and stress dissipation. *J. Fluid Mech.*, 459, 139–166.

57 Jakirlić, S., Eisfield, B., Jester-Züker, R., and Kroll, N. (2007) Near-wall Reynolds-stress model calculations of transonic flow configurations relevant to aircraft aerodynamics. *Int. J. Heat Fluid Flow*, 28, 602–615.

58 Craft, T.J., Launder, B.E., and Suga, K. (1995) A non-linear eddy viscosity model including sensitivity to stress anisotropy. *Proc. Turbulent Shear Flows 10*, 3, 23.19–23.24.

59 Speziale, C.G. and So, R.M.C. (2016) Turbulence modeling and simulation, in *The Handbook of Fluid Dynamics* (ed. R.W. Johnson), CRC Press, 2nd edn.

60 Rodi, W. (1976) A new algebraic relation for calculating Reynolds stress. *Z. Angew. Math. Mech.*, 56, 331–340.

61 Girimaji, S.S. (1996) Fully explicit and self-consistent algebraic Reynolds stress model. *Theor. Comp. Fluid Dyn.*, 8, 387–402.

62 Fröhlich, J. and von Terzi, D. (2008) Hybrid LES/RANS methods for the simulation of turbulent flows. *Prog. Aerospace Sci.*, 44, 349–377.

63 Graham, J., Kanov, K., Yang, X., Lee, M., Malaya, N., Lalescu, C., Burns, R., Eyink, G., Szalay, A., Moser., R., and Meneveau, C. (2016) A web services-accessible database of turbulent channel flow and its use for testing a new integral wall model for les. *J. Turbulence*, 17, 181–215.

Problems

9.1 Show that the two expressions of the Baldwin–Lomax model in Eqs. (9.20) and (9.21) are equivalent.

9.2 Compute ν_t using channel flow data for the Cebeci–Smith model by substituting the channel half-width h for δ, and centerline velocity for U_∞.

9.3 Use channel flow data to calculate ν_t from Eqs. (9.21) and (9.23). Compare this to ν_t predicted using Eqs. (9.18) and (9.19).

9.4 Derive Eq. (9.53) by expressing K and ϵ in Taylor series about the wall location.

9.5 Derive Eq. (9.54) by carrying out the differentiation and using Taylor series expansions.

9.6 Compute the coefficients in the ϵ equation as it would appear in the SST closure away from boundaries.

9.7 Prove that $|b_{ij}| \leq 1$ for all i, j by using the definition of K and applying the Schwarz inequality to R_{ij}.

9.8 Prove Eq. (9.131) by relating the eigenvalues of R_{ij} to those of b_{ij} and using the fact that R_{ij} is real, symmetric, and positive semi-definite.

9.9 Use channel flow data available on the internet (e.g., [63], or at http://turbulence.ices.utexas.edu) to evaluate and plot C_μ. See how it changes with Reynolds number and compare it to the standard value of 0.09.

9.10 Use data for a channel flow to evaluate and compare ν_t for the K–ϵ closure, with and without the near-wall model in Eqs. (9.59) and (9.63). Include a plot of ν_t that corresponds to having Eq. (9.8) exactly satisfied.

9.11 Show that the Coriolis term in Eq. (9.105) can be absorbed in the pressure term for 2D turbulence.

9.12 Show that $\overline{U}(\mathbf{y}, s) - \overline{U}(\mathbf{x}, s)$ satisfies extended Galilean invariance.

9.13 Prove the identities in Eq. (9.134) by applying Eq. (4.8) to the definition in Eq. (9.127) assuming homogeneous turbulence.

9.14 Following the steps outlined in the text, derive Eqs. (9.140).

9.15 Compute streamlines or particle paths corresponding to the mean velocity field satisfying Eqs. (9.147) and (9.148) for different values of S and Ω.

9.16 By contracting indices in the Reynolds stress equation the transport models given in Eqs. (9.153)–(9.155) become models for the transport term in the K equation. Determine the form of these models in channel flow and show that they are all different and none of them are the same as the transport model Eq. (9.33) incorporated in (9.34).

9.17 Write a code (e.g., in MATLAB) that can solve for fully developed channel flow using one or more of the EVM given in the text in coordination with Eqs. (7.6) and (9.8). For example (a) the mixing length model in Eqs. (9.12) and (9.14), (b) the mixing length model in Eq. (9.12) and (9.15), (c) the Cebeci–Smith model in Eqs. (9.18) and (9.19), and (d) the Baldwin–Lomax model in Eqs. (9.21)–(9.25). Compare the solutions for $\overline{U}$ to channel flow data available on the internet (e.g., [63]).

9.18 Write a code (e.g., in MATLAB) that can solve for fully developed channel flow using the low Reynolds number form of the K–ϵ closure given in Eqs. (7.6), (9.8), (9.34), and (9.62)–(9.65). Compare the solutions for $\overline{U}, K$, and ϵ to the channel flow data available on the internet (e.g., [63]).

10

Large Eddy Simulations

RANS modeling as considered in the previous chapter is built on a variety of strategies for deciphering the relationship between the Reynolds stresses and the complex flow physics that produce them. To some extent the difficulty of RANS modeling arises from the need to model the average effect of flow structure without following the details of the structures themselves. For example, the net Reynolds shear stress arises from the action of a relatively small population of individual vortical events – as seen previously in Chapter 8 – whose average effect on transport is not easily related to a functional of mean flow statistics. Methodologies for flow prediction that are capable of resolving and computing the dynamics of vortical structures should therefore, in principle, have a greater likelihood of capturing the essential aspects of the transport mechanism that is fundamental to turbulent motion.

It is in the nature of DNS that all Reynolds stress producing events in the flow field are well accounted for, so Reynolds stress predictions are accurate. While DNS is prohibitively expensive or infeasible for many flows, it is reasonable to expect that the high resolution of a DNS may not be necessary in all cases to obtain useful predictions of turbulent flow quantities since it is primarily the large-scale motions that underlie momentum transport and thus account for the Reynolds shear stress. Methodologies such as LES that directly predict large-scale motions without fully resolving small-scale motions have thus become the best bet for eventually maturing into an affordable and relatively accurate means for studying turbulent flows.

The central difficulty with LES, which continues to be the object of study, concerns the accuracy with which the action of small-scale structure on large can be modeled. This is of crucial importance since errors in computing flow structure translate into errors in modeling transport. In regions of the flow where there are ample meshpoints available to resolve stress-producing structures one can expect a reasonably accurate description of the flow field. Near boundaries where essential structures tend to shrink in size, the demand for fine mesh resolution, particularly for high Reynolds number flows, is enough to bring the cost of LES close that of DNS and hence is uneconomical and impractical. In essence, the concept of a subgrid-scale becomes unrealistic in regions where the smallest events are also the energy-containing events, so subgrid modeling is not feasible near solid boundaries. The consequence for not adequately resolving structure and/or not providing for their accurate modeling leads to the appearance of numerical diffusion that prematurely dissipates dynamically important vortices, leading to significant error.

From this discussion it follows that to a large extent it is the performance of LES in near-wall regions that is the pacing item for the development of LES into a practical tool.

Turbulent Fluid Flow, First Edition. Peter S. Bernard.

Companion website: www.wiley.com/go/Bernard/Turbulent_Fluid_Flow

Even though the wall region may be a small part of the overall flow field, it is the source of vorticity and must be accurately accounted for. In some situations, principally attached boundary layers, RANS modeling has a proven record of providing reasonably accurate mean velocity fields. This has led to the idea that LES schemes might be enhanced by combining them with selective use of RANS techniques in the near-wall flow region. A number of alternative ideas for accomplishing this have been implemented and form the category of methods known as hybrid LES/RANS schemes. These methods allow the computed solution to shift between LES and RANS formulations depending on a variety of criteria, some using local information about the mesh and flow statistics and others based on regional considerations.

In this chapter we present some of the main approaches to formulating LES and then consider some of the relatively recent schemes that attempt to make the LES approach a more robust alternative for real-world turbulent flow prediction. Also fitting within this genre of methodologies are the URANS and other aligned RANS techniques that include transient structure in their computed flow fields and tend to operate in a hybrid sense between RANS and URANS.

10.1 Mathematical Basis of LES

Large eddy simulation (LES) is based on finding a middle ground between averaging and not averaging the flow field. Formally one *filters* the velocity, pressure, and other fields of interest to produce quantities that are still random but are generally free of motions at scales smaller than a chosen resolution threshold. Assuming that the calculations will be done on a numerical mesh, it is the mesh spacing that generally provides a lower limit on resolution. This scale does not necessarily have to be the smallest resolved scale in a LES calculation, but it is generally taken to be so in practice. In fact, as a general rule one would like to maximize the use of computational resources, which means not providing greater resolution than is needed in modeling flow events at a chosen smallest scale.

A filter in an LES scheme is a function of the form $G(\mathbf{x}, \mathbf{y})$ that is used to produce the filtered velocity field via

$$\langle \mathbf{U} \rangle(\mathbf{x}, t) \equiv \int_{\Re^3} G(\mathbf{x}, \mathbf{y}) U(\mathbf{y}, t) d\mathbf{y}. \tag{10.1}$$

By requiring that

$$\int_{\Re^3} G(\mathbf{x}, \mathbf{y}) d\mathbf{y} = 1, \tag{10.2}$$

the filtering of a constant field returns the same constant. After the operation on the right-hand side of Eq. (10.1) $\langle \mathbf{U} \rangle(\mathbf{x}, t)$ can still be expected to be random. The unresolved part of the velocity field is denoted as $\mathbf{u}'$ and is defined according to

$$\mathbf{u}' \equiv \mathbf{U} - \langle \mathbf{U} \rangle. \tag{10.3}$$

Since, unlike traditional averaging, it is not necessarily true for all filters that $\langle \langle \mathbf{U} \rangle \rangle = \langle \mathbf{U} \rangle$, it also may not be true that $\langle \mathbf{u}' \rangle = 0$.

For flows that are homogeneous a function of a single position vector $\mathbf{r}$, say $G^*(\mathbf{r})$, can be introduced so that

$$G(\mathbf{x}, \mathbf{y}) = G^*(\mathbf{x} - \mathbf{y}), \tag{10.4}$$

in which case Eq. (10.1) gives

$$\langle \mathbf{U}\rangle(\mathbf{x},t) = \int_{\Re^3} G^*(\mathbf{x}-\mathbf{y})U(\mathbf{y},t)d\mathbf{y}. \tag{10.5}$$

Assuming further that $G^*(\mathbf{r})$ has bounded support then it may be shown (see Problem 10.1) that the filtering operation, just like the averaging operation, has the helpful property of commuting with differentiation [1], so that

$$\left\langle \frac{\partial U_i}{\partial x_j} \right\rangle = \frac{\partial \langle U_i \rangle}{\partial x_j}. \tag{10.6}$$

This property is particularly helpful in simplifying the governing equations for filtered velocity, as will be seen below.

For the general case of non-homogeneous flows, it may be anticipated that the filter needs to reflect local changes in the scales that characterize the local turbulence. For example, as a boundary is approached filtering should be performed over a smaller region that takes into account the details of the wall. Filters in the form of Eq. (10.4) may not always be compatible with changing conditions from one flow region to another so that in such cases an inhomogeneous filter would have to be used to provide the desired flexibility. It is usual practice, however, to assume the validity of Eq. (10.6) in all circumstances with the justification that the errors caused by its use will typically be of the same order of magnitude as the errors associated with numerically solving the governing equations. On the other hand, if a higher order numerical scheme is used to solve the LES equations, then the errors generated by Eq. (10.6) may have to be given analytic form and evaluated as part of the numerical scheme [2].

It is also common practice to consider filters which can be written as products of 1D filters, say $H(r)$, so that, for example,

$$G^*(\mathbf{x}-\mathbf{y}) = H(x_1 - y_1)H(x_2 - y_2)H(x_3 - y_3). \tag{10.7}$$

Assuming that the integral of H over the real line is unity, Eq. (10.2) is satisfied. It is also frequently the case that different 1D filters are used in different directions, the goal being to take advantage of flow symmetries that may be present. Moreover, the filters may be homogeneous in some directions and inhomogeneous in others. For example, in a channel flow, the streamwise and spanwise directions are homogeneous and the wall-normal direction is not, so filters reflecting this property can be applied. In fact, the inhomogeneous direction in this case is often not given an explicit filter, with reliance instead placed on a de facto implicit filter that comes about from applying discrete approximations to the differential expressions in the inhomogeneous direction.

There are three particular 1D filters that are commonly used in LES. The first is the so-called *Gaussian filter*

$$H(r) = \sqrt{\frac{6}{\pi h^2}}\, e^{-6r^2/h^2} \tag{10.8}$$

in which h is a length scale taken to be indicative of the boundary between resolved and unresolved parts of the turbulent field. The *top hat filter* is defined by

$$H(r) = \begin{cases} 1/h & \text{if } |r| \leq h/2 \\ 0 & \text{otherwise} \end{cases} \tag{10.9}$$

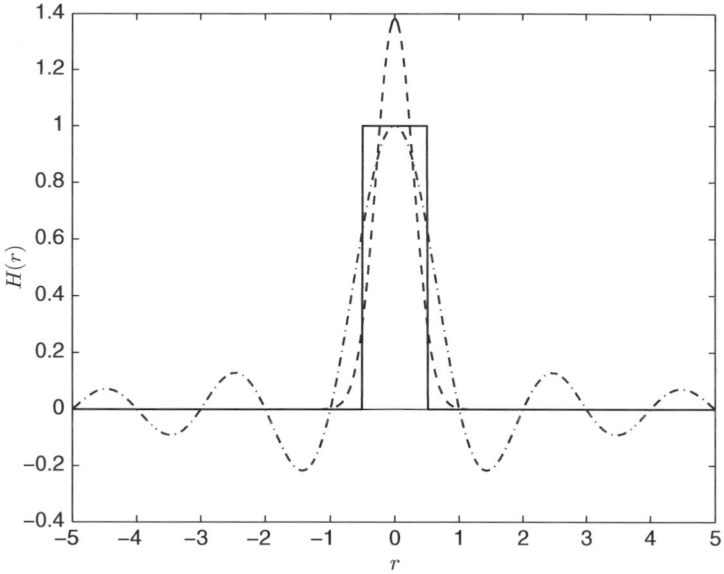

Figure 10.1 Filters: —, top hat; −−, Gaussian; − · −, sharp Fourier cut-off.

while the *sharp Fourier cut-off filter* is given by

$$H(r) = \frac{\sin(\pi r/h)}{\pi r}. \tag{10.10}$$

The differences between these functions is evident from their depictions in Figure 10.1.

The top hat filter is tantamount to equating filtering with averaging over a local region of fixed size surrounding each point. The Gaussian filter is a similar idea except that the averaging produced by the use of Eq. (10.8) is weighted toward the velocity closest to the point of interest. Equation (10.10) is called the sharp Fourier cut-off filter because it is the inverse Fourier transform of what amounts to a "top hat" function in Fourier space. In other words, this is a filter that cuts off contributions from all wave numbers with magnitudes greater than a certain value, in fact, π/h. To see this, note that the Fourier transform of H is defined by

$$\hat{H}(k) \equiv \int_{\Re} e^{\imath rk} H(r) dr, \tag{10.11}$$

which has the corresponding inverse transform

$$H(r) = \frac{1}{2\pi} \int_{\Re} e^{-\imath rk} \hat{H}(k) dk. \tag{10.12}$$

Evaluation of Eq. (10.11) after substituting Eq. (10.10) can be shown to yield

$$\hat{H}(k) = \begin{cases} 1 \text{ if } |k| \le \pi/h \\ 0 \text{ otherwise} \end{cases} \tag{10.13}$$

so Eqs. (10.10) and (10.13) form a Fourier transform pair.

The significance of Eq. (10.13) can be illustrated by considering the case of turbulence in a cubical domain with periodic boundary conditions enforced in all directions. In this case, $\mathbf{U} = \mathbf{u}$ is given by Eq. (2.50) and it is straightforward to show that taking a Fourier transform of Eq. (10.5) and then applying Eqs. (10.7) and (10.10) gives

$$\widehat{\langle U_i \rangle}(\mathbf{k}, t) = \widehat{H}(k_1)\widehat{H}(k_2)\widehat{H}(k_3)\widehat{U}_i(\mathbf{k}, t). \tag{10.14}$$

After utilizing Eq. (10.13), (10.14) yields

$$\langle U_i \rangle(\mathbf{x}, t) = \sum_{|k_1|<\pi/h} \sum_{|k_2|<\pi/h} \sum_{|k_3|<\pi/h} \hat{U}_i(\mathbf{k}, t) e^{\imath \mathbf{k}\cdot\mathbf{x}}. \tag{10.15}$$

Thus, the sharp Fourier cut-off filter is equivalent to the act of truncating a Fourier expansion of the velocity field.

The sharp Fourier cut-off filter best satisfies our intuitive notion of what a filter should do in the sense that it exactly eliminates all modes with wave numbers larger than some cut-off. In this case resolved and unresolved scales have a precise meaning. Moreover, for the sharp Fourier cut-off filter $\langle u_i' \rangle = 0$, since $\langle \langle U_i \rangle \rangle = \langle U_i \rangle$ in this case. One of the drawbacks of the sharp Fourier cut-off filter stems from the fact that the Fourier representation of a spatially localized structure such as a discrete vortex is highly inefficient since many modes in the expansion in Eq. (10.15) will be non-zero and thus necessary for representing the eddy. In contrast, filters such as the top hat and Gaussian are better positioned to well represent coherent vortical structures since the filtered object will be a smooth version of the original self.

To derive an equation governing the filtered velocity field, the operation in Eq. (10.5) is applied to the Navier–Stokes equation under the proviso that filtering commutes with differentiation in all terms. The result is

$$\frac{\partial \langle U_i \rangle}{\partial t} + \langle U_j \rangle \frac{\partial \langle U_i \rangle}{\partial x_j} = -\frac{1}{\rho}\frac{\partial \langle P \rangle}{\partial x_i} + \nu \nabla^2 \langle U_i \rangle - \frac{\partial \tau_{ij}}{\partial x_j}, \tag{10.16}$$

where

$$\tau_{ij} \equiv \langle U_i U_j \rangle - \langle U_i \rangle \langle U_j \rangle \tag{10.17}$$

is referred to as the *subgrid-scale stress tensor*. Note that similar to the way that "Reynolds stress" is used interchangeably with R_{ij} and $\sigma_{t_{ij}} = -\rho R_{ij}$, both τ_{ij} and $-\rho\tau_{ij}$ can be referred to as the subgrid-scale stress tensor. Filtering the continuity equation in the same way as done to the Navier–Stokes equation shows that the filtered velocity field is also incompressible and satisfies

$$\frac{\partial \langle U_i \rangle}{\partial x_i} = 0. \tag{10.18}$$

It may be noted that the equations produced by filtering are formally identical to the equations produced by averaging, specifically Eqs. (3.9) and (3.10), though there is a large conceptual difference in how they are to be interpreted. In particular, $\langle U_i \rangle$ is random, while $\overline{U}_i$ is deterministic. τ_{ij} must be modeled for the same reasons that R_{ij} must be modeled in Eq. (3.9). Since τ_{ij} formally represents the influence that the unresolved part of the flow field has on the resolved part, it cannot be interpreted as characterizing the physical process of momentum transport in the same way that the Reynolds stress tensor can. On the other hand, it is expected that τ_{ij} will make a substantially smaller contribution to the dynamics of $\langle \mathbf{U} \rangle$ than R_{ij} does to $\overline{\mathbf{U}}$. If this is true, then it may be imagined that errors in modeling τ_{ij} will have less impact on the quality of predictions than do errors in modeling R_{ij}. This is a premise that motivates much of the interest in LES.

Though a reduced reliance on modeling is a fundamental advantage of LES over RANS, it is only achieved at the price of a great increase in computational expense compared to solving systems of averaged equations. Moreover, numerical solutions of the LES equations give the filtered velocity only. Whatever information they give about the unresolved field is contained in the form of the assumed subgrid model. Consequently, LES cannot provide unambiguous estimates of $\overline{U}_i$ and other averaged quantities such as K. To see this, note that it follows from Eq. (10.3) that

$$\overline{\mathbf{U}} = \overline{\langle \mathbf{U} \rangle} + \overline{\mathbf{u}'}. \tag{10.19}$$

While $\overline{\langle \mathbf{U} \rangle}$ is readily available by averaging the LES solution, $\overline{\mathbf{u}'}$ is not, and it cannot be said with certainty that $\overline{\mathbf{u}'} = 0$ for any particular filter. Thus, in a formal sense the mean velocity is not always knowable from LES.

If it can be argued that the contribution to $\overline{\mathbf{U}}$ of the small scales (as manifested in $\overline{\mathbf{u}'}$) is small then it follows that

$$\overline{\mathbf{U}} \approx \overline{\langle \mathbf{U} \rangle} \tag{10.20}$$

so that a prediction of $\overline{\mathbf{U}}$ is possible. As a matter of course, LES computations assume that this is a legitimate step. If it were not true, then the usefulness of LES would be much restricted. It is also interesting to note that any observed differences between $\overline{\mathbf{U}}$ and $\overline{\langle \mathbf{U} \rangle}$ may possibly be due to non-negligible contributions from $\overline{\mathbf{u}'}$. This is in addition to any errors deriving from subgrid modeling or numerical errors.

The idiosyncrasies of filtering have larger impact on the possibility for obtaining estimates of quantities such as the turbulent kinetic energy, $K = \overline{u_i^2}/2$. Thus, note the identity

$$u_i = [\langle U_i \rangle - \overline{\langle U_i \rangle}] + [u_i' - \overline{u_i'}] \tag{10.21}$$

that follows from Eqs. (10.3) and (10.19). Equation (10.21) says that the velocity fluctuation in the traditional sense is the sum of the velocity fluctuation of the resolved field, say $u_i^r \equiv \langle U_i \rangle - \overline{\langle U_i \rangle}$, plus the velocity fluctuation of the unresolved scales, $u_i^u \equiv u_i' - \overline{u_i'}$. Thus

$$u_i = u_i^r + u_i^u \tag{10.22}$$

and

$$K = \frac{1}{2}\overline{(u_i^r)^2} + \overline{u_i^r u_i^u} + \frac{1}{2}\overline{(u_i^u)^2}. \tag{10.23}$$

This shows that K is the sum of the energy of the resolved and unresolved scales given by the first and third terms on the right-hand side, respectively, plus a correlation term between the scales. Only the first of the three terms on the right-hand side of Eq. (10.23) is computable in a simulation. Assuming that the middle term in Eq. (10.23) is comparatively small since it lacks the obvious source of correlation that the other terms have, the absence of knowledge of $\overline{(u_i^u)^2}/2$ may preempt the possibility of making accurate estimates of K in some circumstances. For a well-resolved LES this is not likely to be an issue since by definition $\overline{(u_i^u)^2}/2$ is small and the computation can be expected to be similar to that of a DNS. In this case the assumption that

$$K \approx \frac{1}{2}\overline{(u_i^r)^2} \tag{10.24}$$

may be reasonable. In contrast, for a poorly resolved LES in which the principal structures are not well resolved, it may be expected that $\overline{(u_i^u)^2}/2$ represents a sizable portion of K, thus calling into question the validity of Eq. (10.24). Particularly near boundaries much of the energy resides in structures that are unresolved unless extraordinary effort is made to increase the mesh density near walls. It is also the case that the subgrid energy is not readily modeled in an LES unless special effort is made to do so. For example, if it is assumed that $\tau_{ii} \approx \overline{(u_i^u)^2}$, as would follow from Eq. (10.17) if $\langle\langle U_i\rangle\rangle = \langle U_i\rangle$ and $\langle\langle U_i\rangle\langle U_i\rangle\rangle = \langle U_i\rangle\langle U_i\rangle$, this still does not provide a useful value for $\overline{(u_i^u)^2}$ since models for τ_{ij} tend to apply to the off-diagonal terms only, as will be seen below. However, by including a modeled equation for the unresolved kinetic energy K in the LES model it is at least theoretically possible to get an estimate of the total kinetic energy.

As a general rule if an LES resolves at least 3/4 of the energy, then it is likely to make credible predictions of K. One final point is that in view of the expectation that dissipation generally takes place at scales that are smaller than those containing energy, dissipation is expected to occur within the subgrid and hence will be largely determined through modeling.

10.2 Numerical Considerations

It is standard practice in computational fluid dynamics (CFD) to insist that numerical solutions to the fluid equations are grid independent. This is a property one expects in DNS as well as in the solution of the RANS equations. In the case of LES, on the other hand, since the mesh resolution usually coincides with the smallest resolved scales, the grid spacing is a modeling parameter and it is inevitable that the numerical solutions for the filtered velocity are grid dependent. In view of Eqs. (10.20) and (10.24), it is likely that even if the subgrid models that one uses were exactly correct, the mean statistics that are calculated as output of the simulation will change with mesh resolution. Since subgrid models are, in fact, at best approximate, mesh resolution tends to have a major influence on LES solutions. Since enhanced resolution tends to bring LES closer to DNS, LES calculations usually are performed with the largest affordable mesh in the hope of reducing the importance of subgrid models and the errors that are associated with them.

The fact that LES computations are grid-dependent via their incorporation of a subgrid model has the potential to mask grid dependencies that arise from the numerical schemes used in solving the filtered equations. For example, whenever the mesh is inadequate to resolve physically appropriate sharp flow features that form in the course of the calculation, numerical diffusion is likely to provide unphysical smoothing that changes the flow physics. While it may be possible to somewhat reduce the effects of numerical diffusion by adopting higher order algorithms, such schemes typically incur much greater costs [3]. In any event, numerical smoothing occurs over and above any unphysical smoothing that derives from the subgrid modeling. All of these sources of smoothing should be absent in a DNS. In some cases, numerical diffusion errors in an LES can exceed the subgrid stresses in magnitude, thereby casting a cloud over the legitimacy of the LES simulation [4–6].

Since numerical diffusion to some degree is almost certain to be present in turbulent flow simulations on under-resolved meshes, regardless of the presence of subgrid models, a simulation technique referred to as *implicit LES* has arisen based on the idea of omitting the explicit use of subgrid modeling in favor of relying exclusively on numerical diffusion to account for subgrid physics [7, 8]. Such schemes, to be useful, have the property of remaining stable in the face of under-resolution. This type of analysis is clearly simpler to implement than traditional LES incorporating a subgrid model. However, for any flow where it is known that under-resolved DNS or traditional LES fail to accurately predict the flow statistics, it cannot be expected that implicit LES will be more successful. Since LES is often problematical next to solid boundaries due to inadequate mesh resolution, implicit schemes suffer the same limitations. It will be seen below that a common strategy for avoiding this problem is to develop boundary conditions in the form of *wall functions* similar to those developed for RANS models. These are applied outside the most sensitive part of the near-wall flow, so that this region does not have to be simulated.

While the ultimate judge as to the accuracy and usefulness of LES schemes comes from their application to engineering flows, the availability of DNS data allows for testing of subgrid models in isolation from the other aspects of the LES schemes. In these so-called *a priori tests* the DNS data allows for evaluation of τ_{ij} as well as the models that are designed to represent it. Such tests do not typically show great accuracy at the local, pointwise level, though they may show reasonably good results over larger regions. This helps to explain some of the positive results than can be had with LES despite the presence of significant errors exposed in a priori testing. How LES schemes perform in engineering calculations is the most important means for evaluating their usefulness.

10.3 Subgrid-Scale Models

Subgrid-scale models tend to adhere to the same assumptions that motivate the Reynolds stress/mean rate-of-strain relation in Eq. (9.1). The analogous model in the context of LES is given by

$$\tau_{ij} - \frac{1}{3}\tau_{kk}\,\delta_{ij} = -2\nu_T\langle S\rangle_{ij}, \tag{10.25}$$

where

$$\langle S\rangle_{ij} = \frac{1}{2}\left(\frac{\partial\langle U_i\rangle}{\partial x_j} + \frac{\partial\langle U_j\rangle}{\partial x_i}\right) \tag{10.26}$$

is the filtered rate-of-strain tensor and ν_T is a subgrid-scale eddy viscosity. Since most LES modeling depends on Eq. (10.25), determination of effective eddy viscosities is the primary focus for modeling the subgrid stress tensor. Much less common are models that pursue non-linear generalizations to Eq. (10.25) that provide a means for modeling anisotropy in the same way that NLEVM and ARM closures provide this generality in RANS modeling. One example is the explicit algebraic sub-grid stess model (EAM) [9] that includes a term equivalent to that with coefficient A_3 in Eq. (9.114). The inclusion of anisotropic effects near solid boundaries is found to be of considerable aid in improving the efficiency and accuracy of LES in some circumstances [10].

For traditional LES modeling based on Eq. (10.25) a substitution into (10.16) gives the equation for $\langle U_i \rangle$ as

$$\frac{\partial \langle U_i \rangle}{\partial t} + \langle U_j \rangle \frac{\partial \langle U_i \rangle}{\partial x_j} = -\frac{\partial}{\partial x_i}(\langle P \rangle/\rho + \tau_{kk}/3) + 2\frac{\partial}{\partial x_j}((\nu + \nu_T)\langle S \rangle_{ij}). \tag{10.27}$$

Apart from the fact that ν_T is not necessarily constant, Eq. (10.27) is formally identical to the Navier–Stokes equation. In fact, in view of the computationally intensive demands of LES there is little leeway for developing more complicated subgrid models that do not take advantage of the fast algorithms already in place for solving equations such as (10.27).

The appearance of τ_{kk} in Eq. (10.27) is in combination with the pressure so that it does not have to be separately evaluated. In fact, a contraction of Eq. (10.25) shows that this relation gives no information about τ_{kk}. The comparable situation was faced in the RANS context where information about K had to come from a separate model of the K equation since it was not available from Eq. (9.1). Equation (10.25) implies that ν_T serves as the proportionality constant relating left- and right-hand sides for six different components of τ_{ij}. The same kind of simplification was shown to be a weakness of Reynolds stress modeling and it is not unexpected that similar problems will also occur for LES based on Eq. (10.25). Indeed, a priori tests show that only a weak correlation exists between the principle axes of the tensors on either side of Eq. (10.25). This means, similar to the RANS case, that there very well might not be any choice of ν_T that can guarantee accurate simulations via LES.

Another difficulty with Eq. (10.25) concerns the sign of ν_T. Since ν_T appears in (10.27) in the guise of a diffusion term, it is generally true that the condition $\nu_T > 0$ is necessary for the stability of numerical schemes that solve the LES equations. On the other hand, there are times when ν_T should be negative, a conclusion that can be reached from considering the equation for the resolved turbulent kinetic energy, $\langle K \rangle \equiv \frac{1}{2}\langle U_i \rangle^2$. To derive this relation, multiply Eq. (10.16) by $\langle U_i \rangle$ and following steps similar to those used in deriving the equation for $\overline{K} = \frac{1}{2}\overline{U}_i^2$ in Eq. (3.24) it is found that

$$\begin{aligned}\frac{\partial \langle K \rangle}{\partial t} + \langle U_j \rangle \frac{\partial \langle K \rangle}{\partial x_j} &= -\frac{1}{\rho}\frac{\partial \langle P \rangle \langle U_j \rangle}{\partial x_j} + \nu \nabla^2 \langle K \rangle \\ &\quad - \frac{\partial \tau_{ij} \langle U_i \rangle}{\partial x_j} - \nu \frac{\partial \langle U_i \rangle}{\partial x_j}\frac{\partial \langle U_i \rangle}{\partial x_j} + \tau_{ij}\langle S \rangle_{ij}.\end{aligned} \tag{10.28}$$

The terms on the right-hand side of Eq. (10.28) account for, respectively, total pressure work, viscous diffusion, subgrid stress diffusion, viscous dissipation, and in the last term, ostensibly the loss of $\langle K \rangle$ to energy in the subgrid where it will be dissipated. This interpretation of $\tau_{ij}\langle S \rangle_{ij}$ is consistent with ν_T being strictly positive since according to Eq. (10.25)

$$\tau_{ij}\langle S \rangle_{ij} = -2\nu_T \langle S \rangle_{ij}\langle S \rangle_{ij} \tag{10.29}$$

and ν_T controls the sign of the expression on the right-hand side. However, DNS has shown that $\tau_{ij}\langle S \rangle_{ij}$ is often positive, suggesting that in such locations this term accounts for the production of resolved energy out of the subgrid motion. This reverse process is known as *backscatter* and is not easily captured using the traditional eddy viscosity formula indicated in Eq. (10.25).

The consequence of adhering to the condition that $\nu_T \geq 0$ is that the physical process of backscatter does not get included in simulations. Somewhat mitigating this limitation of the eddy viscosity model is the fact that regions where $\tau_{ij}\langle S\rangle_{ij}$ is negative tend to be favored over regions where it is positive, so when averaged over a large enough region the sign of $\tau_{ij}\langle S\rangle_{ij}$ tends to be negative. This explains why maintaining $\nu_T > 0$ does not prevent LES calculations from making reasonable average predictions, despite their inevitable distortion of the local physics in some locations.

A model for τ_{ij} that does allow for inclusion of some degree of backscatter can be developed by creating a more active role for the filtering operation. In these *scale similarity models* Eq. (10.3) is used to replace **U** in τ_{ij}, giving

$$\tau_{ij} = [\langle\langle U_i\rangle\langle U_j\rangle\rangle - \langle U_i\rangle\langle U_j\rangle] + [\langle u_i'\langle U_j\rangle\rangle + \langle u_j'\langle U_i\rangle\rangle] + \langle u_i' u_j'\rangle. \tag{10.30}$$

The first term on the right-hand side, which may be evaluated from $\langle U_i\rangle$, is referred to as the *Leonard stress* [1]. It may be thought of as representing the creation of small-scale motions from the resolved scales, as is easily visualized in the case of the sharp Fourier cut-off filter when the only Fourier modes that appear in the Leonard stress are beyond the cut-off wave number. The second term in Eq. (10.30) is referred to as the subgrid-scale cross stress and is imagined to directly connect resolved and unresolved scales and thus may be a source of backscatter. The final term is called the subgrid-scale Reynolds stress and is expected to account for the influence of unresolved scales on resolved scales and thus also may be associated with backscatter.

The scale similarity model assumes that there is value in keeping the Leonard stress as an independent entity, as against adopting a model such as Eq. (10.25) in which the decomposition in Eq. (10.30) is not introduced. An additional assumption that is often made [11] is to assume that the cross stress term in Eq. (10.30) can be modeled as

$$\langle u_i'\langle U_j\rangle + u_j'\langle U_i\rangle\rangle = \langle U_i\rangle\langle U_j\rangle - \langle\langle U_i\rangle\rangle\langle\langle U_j\rangle\rangle, \tag{10.31}$$

with the consequence that

$$\tau_{ij} = \langle\langle U_i\rangle\langle U_j\rangle\rangle - \langle\langle U_i\rangle\rangle\langle\langle U_j\rangle\rangle + \langle u_i' u_j'\rangle. \tag{10.32}$$

To complete the model, the subgrid term in Eq. (10.32) is given a diffusive form yielding

$$\begin{aligned}\tau_{ij} - \frac{1}{3}\tau_{kk}\,\delta_{ij} &= [\langle\langle U_i\rangle\langle U_j\rangle\rangle - \langle\langle U_i\rangle\rangle\langle\langle U_j\rangle\rangle] \\ &\quad - \frac{1}{3}\delta_{ij}[\langle\langle U_k\rangle\langle U_k\rangle\rangle - \langle\langle U_k\rangle\rangle\langle\langle U_k\rangle\rangle] - 2\nu_T\langle S\rangle_{ij}.\end{aligned} \tag{10.33}$$

In this, ν_T can be taken to be any of the forms normally used in Eq. (10.25), though with adjusted model constants. It may be checked that Eq. (10.33) is self-consistent in the sense that both sides satisfy Galilean invariance [12] to the effect that the terms have form invariance for all inertial observers. In fact, a main motivation for the modeling used in obtaining Eq. (10.32) is that with it, Galilean invariance of Eq. (10.33) is assured. Computations show that the scale invariant part of Eq. (10.33), which is the first term on the right-hand side, can make substantial contributions to backscatter [13]. Moreover, because the last term in Eq. (10.33) does not act alone it has less of a role in simulations than it does in models such as Eq. (10.25) [14]. On the other hand, the inclusion of the diffusive term in Eq. (10.33) is considered essential since the scale similarity term by itself does not allow for sufficient energy dissipation. It is generally accepted that "mixed" models such as Eq. (10.33) are superior to either models such as Eq. (10.25) or scale similarity models by themselves.

Models incorporating Eq. (10.25) do not depend on the filter since it never appears in the equations determining $\langle U_i \rangle$. In this case a priori testing of the effect of filtering on the accuracy of the models provides no benefit. In contrast, Eq. (10.33) is filter sensitive and there may be some advantages to performing a priori tests. Numerical experiments show [15] that the Gaussian filter is a good choice to use with the scale similarity model because both model and filter appear to have a similar propensity to accommodate the influence of a relatively broad range of length scales.

In those instances when the filtering of flow variables must be computed, it is necessary to numerically evaluate a filtering operation, and the best way to proceed depends on which filter is being applied, the particular flow field, its symmetries, and the properties of the numerical mesh and algorithm. The sharp cut-off filter is readily applied in spectral space, while the top-hat filter can naturally be implemented in real space by averaging over nearby values in the mesh. The Gaussian filter, after truncation so that it has finite support, can be applied by quadrature in either physical or spectral space. Oftentimes, LES implementations use different filters in different directions, so a combination of techniques for evaluating filters may be necessary.

10.3.1 Smagorinsky Model

The first ideas for modeling the subgrid stress tensor borrow the gradient diffusion model as it appears in the mixing length theory. Thus, as given in the widely used Smagorinsky [16] model,

$$\nu_T = (C_S h)^2 |\langle S \rangle|, \tag{10.34}$$

where C_S is referred to as the "Smagorinsky constant," h is a length scale, and $|\langle S \rangle| \equiv (2\langle S \rangle_{ij} \langle S \rangle_{ij})^{1/2}$. Equation (10.34) is based on assuming that $\nu_T \sim \mathcal{U} L$, as is done for RANS modeling in Eq. (9.7), and then choosing the velocity scale $\mathcal{U} = h|\langle S \rangle|$ and length scale h. Another route to the same result is the recognition that the choice of $\mathcal{U}$ for the Smagorinsky model is what is necessary to equilibrate the loss of resolved energy given by the term $\tau_{ij}\langle S \rangle_{ij}$ in Eq. (10.28) with the modeled subgrid energy dissipation rate $\mathcal{U}^3/h$, a scaling which was previously considered in regard to Eq. (9.28). Note that this argument is ultimately a global one since it ignores the regions of backscatter.

In practice h usually is defined to be a function of the local gridding with the convention that for a uniform mesh h is taken to be twice the grid spacing. For grids whose spacing is not the same in every direction, h may be formed from the geometric mean of the grid spacings or by other formulas. As in all mixing length models, ad hoc adjustment of the Smagorinsky constant C_S and or subgrid scale h is required near boundaries so as to reflect the reduced scale of the local eddies that are responsible for transport. In fact, to be consistent with the vanishing of τ_{ij} at the wall surface, $\nu_T = 0$ must be zero at the same location. This requirement conflicts with Eq. (10.34) so it is customary to apply an ad hoc modification to this formula in order to achieve consistency. A common technique in this regard is to employ van Driest damping as given in Eq. (9.15) for mixing length models so that h is replaced near the boundary by the empirical relation $h(1 - e^{-y^+/25})$. Implementation of this condition requires a method for accurately computing y^+ which means that the surface shear stress must be calculated. This then requires a fine mesh near the wall thus raising the cost of the simulation. For complicated geometries it is likely to be much less obvious how h appearing in Eq. (10.34) should be modified in order to force $\nu_T = 0$ at the surface.

Several alternative models for ν_T that will be discussed below have the property of naturally reducing to zero at the wall surface. Another way to do this is to limit the use of the Smagorinsky model to a region off the wall while a scheme that is better able to accommodate the properties of ν_T near the surface is utilized in the zone adjacent to the surface. Such approaches include the hybrid LES/RANS methods considered in Section 10.4. Another means to this end is to employ wall function boundary conditions as mentioned in Section 10.2. Essentially, wall functions are empirical relations giving the flow variables near solid boundaries, so that the near-wall boundary layer does not have to be explicitly calculated.

While van Driest damping can help improve the physicality of Eq. (10.34) in the region very close to boundaries, the Smagorinsky constant C_S is subject to differing requirements depending on whether or not a region is within the sphere of influence of a solid boundary. In fact, tuning the Smagorinsky model for a far-field flow will lead to an excess of dissipation in the near-wall regions. Modification of the Smagorinsky constant with position in relation to walls is generally necessary as a consequence.

An estimate of C_S in the region away from boundaries can be achieved via a heuristic argument [17, 18]. Thus, assume that turbulence is locally stationary, homogeneous, and isotropic. Assume that $0 < k < k_c$ is the range of resolved wave numbers where k_c lies in the inertial range. In view of Eq. (10.13) it can be assumed that $k_c = \pi/h$. The energy dissipated from the resolved field, assuming that transport is governed by an eddy viscosity ν_T, can be deduced in analogy to the analysis leading to Eq. (4.12) so that

$$\epsilon = 2\nu_T \int_0^{k_c} k^2 E(k)dk. \tag{10.35}$$

Since dissipation is minimal at small wave numbers and k_c extends to the inertial range, it may be assumed that Eq. (4.57) can be used to evaluate the right-hand side of Eq. (10.35), yielding

$$\epsilon = \frac{3}{2}\nu_T C_K \epsilon^{2/3} k_c^{4/3}, \tag{10.36}$$

an equation which can then be solved for ϵ. A second approximation of ϵ can be made from the equilibrium assumption to the effect that ϵ is balanced by the production term $-\tau_{ij}\langle S_{ij}\rangle$. Equating the latter two expressions and assuming that ν_T has the Smagorinsky form given in Eq. (10.34) leads to a formula for C_S [19] (see Problem 10.5). Taking the Kolmogorov constant $C_K = 1.4$, a calculation then gives $C_S = 0.18$.

This estimation of C_S proves to be overly dissipative next to boundaries in flows containing mean shear, such as a boundary layer or channel. A refined version of the methodology for predicting C_S [18] incorporates some improvements in the physical modeling and leads to a smaller value that is more closely aligned with the value $C_S = 0.065$ that is empirically observed to be appropriate for boundary layers [20]. In view of the fact that it is actually C_S^2 that appears in the eddy viscosity, an order of magnitude reduction in the coefficient of ν_T is realized by this change in coefficient.

On the whole, however, the need to modify C_S from one region to the next imposes the same kind of hardships on the implementation of the Smagorinsky model as occurs in the mixing length model used in RANS calculations. In many applications the Smagorinsky model proves to be overly dissipative and tends to smooth turbulence fluctuations, even to the extent of eradicating boundary layer structures. It is also known to prevent transition in very much the same way as happens for equivalent RANS models.

10.3.2 WALE Model

The relative ease with which the Smagorinsky model can be adopted for applications has helped promote its widespread use despite its well-known limitations. The particular form adopted for the eddy diffusivity ν_T in this case as given in Eq. (10.34) is based on imitating the mixing length transport model used in RANS computations. There is no reason why alternative forms should not be considered in the same way that Eqs. (9.18) and (9.20) were used in deriving alternative generalized forms of the mixing length model. All such approaches have in common the use of the mean velocity gradient tensor in developing subgrid-scale models.

One such alternative to the Smagorinsky model that is built from the mean velocity gradient tensor and is readily incorporated into LES codes is the wall-adapting local eddy (WALE) viscosity model [19]. This model is designed specifically to overcome some of the weaknesses of the Smagorinsky model without adding significant new complexities or expense to mean flow simulations.

The WALE model is based on incorporating an invariant of the filtered velocity gradient tensor within the eddy viscosity model that is more responsive to the local physics than the quantity $|\langle S\rangle| = \sqrt{2\langle S\rangle_{ij}\langle S\rangle_{ij}}$ in Eq. (10.34), which depends only on the rate of strain tensor. Among the limitations of the Smagorinsky form of ν_T is that it is not receptive to the influence that rotational motions are likely to have on transport and dissipation. This has implications for how energy dissipation at subgrid scales is modeled in Eq. (10.28). Computations and experiment suggest that rotational motions are often locations of significant dissipation of the turbulent kinetic energy so that failure to accommodate such motions in the Smagorinksy model contributes to its overall marginal performance.

The WALE model is built up from the invariant, traceless, symmetric tensor defined from the filtered velocity gradient tensor as

$$S_{ij}^d = \frac{1}{2}\left(\frac{\partial\langle U_i\rangle}{\partial x_k}\frac{\partial\langle U_k\rangle}{\partial x_j} + \frac{\partial\langle U_j\rangle}{\partial x_k}\frac{\partial\langle U_k\rangle}{\partial x_i}\right) - \frac{1}{3}\delta_{ij}\frac{\partial\langle U_k\rangle}{\partial x_l}\frac{\partial\langle U_l\rangle}{\partial x_k} \tag{10.37}$$

that accommodates both the influence of straining and rotational motions. Substituting

$$\frac{\partial\langle U_i\rangle}{\partial x_j} = \langle S\rangle_{ij} + \langle W\rangle_{ij} \tag{10.38}$$

into Eq. (10.37) and squaring yields

$$S_{ij}^d S_{ij}^d = \frac{1}{6}(S^2S^2 + \Omega^2\Omega^2) + \frac{2}{3}S^2\Omega^2 + 2I(\langle S\rangle_{ik}\langle S\rangle_{kj}\langle W\rangle_{jl}\langle W\rangle_{li}), \tag{10.39}$$

where $S^2 \equiv \langle S \rangle_{ij} \langle S \rangle_{ij}$ and $\Omega^2 \equiv \langle W \rangle_{ij} \langle W \rangle_{ij}$. Clearly, both strain and rotation play a role in determining the magnitude of $S^d_{ij} S^d_{ij}$. The quantity $S^d_{ij} S^d_{ij}$ has a number of useful properties that can be harnessed in developing a subgrid eddy viscosity that is more responsive to the physics of turbulent flow than the Smagorinsky model.

Among the desirable properties of $S^d_{ij} S^d_{ij}$ is that it is an invariant of the motion. Moreover, in a unidirectional flow, such as laminar flow in a channel where $\langle \mathbf{U} \rangle$ is non-random and has components $(U(y), 0, 0)$, the only non-zero components of $\langle S \rangle_{ij}$ and $\langle W \rangle_{ij}$ are

$$\langle S \rangle_{12} = \langle S \rangle_{21} = \langle W \rangle_{12} = -\langle W \rangle_{21} = (dU/dy)/2. \tag{10.40}$$

A calculation in this case (see Problem 10.4) shows that $S^d_{ij} S^d_{ij} = 0$. This may be contrasted with the invariant S in the Smagorinsky model which for the same conditions is non-zero. For flows that are approximately unidirectional, such as a Blasius boundary layer, it can be expected that $S^d_{ij} S^d_{ij}$ is much diminished in magnitude. This property implies that unlike the traditional Smagorinsky model, models such as the WALE model that are proportional to Eq. (10.39) will naturally diminish in regions of laminar flow. This also means that this kind of model does not tend to erroneously impose the presence of turbulent flow in upstream regions prior to the natural appearance of turbulent motion after transition. This is a major benefit in comparison to the Smagorinsky model among others.

The way in which the WALE model incorporates Eq. (10.39) is designed to take advantage of an additional property of $S^d_{ij} S^d_{ij}$, specifically the fact that near a solid boundary at $y = 0$

$$S^d_{ij} S^d_{ij} \sim y^2. \tag{10.41}$$

Consequently, since

$$\nu_t \sim y^3, \tag{10.42}$$

next to a boundary, according to Eq. (10.25) consistency with Eq. (10.42) is achieved if

$$\nu_T \sim (S^d_{ij} S^d_{ij})^{3/2}. \tag{10.43}$$

This is a limiting condition that cannot be attained with S in the Smagorinsky model since it is $O(1)$ next to solid boundaries.

Following a similar construction as in the Smagorinksy model, the eddy viscosity in the WALE model is derived from the product of a squared length scale and a term with time dependence t^{-1}. Since ν_T is to be proportional to $(S^d_{ij} S^d_{ij})^{3/2}$ (a factor that has time dependency t^{-6}) in order to arrive at an expression for ν_T with time unit proportional to t^{-1}, this term must be divided by a term with dimension t^{-5}. This leads to the expression

$$\nu_T = (C_w \Delta)^2 \frac{(S^d_{ij} S^d_{ij})^{3/2}}{(S^d_{ij} S^d_{ij})^{5/4} + (\langle S \rangle_{ij} \langle S \rangle_{ij})^{5/2}} \tag{10.44}$$

where C_w is a fixed constant everywhere and Δ is a given length scale. The particular choice of terms in the denominator prevents it from being zero at all locations in a flow. In fact, the term containing $\langle S \rangle_{ij}$ remains finite at solid surfaces in the presence of shearing, while at locations where $\langle S \rangle_{ij}$ itself might be zero it is unlikely that the term containing S^d_{ij} will also be zero. This prevents singularities in the eddy viscosity prediction.

It is evident from the form of Eq. (10.44) that codes for which the Smagorinsky model has been implemented will readily accommodate the WALE model. Moreover, with the WALE model there is no need to include ad hoc measures to reduce the eddy viscosity in the wall vicinity nor in upstream laminar flow regions. Applications of the WALE approach are now widespread and tend to confirm its capacity to offer improvement in comparison to the Smagorinksy model. As in the case of all eddy viscosity models, however, the WALE model cannot supply backscatter to the simulations, thus depriving them of an important degree of physicality.

10.3.3 Alternative Eddy Viscosity Subgrid-Scale Models

Very much in the same way that alternatives to the mixing length model have been developed that depend on such quantities as K and ϵ, so too subgrid models have been derived that depart from the Smagorinsky form to use different ideas about the fundamental scales upon which a subgrid-scale viscosity might depend. One example of this [21, 22] uses the local subgrid kinetic energy

$$K_{sgs} = \frac{1}{2}\tau_{kk}, \tag{10.45}$$

as the basis for the velocity scale of the eddy viscosity, so that

$$\nu_T = C_k h\sqrt{K_{sgs}}. \tag{10.46}$$

An equation to determine K_{sgs} can be developed by adopting the traditional modeling of the K equation for this purpose. For example, in a typical formulation the K_{sgs} equation takes the form

$$\begin{aligned}\frac{\partial K_{sgs}}{\partial t} + \langle U_j\rangle\frac{\partial K_{sgs}}{\partial x_j} &= 2hC_k\sqrt{K_{sgs}}\langle S\rangle_{ij}\langle S\rangle_{ij} - C_\epsilon\frac{K_{sgs}^{3/2}}{h}\\ &+\frac{\partial}{\partial x_j}\left((\nu + C_k h\sqrt{K_{sgs}})\frac{\partial K_{sgs}}{\partial x_j}\right),\end{aligned} \tag{10.47}$$

where the terms on the right-hand side represent production, dissipation, and transport, respectively. Note that since this is a one-equation model, the dissipation takes the characteristic form in Eq. (9.28) with C_ϵ a model parameter. Usual values of the parameters are $C_k = 0.07$ and $C_\epsilon = 1.05$. The length scale h is related to the grid spacing. Originally it was formulated to depend on the local mesh spacings in orthogonal directions as $h = (\Delta x\Delta y\Delta z)^{1/3}$. The choice $h = \min(\Delta x, \Delta y, \Delta z)$ has been found to be more effective in some circumstances [23].

Another approach toward eddy viscosity modeling [24, 25] has its origins in spectral theory as it was employed in deriving Eqs. (10.35) and (10.36). Specifically, Eq. (10.36) can be rewritten as

$$\nu_T = \frac{2}{3}C_K^{-1}\epsilon^{1/3}k_c^{-4/3}. \tag{10.48}$$

Using Eq. (4.57) this can be expressed alternatively as

$$\nu_T = \frac{2}{3}C_K^{-3/2}\left(\frac{E(k_c, t)}{k_c}\right)^{1/2}, \tag{10.49}$$

which is tantamount to taking the velocity scale $\mathcal{U} = \sqrt{E(k_c, t)k_c}$ and length scale $\mathcal{L} = k_c^{-1}$ in the definition of the eddy viscosity. Previously, the second-order structure function was connected to ϵ via Eq. (4.99). Replacing ϵ in Eq. (10.48), taking an average over $|r| = h$, assuming that $k_c = \pi/h$, and finally entering the values of the parameters yields

$$\nu_T = 0.063\, h\sqrt{(|\langle U_i\rangle(\mathbf{x}+\mathbf{r},t) - \langle U_i\rangle(\mathbf{x},t)|)^2_{|\mathbf{r}|=h}} \tag{10.50}$$

where the filtered velocity difference in the formula includes an average over the values of $|\mathbf{r}| = h$. A number of variants of this approach have been formulated which attempt to add additional physics into the determination of the eddy viscosity.

10.3.4 Dynamic Models

To the extent that LES can be trusted to faithfully represent the physics of the turbulent velocity field, it is generally the case that the smallest simulated scales contain useful information about the largely inviscid subgrid motions within the inertial range. Dynamic subgrid models are a formal means of estimating the coefficients in the subgrid stress tensor by leveraging information taken from the smallest resolved scales. The dynamic procedure has most often been used in the context of Eq. (10.25) incorporating (10.34) as well as with the scale similarity model Eq. (10.33), though it can be applied to other models. For concreteness the present discussion is framed in terms of the specific example of the Smagorinsky model.

The filtering operation in an LES determines the boundary between resolved and unresolved scales. The filter that is tied to the mesh used in a LES, referred to as the *grid* filter, is assumed to smooth away the action of smaller scale motions occurring in between the mesh points. Their influence on the computation is relegated to the subgrid-scale tensor defined in Eq. (10.25). If one considers a filter based on a somewhat larger scale than is used in the grid filter – but still, ideally, within the inertial range – the so-called *test filter*, then from the perspective of test filtering the flow at the scale of the grid filter is within its "invisible" subgrid range. Thus, unlike the grid filter the test filter has some information about its subgrid motion that it may be able to use to determine local values of its parameters. The dynamic model is built by taking advantage of this special knowledge to help model τ_{ij} associated with the grid filter.

To formally develop the dynamic model, assume that h is the scale of the grid filter and h_t is that of the test filter, so by definition $h_t > h$. For the purpose of the following discussion, test filtering is denoted by curly brackets as in $\{\mathbf{U}\}$ for the test filtered velocity field. Frequently in what follows both test and grid filters are applied, first the grid then the test filter, leading to notation such as $\{\langle U_i\rangle\}$. In the case of the sharp Fourier cut-off filter it is clear that $\{\langle U_i\rangle\} = \{U_i\}$, but this does not necessarily hold for other filters.

If the test and grid filters are applied consecutively to the Navier–Stokes equation, there appears the subgrid stress term

$$T_{ij} \equiv \{\langle U_i U_j\rangle\} - \{\langle U_i\rangle\}\{\langle U_j\rangle\} \tag{10.51}$$

in very much the same way that τ_{ij} appears in Eq. (10.16). In fact, it is not hard to show that τ_{ij} and T_{ij} are related to each other through the identity [26]

$$T_{ij} = \{\tau_{ij}\} + \mathcal{L}_{ij} \tag{10.52}$$

where

$$\mathcal{L}_{ij} \equiv \{\langle U_i\rangle\langle U_j\rangle\} - \{\langle U_i\rangle\}\{\langle U_j\rangle\} \tag{10.53}$$

closely resembles the Leonard stress that appeared in Eq. (10.30).

The dynamic form of the Smagorinsky model is derived by assuming that both τ_{ij} and T_{ij} satisfy Eqs. (10.25) and (10.34), in which case it can be asserted that

$$\tau_{ij} - \frac{1}{3}\tau_{kk}\,\delta_{ij} = -2Ch^2|\langle S\rangle|\langle S\rangle_{ij} \tag{10.54}$$

and

$$T_{ij} - \frac{1}{3}T_{kk}\,\delta_{ij} = -2Ch_t^2|\{\langle S\rangle\}|\{\langle S\rangle_{ij}\}, \tag{10.55}$$

where C is assumed to be a function of position and time, but independent of h or any other flow variables. C has been introduced in place of C_S^2 appearing in Eq. (10.34) in order to allow for the possibility of a negative coefficient in the subgrid model. When $C < 0$ there is local backscatter, so the idea in this case is to allow for inclusion of more physics than the traditional Smagorinsky model.

The scale similarity assumption to the effect that the identical parameter C may be utilized at the different scales associated with Eqs. (10.54) and (10.55) cannot be expected to be universally true. For example, near boundaries one or both of the test and grid filters may encompass scales larger than the inertial range in which case some scale dependence of C might be expected. In situations like this, the legitimacy of assuming that there is no variation in C between scales is called into question. Generalizations of the dynamic model have been developed [27] in which accommodation is made for variations in C between the grid and test filters. This can be an advantage in improving the prediction of near-wall dissipation that would otherwise be underestimated if C were assumed to be insensitive to differences in scale.

To determine C under normal circumstances, substitute Eqs. (10.54) and (10.55) into Eq. (10.52) yielding

$$\mathcal{L}_{ij} - \frac{1}{3}\delta_{ij}\mathcal{L}_{kk} = 2CM_{ij} \tag{10.56}$$

where

$$M_{ij} \equiv h^2\{|\langle S\rangle|\langle S\rangle_{ij}\} - h_t^2|\{\langle S\rangle\}|\{\langle S\rangle_{ij}\}. \tag{10.57}$$

The appearance of C in front of the quantity M_{ij} in Eq. (10.56) is a result of the inconsistent assumption that C is a constant in space so that it can be taken outside of the coarse filtering that is applied to τ_{ij} in Eq. (10.52). If this assumption is not made, then Eq. (10.52) leads to an integral equation for $C(\mathbf{x}, t)$ instead of the simpler algebraic form in Eq. (10.56). Models which pursue a solution for C from an integral equation are referred to as dynamic localization models [28]. In practice, the considerable extra cost of this approach does not produce a comparable benefit so such models are not generally employed.

Since Eq. (10.56) consists of an overdetermined system of five independent relations for the scalar quantity C, not all of the relations can be satisfied. Instead, it is reasonable to pursue a strategy in which Eq. (10.56) is satisfied only in a least square sense [29], wherein the squared error

$$E_r \equiv \left(\mathcal{L}_{ij} - \frac{1}{3}\delta_{ij}\mathcal{L}_{kk} - 2CM_{ij}\right)^2 \tag{10.58}$$

with respect to C is minimized. Thus, setting $\partial E_r/\partial C = 0$ yields the requirement that

$$C = \frac{1}{2}\frac{\mathcal{L}_{ij}M_{ij}}{M_{ij}^2}, \tag{10.59}$$

which is a minimum since $\partial^2 E_r/\partial C^2 = 8M_{ij}^2 > 0$. The ratio of length scales, h_t/h, remains as a free parameter in Eq. (10.59) that must be determined. Since h is given, the question is tantamount to deciding how much coarser than h should one chose the test filter, and numerical experiments suggest that an optimal value is $h_t = 2h$.

In practice it is found that C determined from Eq. (10.59) has an unacceptably rapid spatial variation that can lead to instability in the numerical computation of the resolved field. Evidently, the numerator and denominator in Eq. (10.59) tend to vary rapidly throughout the flow, causing their ratio to be unacceptably noisy. The numerator can, and does in practice, change sign. In and of itself, this is an advantage since it opens up the door to modeling backscatter. However, it is a distinct obstacle to achieving stable numerical solutions to the governing equations, since according to Eq. (10.54) it is tantamount to imposing negative viscosity. To bring the dynamic model to a useful form it is necessary to prevent the occurrence of negative C as well as curtail the rapidity of its variation throughout the flow. A common means of satisfying these requirements is to determine C from minimizing the error over a finite region. Thus Eq. (10.58) may be generalized to

$$E_r \equiv \int_{\mathcal{V}} (\mathcal{L}_{ij} - \frac{1}{3}\delta_{ij}\mathcal{L}_{kk} - 2CM_{ij})^2 d\mathcal{V}, \tag{10.60}$$

where the domain, $\mathcal{V}$, is often chosen to conform to particular attributes of the flow of interest. For example, in a channel flow $\mathcal{V}$ is chosen to encompass planes parallel to the boundary, so that C is only a function of the inhomogeneous direction. Minimizing Eq. (10.60) with respect to C then gives

$$C(\mathbf{x}, t) = \frac{1}{2}\frac{\int_{\mathcal{V}} \mathcal{L}_{ij}M_{ij} d\mathcal{V}(\mathbf{x})}{\int_{\mathcal{V}} M_{ij}^2 d\mathcal{V}}, \tag{10.61}$$

where $\mathcal{V}(\mathbf{x})$ is a region containing the point $\mathbf{x}$. This procedure not only smoothes out rapid variations in C, but also prevents C from becoming negative. In a channel flow with $\mathcal{V}$ representing the x, z plane, Eq. (10.61) yields $C(y, t)$ with y the coordinate normal to the boundary. A particularly attractive property of Eq. (10.61) in this case is that it provides a natural damping near the wall so that $C = 0$ at the surface and van Driest damping is unnecessary.

The reduction in C as the boundary is approached in the dynamic model can be attributed to the growing presence of regions where the local values of C are negative so that they cancel with positive values. To some extent this is fortuitous because it occurs regardless of whether or not an LES has sufficient mesh points to resolve the boundary layer structure.

In complex flow geometries where symmetries are absent, the averaging in Eq. (10.61) cannot be done over planes such as has proven to be effective in channel flow. Local averaging over small volumes surrounding each point can be done, but this is not likely to smooth C as much as is desired. For example, despite such averaging C may still be negative. It is not uncommon for applications faced with this limitation [14] to only accept negative C so long as $\nu + \nu_T > 0$ so that instability of the diffusion term in the numerical calculation is prevented. If C is so negative as to violate this condition then it is forcibly increased to prevent its occurrence, a practice known as *clipping*. Fortunately, it is usually the case that this kind of ad hoc intervention only has to be imposed at a modest number of locations in the course of a typical calculation.

Another approach to smoothing the dynamic coefficient for non-homogeneous flows is based on a weighted average over fluid particle paths [30]. In this, for a given point $\mathbf{x}$ in the flow at time t the spatial averaging in Eq. (10.60) is replaced by a weighted integration over the particle paths arriving at $\mathbf{x}$ at time t. Specifically, the error to minimize is:

$$E_r \equiv \int_{-\infty}^{t} (\mathcal{L}_{ij}(\mathbf{X}(s), s) - \frac{1}{3}\delta_{ij}\mathcal{L}_{kk}(\mathbf{X}(s), s) \\ -2C(\mathbf{x}, t)M_{ij}(\mathbf{X}(s), s))^2 W(t-s)ds, \tag{10.62}$$

where $\mathbf{X}(s)$ is the path of a fluid particle satisfying $\mathbf{X}(t) = \mathbf{x}$ and $W(t)$ is a weighting function designed to force greater contributions from the immediate past than the distant past. By the same steps as led to Eq. (10.61) it follows that

$$C(\mathbf{x}, t) = \frac{1}{2}\frac{N}{D} \tag{10.63}$$

where

$$N \equiv \int_{-\infty}^{t} \mathcal{L}_{ij}(\mathbf{X}(s), s) M_{ij}(\mathbf{X}(s), s) W(t-s)ds \tag{10.64}$$

and

$$D \equiv \int_{-\infty}^{t} M_{ij}(\mathbf{X}(s), s)^2 W(t-s)ds. \tag{10.65}$$

The evaluation of N and D is made less onerous if the weighting function is chosen to be an exponential of the form

$$W(t) = \frac{1}{T}e^{-t/T}, \tag{10.66}$$

where T is a time scale. Empirical study suggests that the latter may be taken to be

$$T = 1.5h(ND)^{-1/8}. \tag{10.67}$$

With this assumption it may be shown that N and D satisfy relaxation-transport equations in the form

$$\frac{\partial N}{\partial t} + \langle \mathbf{U} \rangle \cdot \nabla N = \frac{1}{T}(\mathcal{L}_{ij}M_{ij} - N) \tag{10.68}$$

and

$$\frac{\partial D}{\partial t} + \langle \mathbf{U} \rangle \cdot \nabla D = \frac{1}{T}(M_{ij}^2 - D). \tag{10.69}$$

These equations are readily solved in the course of an LES so that the expensive path averaging in Eqs. (10.64) and (10.65) can be avoided.

Some justification for the Lagrangian model comes from test calculations showing that the effect of path averaging is to make C relatively scale independent. This helps justify the use of the same C for the test and grid filters. Calculations have also shown that the Lagrangian model responds well to the presence of large-scale structures. In fact, the Lagrangian model has the benefit of being responsive to particular events in the homogeneous directions that would ordinarily be averaged out in the typical dynamic model formulation.

The art of subgrid modeling continues to receive considerable attention with many new variations of the familiar approaches being developed, as well as new means of including dynamic behavior [31]. It can therefore be imagined that some improvements beyond current models will appear in the future.

10.4 Hybrid LES/RANS Models

The constraint of affordability limits the size of numerical grids that can be utilized in practical LES computations. In particular, the need to resolve near-wall flow structures at a resolution rivaling that of a DNS means that grid spacings in the neighborhood of $\Delta x^+ = 100$, $\Delta y^+ = 1$, and $\Delta z^+ = 15$ are required near the boundary and this means a generally large mesh with a costly evaluation of the LES equations. In contrast to LES, RANS calculations in a similar application to an attached boundary layer can relax the Δx^+ and Δz^+ requirements while maintaining the $\Delta y^+ = 1$ condition near the wall. This makes such computations much more affordable than LES in this situation.

If a proper near-wall mesh cannot be used for LES, then the subgrid modeling cannot resolve the energy-containing wall structures and the predictions will be diminished in accuracy. A remedy for this failing is essential for the future success of LES since otherwise calculations at high Reynolds numbers near walls will tend to be governed by excessive diffusion deriving from numerical viscosity.

In early work with LES applied to flows with attached, thin boundary layers, the lack of near-wall mesh resolution was overcome by the use of wall-function boundary conditions derived from an assumed log law mean velocity in the wall vicinity. In this, the velocities at locations just outside the boundary layer are assigned from empirical formulas so there is no need to compute the flow closer to the boundary. This approach assumes that the entire boundary layer fits within the first row of mesh points.

The wall function approach to some extent can be generalized to include a wider range of flows, for example those including acceleration of the outer flow in boundary layers and other effects. However, for the many applications where boundary layers separate from the solid body or contain other complex effects such as system rotation, the wall-function approach is destined to be of questionable accuracy. For example, a log law is inapplicable in separated flow regions.

One means of obtaining somewhat greater generality and accuracy than wall functions, yet at less cost than a well-resolved LES, is to solve for the near-wall flow using a RANS model. As noted previously, the meshing required for implementing RANS near boundaries is much less demanding than that needed in LES or DNS so that calculations with LES in this case are feasible. This suggests that another potentially fruitful means of modelling near-wall flows is to use a hybrid LES/RANS method in which RANS calculations are used in attached boundary layers, while the LES model is used in the remainder of the flow. The success of such methods depends on the accuracy of the RANS scheme in capturing the near-wall flow, the criteria for deciding where the calculation shifts between RANS and LES, the mechanism for melding the different schemes together, and on the accuracy of the LES modeling.

The demarcation of separate RANS and LES zones in a calculation can be achieved in a variety of ways. They may be determined dynamically by the flow or fixed a priori by the user. In the case of the former, the changeover from one mode to the other can be based on mesh resolution or the values of a variety of physical parameters. All such methods have to contend with the fact that the RANS solutions are deterministic while the LES solutions are random, so that special care has to be taken in uniting the two solutions. There is an abundance of schemes devised for implementing the hybrid LES/RANS idea that have been demonstrated in test calculations to offer some potential benefit in flow prediction. At the same time, it is also not possible to point to one approach that has become dominant in applications or enjoys a stronger theoretical basis than other schemes. Our discussion here will consider several of the more commonly applied methodologies.

10.4.1 Detached Eddy Simulation

A widely applied methodology for hybrid LES/RANS computation known as detached eddy simulation (DES) is based on a modification to the Spalart–Allmaras (SA) eddy viscosity model (considered in Section 9.2.5) that promotes LES behavior in the region away from solid boundaries [32]. The essential idea is to alter the dissipation term in the $\widetilde{\nu}$ equation (9.93) in such a way that it maintains its RANS form near the wall and changes to a form more appropriate to LES elsewhere. The desired LES expression can be derived by considering the form that the production and dissipation terms in the $\widetilde{\nu}$ equation should have if they are to be used in the context of a LES. In particular, a balance of production and dissipation gives, in essence,

$$\widetilde{S}\widetilde{\nu} \propto \left(\frac{\widetilde{\nu}}{d}\right)^2, \tag{10.70}$$

with d the distance to the closest wall, in which case

$$\widetilde{\nu} \propto \widetilde{S}d^2. \tag{10.71}$$

Comparing this with Eq. (10.34) for the Smagorinsky model suggests that

$$d \sim h \tag{10.72}$$

where h is a measure of the local grid spacing, so that changing d to a value proportional to h brings the production/dissipation balance in the SA model into line with an LES. This then leads to selecting

$$\widetilde{d} = \min(d, C_{DES}h) \tag{10.73}$$

as a replacement for d in the dissipation term in Eq. (9.93). h is generally taken to be the maximum grid spacing among the three directions and $C_{DES} = 0.65$ is found by applying DES to decaying isotropic turbulence [33]. In practice, the substitution for d has the effect of raising dissipation in the region $d > C_{DES}h$, which lowers the eddy viscosity and promotes the instabilities leading to the random flow characteristic of a LES. The method is particularly attractive since it requires no user input apart from that of an appropriate mesh. Moreover, the shift between RANS and LES behavior is smooth since the eddy viscosity varies only due to changes to a source-like term in its governing equation. The random flow in the LES region promotes some unsteadiness in the RANS region as well so the RANS calculation behaves to some extent like a URANS.

DES can be an effective approach for accommodating flows for which large separated regions are a central aspect of the dynamics. When attached boundary layers are present and of importance, then DES is found to be susceptible to the presence of what are referred to as *gray regions* where the RANS eddy viscosity drops in magnitude prior to the appearance of a robust random field belonging to the LES. Within the gray region may appear anomalous structural features including vortices and "super streaks." Artificial means of exciting the flow in this region [34] can be beneficial in improving accuracy. A more general means of improving results is to prevent a premature shift away from the RANS solution so as to give the LES solution an opportunity to become fully developed. This is accomplished in what is known as the delayed detached eddy simulation (DDES) scheme [35].

For DDES the assumption in Eq. (10.73) is replaced by

$$\widetilde{d} = d - f_d \max(0, d - C_{DES}h) \tag{10.74}$$

where

$$f_d = 1 - \tanh[(8r_d)^3] \tag{10.75}$$

and

$$r_d = \frac{\widetilde{\nu}}{S\kappa^2 d^2}. \tag{10.76}$$

When $d < C_{DES}h$ in the RANS region, $\widetilde{d} = d$ as before. When $d > C_{DES}h$ in the LES region, Eq. (10.74) shows that $\widetilde{d}$ remains close or equal to d as long as f_d is small and this occurs when r_d is large. Thus Eq. (10.74) is an empirically derived relation that can force the ν_t equation to operate in the RANS mode for conditions that do not depend on h.

An improved version of DDES has also been developed (IDDES) [36] that provides greater flexibility in accommodating a variety of flow scenarios. It combines DDES with a wall model LES approach together with a set of empirical functions that allow each aspect of the scheme to act as designed as well as to work together where it is warranted by virtue of the initial and inflow conditions. The IDDES model includes a subgrid length scale that depends explicitly on wall distance, unlike the typical LES and DES schemes, which only depend on local grid spacing.

10.4.2 A Hybrid LES/RANS Form of the Menter SST Model

Another means of shifting between RANS and LES modes is contained in a modification of the Menter SST model considered in Section 9.2.4. In particular, where the SST RANS model incorporates a shift between the K–ω model near the wall to a K–ϵ formulation away from the wall, this approach can be converted to a hybrid LES/RANS scheme by replacing the K–ϵ part of the SST model with a LES scheme [37]. In essence, this method builds upon the capacity of the SST model to automatically sense the end of near-wall conditions so that the modeling can shift into a different modeling regime at points sufficiently far from the wall.

Similar to the Menter SST model a blended eddy viscosity is computed via the expression

$$\nu_t = f\frac{K}{\omega} + (1-f)C_s\sqrt{K}h \tag{10.77}$$

together with a blended dissipation rate given by

$$\epsilon = f\beta^* K\omega + (1-f)C_s K^{3/2}/h. \tag{10.78}$$

The constant $C_s = 0.01$ and f is similar to Eq. (9.82) and is given by

$$f = \tanh(\eta^4) \tag{10.79}$$

with

$$\eta = \frac{1}{\omega}\max\left(\frac{500\nu}{d^2}, \frac{\sqrt{K}}{C_\mu d}\right). \tag{10.80}$$

β^* is given as in Eq. (9.83) and ω appearing in Eq. (10.80) may either be the value computed from its own equation or a blended value.

It is evident from Eq. (10.80) that near the wall η is large so that $f = 1$ and the usual K–ω model is implemented. Further from the wall as η decreases the model shifts to an eddy viscosity associated with an LES scheme. A problem observed in implementations of this approach is that the blended eddy viscosity outside the RANS region, which accounts for subgrid effects, depends on having the resolved stress equal to its expected magnitude. To the extent that the latter condition is not achieved because the simulated flow is in the process of shifting from RANS to a LES mode, the subgrid eddy viscosity will be under-represented. In the literature this is sometimes referred to as modeled stress depletion (MSD). This deficit shows up in the appearance of unphysical turbulent flow features including intense vortices similar to the gray areas observed in the DES model.

10.4.3 Flow Simulation Methodology

The flow simulation methodology (FSM) [38, 39] is a simple procedure for developing hybrid LES/RANS methods from any given RANS model by taking advantage of the close similarity between the RANS equation (3.9) and the LES equation (10.16). In this, a shift between a given RANS model and its functioning as a LES subgrid model is achieved by expressing the governing mean momentum equation as

$$\frac{\partial\langle U_i\rangle}{\partial t} + \langle U_j\rangle\frac{\partial\langle U_i\rangle}{\partial x_j} = -\frac{1}{\rho}\frac{\partial\langle P\rangle}{\partial x_i} + \nu\nabla^2\langle U_i\rangle - \frac{\partial\tau_{model_{ij}}}{\partial x_j} \tag{10.81}$$

and setting

$$\tau_{model} = f(h/\eta)R_{model}, \tag{10.82}$$

where h is a length associated with the local mesh, η is the Kolmogorov length scale defined in Eq. (4.39), $0 \le f(h/\eta) \le 1$, and f is a function that controls the degree to which the RANS model is used locally. The smaller the ratio h/η the better the flow is resolved, and the numerical scheme should be more like a DNS, which is exactly the case if $f = 0$. On the other hand, for relatively large h/η the solution needs to operate as a RANS model, which it does exactly when $f = 1$. When f is between its limits the turbulent transport represented by the R_{model} has the capacity to become sufficiently diminished so that the flow field becomes more responsive to instability and behaves in the manner of an LES.

The function $f(h/\eta)$ is chosen largely through empirical means and a variety of forms have been used. One model that has been used is

$$f = (1 - e^{-\beta h/\eta})^n \tag{10.83}$$

where n controls how fast the shift between DNS and RANS takes place and β helps establish the grid density at which the DNS regime might take over. This f vanishes for small h/η and is 1 for large ratios. In another study [40] the choice is made that

$$f = 1 - e^{-\beta h^2}. \tag{10.84}$$

It can be seen that there is great latitude in the selection of f with the accumulated experience in applications representing the best way to determine an optimal form.

A conceptual problem with the FSM approach is that the RANS models upon which it is based are expressed in terms of the mean velocity and other quantities such as K and ϵ. However, when the scheme is operating in the LES mode, the mean velocity is not generally available unless an assumption to the effect that Eq. (10.20) is satisfied is made and that the average of the LES field is calculated at the same time that the solution is being computed. This can pose difficulties in complex flow situations. In practice, the FSM tends to be applied using the filtered velocity field without regard to the condition that it be averaged. The performance of the method in this case has been shown to be acceptable [41].

10.4.4 Example of a Zonal LES/RANS Formulation

While methods such as DES strive to more or less continuously shift between RANS and LES modeling, other approaches have been constructed using separate RANS and LES models in different regions, with a scheme for communication between the regions built into the methodology. One method of this sort that has been subject to ongoing development and testing is that of Davidson and Peng [23] in which the K–ω RANS model is solved within a zone near the boundary and an LES calculation based on the Yoshizawa [22] one-equation model is solved elsewhere for the unresolved kinetic energy K_{sgs}. The unknowns in the RANS side of the calculation are averaged quantities while those on the LES side are filtered. The demarcation line between the solutions is given by a pre-assigned value of y^+. In various test calculations this varied between $y^+ = 120$ to 1120.

In its original formulation the RANS domain utilizes the equation for K given by

$$\frac{\partial K}{\partial t} + \overline{U}_j \frac{\partial K}{\partial x_j} = \mathcal{P} - C_K f_K K\omega + \frac{\partial}{\partial x_i}\left[\left(\nu + \frac{\nu_t}{\sigma_K}\right)\frac{\partial K}{\partial x_i}\right], \tag{10.85}$$

with the value $C_K = 0.09$, $\sigma_K = 0.8$, and low Reynolds number damping function

$$f_K = 1 - 0.722 e^{-(R_T/10)^4} \tag{10.86}$$

where $R_T = K/(\nu\omega)$ is a Reynolds number. The equation satisfied by ω is

$$\frac{\partial \omega}{\partial t} + \overline{U}_j \frac{\partial \omega}{\partial x_j} = C_{\omega_1} f_\omega \frac{\omega}{K} \mathcal{P} - C_{\omega_2} \omega^2 + \frac{\partial}{\partial x_i}\left[\left(\nu + \frac{\nu_t}{\sigma_\omega}\right)\frac{\partial \omega}{\partial x_i}\right] + C_\omega \frac{\nu_t}{K}\frac{\partial K}{\partial x_j}\frac{\partial \omega}{\partial x_j} \tag{10.87}$$

where $C_{\omega_1} = 0.42$, $C_{\omega_2} = 0.075$, $\sigma_w = 1.35$, and damping function

$$f_\omega = 1 + 4.3e^{-(R_T/1.5)^{1/2}}. \tag{10.88}$$

The eddy viscosity used in the model is

$$\nu_t = f_\mu \frac{K}{\omega} \tag{10.89}$$

with the damping function

$$f_\mu = 0.025 + (1 - e^{-(R_T/10)^{3/4}})\left(0.975 + \frac{0.001}{R_T} e^{-(R_T/200)^2}\right). \tag{10.90}$$

In the LES zone K_{SGS} is determined as the solution to the model equation [22]

$$\frac{\partial K_{SGS}}{\partial t} + \overline{U}_j \frac{\partial K_{SGS}}{\partial x_j} = \mathcal{P}_{SGS} - C_\epsilon \frac{K_{SGS}^{3/2}}{h} + \frac{\partial}{\partial x_i}\left((\nu + \nu_{SGS})\frac{\partial K_{SGS}}{\partial x_i}\right), \tag{10.91}$$

where the eddy viscosity is

$$\nu_{SGS} = C_K h \sqrt{K_{SGS}} \tag{10.92}$$

and $C_K = 0.07$, $C_\epsilon = 1.05$, and h is the minimum of the local orthogonal grid spacing. No explicit conditions for K and K_{SGS} are imposed on the interface while ω is given a zero normal slope at this location.

Subsequent enhancements of the basic model include adopting a K_{SGS}, ω model in the LES region [42] and otherwise addressing issues that are common to the hybrid LES/RANS approach. Among these are the gray region problem as mentioned previously in regard to DES. Another frequently encountered issue in hybrid LES/RANS methods that is relevant to the present case is a log-law mismatch at the boundary between RANS and LES calculations associated with a rapid change in the length scale as the modeling shifts [43–47]. Generalizations of the approach in Eqs. (10.89) to (10.92) attempt to alleviate the gray region and log mismatch problems by such techniques as adding a degree of empiricism to the definition of h in Eq. (10.92) similar to the way in which the scale $\tilde{d}$ is modified according to Eq. (10.74) in the DDES model. By this step it is possible to suppress the premature drop in eddy viscosity that leads to a gray region appearing prior to the full establishment of the LES field. Additional modifications include the use of artificial stimulation [46] of the fluid on the LES side of the boundary, as well as source terms that attempt to ameliorate differences between the RANS and LES formulations at the interface [48]. Another approach depends on providing filtered LES data to the wall model in control volumes adjacent to the boundary [47]. The result of such efforts is to reduce and/or eliminate the gray region and log-law mismatch.

10.4.5 Partially Averaged Navier–Stokes

A method [49] that is referred to as partially averaged Navier–Stokes (PANS) creates what are essentially URANS simulations from arbitrary RANS models via manipulation of the model parameters. To take a concrete case as illustration, in a PANS formulation of the $K{-}\epsilon$ closure, the equations

$$\frac{\partial K_u}{\partial t} + \overline{U}_j \frac{\partial K_u}{\partial x_j} = \mathcal{P}_u - \epsilon_u + \nabla \cdot ((\nu + \nu_u/\sigma_{ku})\nabla K_u) \tag{10.93}$$

and

$$\frac{\partial \epsilon_u}{\partial t} + \overline{U}_j \frac{\partial \epsilon_u}{\partial x_j} = C^*_{\epsilon_1} \frac{\epsilon_u}{K_u} \mathcal{P}_u - C^*_{\epsilon_2} \frac{\epsilon_u^2}{K_u} + \frac{\partial}{\partial x_i}\left[\left(\nu + \frac{\nu_u}{\sigma_{\epsilon u}}\right)\frac{\partial \epsilon_u}{\partial x_i}\right], \tag{10.94}$$

where

$$\mathcal{P}_u = \nu_{tu} \frac{\partial \overline{U}_i}{\partial x_j}\left(\frac{\partial \overline{U}_i}{\partial x_j} + \frac{\partial \overline{U}_j}{\partial x_i}\right) \tag{10.95}$$

is the turbulent kinetic energy production term, are solved for the unresolved parts of the kinetic energy and dissipation rate referred to, respectively, as K_u and ϵ_u. Here,

$$\nu_t = C_\mu \frac{K_u^2}{\epsilon_u} \tag{10.96}$$

$$\sigma_{ku} = \frac{f_k^2 \sigma_k}{f_\epsilon} \tag{10.97}$$

$$\sigma_{\epsilon u} = \frac{f_k^2 \sigma_\epsilon}{f_\epsilon} \tag{10.98}$$

$$C^*_{\epsilon_1} = C_{\epsilon_1} \tag{10.99}$$

and

$$C^*_{\epsilon_2} = \frac{f_k}{f_\epsilon}(C_{\epsilon_2} - C_{\epsilon_1}) + C_{\epsilon_1}. \tag{10.100}$$

The standard $K{-}\epsilon$ model coefficients $C_{\epsilon_1} = 1.44, C_{\epsilon_2} = 1.92, \sigma_k = 1$, and $\sigma_\epsilon = 1.3$ are used in the model. $f_k = K_u/K$ and $f_\epsilon = \epsilon_u/\epsilon$ are the ratios of unresolved to fully resolved kinetic energy and dissipation, respectively. For any given application of PANS, values are assigned to f_k and f_ϵ. When $f_k = f_\epsilon = 0$, the eddy viscosity in Eq. (10.96) vanishes and the method is forced to be a DNS while if $f_k = f_\epsilon = 1$, the method is exactly the $K{-}\epsilon$ closure. When $f_k < 1$ and $f_\epsilon = 1$ the method is a traditional URANS approach since all the dissipation is modeled while the kinetic energy is divided between a resolved, time-dependent part and a modeled part. Through the selection of f_k and f_ϵ the user controls what degree of approach to either limit one wants to achieve.

For sufficiently small values of f_k and f_ϵ, solutions to the PANS equation contain random vortical structure. In fact, such structures, which are reminiscent of those seen in LES, increase in prominence as the parameters are lowered. Applications of PANS show some promise as well as areas where further development is needed.

10.4.6 Scale-Adaptive Simulation

The scale-adaptive simulation (SAS) strategy developed over a number of years by Menter, Egorov, and associates [50] is a hybrid approach that shifts smoothly between RANS and LES/URANS mode depending on a physical scale that is largely independent of the mesh spacing. With the intention of allowing for the inclusion of structural ideas about the turbulent motion, the starting point for developing SAS is a modeled form of an exact equation governing two-point velocity correlations developed by Rotta [51]. In its recent manifestations, a more careful consideration of the source term in Rotta's velocity correlation equation has led to its being modeled in such a way as to include the von Kármán length scale

$$L_{vK} = \kappa \left| \frac{\partial U/\partial y}{\partial^2 U/\partial y^2} \right|, \tag{10.101}$$

which is representative of the largest turbulent eddies. The appearance of L_{vK} represents an additional length scale in the modeling besides that naturally appearing within the eddy viscosity. A new source term in the velocity correlation equation containing L_{vK} is added so as to supply a means by which the turbulent viscosity is modulated so that vortical structures can be resolved and not smoothed away. In essence the eddy viscosity can be reduced locally to allow for the appearance of resolved turbulence whose scale is consistent with that of the local structure. In this way local conditions determine where and when the computed field shifts between RANS and LES/URANS mode.

While the approach has been developed using several RANS models it is most often encountered in the form of a two-equation model for K and a variable $\Phi \equiv \sqrt{KL}$ where L is an integral scale of the turbulence. The velocity field, denoted U_i, is meant to represent the averaged RANS field in the steady parts of the domain, and an unsteady URANS field with the behavior and characteristics of a resolved LES field in those regions where the flow is unsteady. Which mode is activated depends on the eddy viscosity given by

$$\nu_T = C_\mu^{1/4} \Phi, \tag{10.102}$$

which is determined as part of the two-equation closure consisting of the K equation

$$\frac{\partial K}{\partial t} + \overline{U}_j \frac{\partial K}{\partial x_j} = \mathcal{P} - c_\mu^{3/4} \frac{K^2}{\Phi} + \frac{\partial}{\partial x_j} \left(\left(\nu + \frac{\nu_T}{\sigma_k} \right) \frac{\partial K}{\partial x_j} \right) \tag{10.103}$$

and the Φ equation

$$\frac{\partial \Phi}{\partial t} + \overline{U}_j \frac{\partial \Phi}{\partial x_j} = \frac{\Phi}{K} \mathcal{P} \left(\zeta_1 - \zeta_2 \left(\frac{L}{L_{vK}} \right)^2 \right) - \zeta_3 K + \frac{\partial}{\partial x_j} \left(\left(\nu + \frac{\nu_T}{\sigma_\Phi} \right) \frac{\partial \Phi}{\partial x_j} \right). \tag{10.104}$$

For these equations to apply to general flows, the definition of the von Kármán length scale is set to

$$L_{vK} = \kappa \left| \frac{U'}{U''} \right| \tag{10.105}$$

where

$$U' = S = \sqrt{2 S_{ij} S_{ij}} \tag{10.106}$$

S_{ij} is the rate of strain tensor, and

$$U'' = \sqrt{\frac{\partial^2 U_i}{\partial x_j^2}\frac{\partial^2 U_i}{\partial x_k^2}}. \quad (10.107)$$

After calibration with decaying turbulence and the log law the model parameters are assigned the values $\zeta_1 = 0.8, \zeta_2 = 1.47, \zeta_3 = 0.0288, \sigma_K = 2/3, \sigma_\Phi = 2/3, C_\mu = 0.09$, and $\kappa = 0.41$.

Validation studies for the SAS model show that there is a natural tendency for the range of scales subject to the URANS/LES mode to increase in wave number as far as the local mesh will allow. This represents a symbiotic reduction in the scale $L_{\nu K}$ coordinated with the appearance of smaller and smaller structure. The observed end result is an unphysical accumulation of energy at high wave numbers that suggests that a required damping process is missing from the model. To correct this, a simple approach is to generalize Eq. (10.102) to the form

$$\nu_T = \max(C_\mu^{1/4}\Phi, \nu_{WALE}) \quad (10.108)$$

where it is seen that the eddy viscosity given by the WALE model in Eq. (10.44) is used to establish a minimum. Thus, if Φ is reduced below the WALE expression, then the WALE viscosity takes over. Moreover, in those regions where ν_{WALE} would naturally be zero, the SAS model is able to function as a RANS model.

10.5 Chapter Summary

By modeling the influence of subgrid scales on the resolved flow, LES attempts to capitalize on the increasing capabilities of numerical computation to provide credible solutions to complex engineering flow fields that are well beyond the practical limitations of DNS. The popularity of LES attests to the belief that it supplies a way forward beyond the predictive capabilities of RANS modeling and is well positioned to take advantage of each new milestone in the development of faster and larger computers. The question of how to best capture subgrid-scale effects in LES has received a great deal of attention, and significant progress beyond the classical Smagorinsky model has been achieved. Various strategies for developing subgrid models that better reflect local conditions have appeared, including the WALE model and the dynamic procedure. Limiting the advancement of LES as a technology is the need for mesh refinement next to boundaries of a density approaching that of DNS so that the technique becomes economically infeasible. This has led to a great interest in developing hybrid schemes that hope to combine the best attributes of LES and RANS approaches for treating near-wall flows. Numerous approaches for accommodating both deterministic RANS calculations and random LES calculations have been devised. It remains uncertain at present which of these approaches offers a clear advantage going in to the future.

References

1 Leonard, A. (1974) Energy cascade in large-eddy simulations of turbulent fluid flows. *Adv. Geophys.*, 18A, 137–148.

2 Ghosal, S. and Moin, P. (1995) The basic equations for the large eddy simulation of turbulent flows in complex geometry. *J. Comp. Phys.*, 118, 24–37.

3 Kravchenko, A.G., Moin, P., and Moser, R. (1996) Zonal embedded grids for numerical simulations of wall-bounded turbulent flows. *J. Comp. Phys.*, 127, 412–423.

4 Ghosal, S. (1996) An analysis of numerical errors in large eddy simulation of turbulence. *J. Comp. Phys.*, 125, 187–206.

5 Kravchenko, A.G. and Moin, P. (1997) On the effect of numerical errors in large eddy simulations of turbulent flows. *J. Comput. Phys.*, 131, 310–322.

6 Vreman, B., Geurts, B., and Kuerten, J. (1996) Comparison of numerical schemes in large-eddy simulation of the temporal mixing layer. *Intl. J. Num. Meth. Fluids*, 22, 297–311.

7 Boris, J.P., Grinstein, F.F., Oran, E.S., and Kolbe, R.L. (1992) New insights into large eddy simulation. *Fluid Dyn. Res.*, 10, 199–228.

8 Kawamura, T. and Kuwahara, K. (1984) Computation of high Reynolds number flow around a circular cylinder with surface roughness. *AIAA Paper 1984-0340*.

9 Marstorp, L., Brethouwer, G., Grundestam, O., and Johansson, A.V. (2009) Explicit algebraic subgrid stress models with application to rotating channel flow. *J. Fluid Mech.*, 639, 403–432.

10 Montecchia, M., Brethouwer, G., Johansson, A.V., and Wallin, S. (2017) Taking large-eddy simulation of wall-bounded flows to higher Reynolds numbers by use of anisotropy-resolving subgrid models. *Phys. Rev. Fluids*, 2, 034 601.

11 Bardina, J., Ferziger, J.H., and Reynolds, W.C. (1980) Improved subgrid scale models for large-eddy simulation. *AIAA Paper 80-1357*.

12 Speziale, C.G. (1985) Galilean invariance of subgrid scale models in the large-eddy simulation of turbulence. *J. Fluid Mech.*, 156, 55–62.

13 Horiuti, K. (1989) The role of the Bardina model in large eddy simulation of turbulent channel flow. *Phys. Fluids A*, 1, 426–428.

14 Zang, Y., Street, R.L., and Koseff, J.R. (1993) A dynamic mixed subgrid-scale model and its application to turbulent recirculating flows. *Phys. Fluids A*, 5, 3186–3196.

15 Piomelli, U., Moin, P., and Ferziger, J. (1988) Model consistency in LES of turbulent channel flows. *Phys. Fluids*, 31, 1884–1891.

16 Smagorinsky, J. (1963) General circulation experiments with the primitive equations. *Monthly Weather Rev.*, 91, 99–165.

17 Lilly, D.K. (1967) The representation of small-scale turbulence in numerical simulation experiments, in *Proc. IBM Scientific Computing Symposium on Environmental Sciences*, IBM, White Plains, NY.

18 Canuto, V.M. and Cheng, Y. (1997) Determination of the Smagorinsky-Lilly constant C_S. *Phys. Fluids*, 9, 1368–1378.

19 Nicoud, F. and Ducros, F. (1999) Subgrid-scale stress modelling based on the square of the velocity gradient tensor. *Flow, Turb. Combustion*, 62, 183–200.

20 Moin, P. and Kim, J. (1982) Numerical investigation of turbulent channel flow. *J. Fluid Mech.*, 118, 341–377.

21 Schumann, U. (1975) Subgrid scale model for finite difference simulations of turbulent flows in plane channel and annuli. *J. Comp. Phys.*, 18, 376–404.

22 Yoshizawa, A. and Horiuti, K. (1985) A statistically-derived subgrid-scale kinetic energy model for the large-eddy simulation of turbulent flows. *J. Phys. Soc. Japan*, 54, 2834–2839.

23 Davidson, L. and Peng, S.H. (2003) Hybrid LES-RANS modelling: a one-equation SGS model combined with a $K-\omega$ model for predicting recirculating flows. *Int. J. Num. Meth. Fluids*, 43, 1003–1018.

24 Métais, O. and Lesieur, M. (1992) Spectral large-eddy simulation of isotropic and stably stratified turbulence. *J. Fluid Mech.*, 239, 157–194.

25 Lesieur, M. and Métais, O. (1996) New trends in large-eddy simulations of turbulence. *Ann. Rev. Fluid Mech.*, 28, 45–82.

26 Germano, M., Piomelli, U., Moin, P., and Cabot, W.H. (1991) A dynamic subgrid-scale eddy viscosity model. *Phys. Fluids A*, 3, 1760–1765.

27 Agel, F., Meneveau, C., and Parlange, M.B. (2000) A scale-dependent dynamic model for large-eddy simulation: application to a neutral atmospheric boundary layer. *J. Fluid Mech.*, 415, 261–284.

28 Ghosal, S., Lund, T., Moin, P., and Akselvoll, K. (1995) A dynamic localization model for large-eddy simulation of turbulent flow. *J. Fluid Mech.*, 286, 229–255.

29 Lilly, D.K. (1992) A proposed modification of the Germano subgrid scale closure method. *Phys. Fluids A*, 4, 633–635.

30 Meneveau, C., Lund, T.S., and Cabot, W.H. (1996) A Lagrangian dynamic subgrid-scale model of turbulence. *J. Fluid Mech.*, 319, 353–385.

31 Meneveau, C. (2012) Germano identity-based subgrid-scale modeling: A brief survey of variations on a fertile theme. *Phys. Fluids*, 24, 121 301.

32 Spalart, P.R. (2009) Detached-eddy simulation. *Ann. Rev. Fluid Mech.*, 41, 181–202.

33 Shur, M., Spalart, P.R., Strelets, M., and Travin, A. (1999) Detached-eddy simulation of an airfoil at high angle of attack, in *Engineering Turbulence Modeling and Experiments*, vol. 4 (eds W. Rodi and D. Laurence), Elsevier, Amsterdam, pp. 669–678.

34 Piomelli, U., Balaras, E., Pasinato, H., Squires, K., and Spalart, P.R. (2003) The inner-outer layer interface in large-eddy simulations with wall-layer models. *Int. J. Heat Fluid Flow*, 24, 538–550.

35 Spalart, P.R., Deck, S., Shur, M.L., and Travin, A. (2006) A new version of detached-eddy simulation, resistant to ambiguous grid densities. *Theor. Comp. Fluid Dynamics*, 20, 181–195.

36 Shur, M.L., Spalart, P.R., Strelets, M.K., and Travin, A.K. (2008) A hybrid RANS-LES approach with delayed-DES and wall-modelled LES capabilities. *Int. J. Heat Fluid Flow*, 29, 1638–1649.

37 Fan, T.C., Edwards, J.R., Hassan, H.A., and Baurle, R.A. (2001) Validation of a hybrid Reynolds-averaged/large-eddy simulation method for simulating cavity flameholder configurations. *AIAA Paper 2001-2929*.

38 Speziale, C.G. (1998) Turbulence modeling for time-dependent RANS and VLES: A review. *AIAA J.*, 36, 173–184.

39 Fasel, H.F., Seidel, I., and Wernz, S. (2002) A methodology for simulation of complex turbulent flows. *J. Fluids Eng.*, 124, 933–942.

40 Hussaini, M.Y., Thangam, S., Woodruff, S.L., and Zhou, Y. (2006) Development of a continuous model for simulation of turbulent flows. *J. Appl. Mech.*, 73, 441–448.

41 Fasel, H.F., von Terzi, D.A., and Sandberg, R.D. (2006) A methodology for simulating compressible turbulent flows. *J. Appl. Mech.*, 73, 405–412.

42 Arvidson, S., Davidson, L., and Peng, S.H. (54) Hybrid Reynolds-averaged Navier–Stokes/large-eddy simulation modeling based on a low-Reynolds-number $K - \omega$ model. *AIAA J.*, 2016, 4032–4037.
43 Hamba, F. (2009) Log-layer mismatch and commutation error in hybrid RANS/LES simulation of channel flow. *Int. J. Heat Fluid Flow*, 30, 20–31.
44 Kawai, S. and Larsson, J. (2013) Wall-modeling in large eddy simulation: length scales, grid resolution and accuracy. *Phys. Fluids*, 24, 015 105.
45 Wu, P. and Meyers, J. (2013) A constraint for the subgrid-scale stresses in the logarithmic region of high Reynolds number turbulent boundary layers: A solution to the log-layer mismatch problem. *Phys. Fluids*, 25, 015 104.
46 Larsson, J., Kawai, S., Bodart, J., and Bermejo-Moreno, I. (2016) Large eddy simulation with modeled wall-stress: recent progress and future directions. *Bull. JSME, Mech. Engrg. Review*, 3, 15–00 418.
47 Yang, X.I.A., Park, G.I., and Moin, P. (2017) Log-layer mismatch and modeling of the fluctuating wall stress in wall-modeled large-eddy simulations. *Phys. Rev. Fluids*, 2, 104 601.
48 Davidson, L. (2015) Two-equation hybrid RANS-LES models: Novel way to treat k and ω at the inlet, in *Turbulence, Heat, and Mass Transfer, THMT-15* (eds K. Hanjalic, T. Miyauchi, D. Borello, M. Hadziabdic, and P. Venturini), Begell House, New York, pp. 1–13.
49 Girimaji, S.S. (2006) Partially-averaged Navier–Stokes model for turbulence: A Reynolds-averaged Navier–Stokes to direct numerical simulation bridging method. *J. Applied Mech.*, 73, 413–421.
50 Menter, F.R. and Egorov, Y. (2010) The scale adaptive simulation method for unsteady turbulent flow predictions. Part 1: Theory and model description. *Flow, Turb. Combustion*, 85, 113–138.
51 Rotta, J.C. (1972) *Turbulente Strömumgen*, BG Teubner, Stuttgart.

Problems

10.1 Prove that Eq. (10.6) holds for a homogeneous filter satisfying Eq. (10.4).

10.2 Show that Eqs. (10.10) and (10.13) form a Fourier transform pair.

10.3 Consider the x component of velocity U due to a point vortex situated at the origin. Along the line $y = y_0$, $U(x) = -\Gamma/(2\pi)y_0/(x^2 + y_0^2)$. For $y_0 = 1$ and 0.25, evaluate $U(x)$ on a numerical grid spanning the interval $|x| \leq 10$ and then compute its discrete Fourier transform (e.g., using *fft* in MATLAB). Compare plots of $U(x)$ for the two cases as well as a stem plot of the magnitudes of the Fourier coefficients. Assuming that $U(x)$ is a model of the velocity field due to an eddy, comment on how efficient the Fourier transform is in providing a description of the eddy.

10.4 Show that $S_{ij}^d S_{ij}^d \equiv 0$ in a unidirectional flow $\overline{U}(y)$.

10.5 Derive Eq. (10.36) and then show that

$$C_S = \frac{1}{\pi}\left(\frac{3C_K}{2}\right)^{-3/4} \tag{10.109}$$

by equating Eq. (10.36) to $-\tau_{ij}\langle S_{ij}\rangle$ after using the Smagorinsky model Eq. (10.34).

10.6 Show that Eq. (10.25) cannot be solved for the normal Reynolds stresses.

11

Properties of Turbulent Free Shear Flows

In many instances turbulent flows evolve without the confining influence of solid boundaries. Such *free shear flows* tend to originate through contact with solid surfaces, as in the flow in the wake of a solid body or a jet emanating from a nozzle. To some extent the particular free shear flows of interest here represent the aftermath – without boundaries – of flows that were considered in the previous chapters.

This chapter will focus on describing some of the fundamental properties of wake and jet flows as well as mixing layers in which streams at two different velocities are brought together behind a splitter plate. Figure 11.1 illustrates the essential aspects of the mean velocity profiles that occur at any transverse cut through these flows. Wake flows develop downstream of either streamlined or bluff bodies. In such cases there will be a deficit in the mean velocity behind the body that gradually returns to that of the outlying fluid. In a turbulent jet, fluid at a speed higher than ambient exits a nozzle or orifice and expands into a larger domain. In a mixing layer, the mean velocity monotonically falls from the higher speed free stream to the lower speed free stream over a lateral distance that increases with downstream distance.

Each of the flows illustrated in Figure 11.1 tends to show some variation in properties as the Reynolds number increases. Besides the general transition from a laminar to a turbulent state there also may be changes in the nature of the structural aspects of the flow. In this way free shear flows can be a rich source of interesting physical phenomena. We will consider here some aspects of each type of flow, mainly concentrating on their planar version, though each case also has a 3D analogue. For example, jets may be planar where fluid leaves a slot or round if they leave a circular orifice.

11.1 Thin Flow Approximation

A general property of the free shear flows to be considered here is that the streamwise (x) variation of the mean velocity is small compared to that in the cross-flow direction (y). This assumption yields a simplification to the governing equations similar to that of a boundary layer flow and, as in that case, leads to an analysis that allows for similarity solutions to the mean field. To apply the thin layer approximation to the different free shear flows requires defining an appropriate velocity scale, $\mathcal{U}$, as well as a length scale in the crossflow direction, ℓ. To satisfy the thin flow approximation the latter must generally satisfy $\ell(x)/x << 1$. In the case of wake flows the maximum velocity defect $U_e - \overline{U}_{min}$ is a natural characteristic velocity scale at each streamwise location, where

Turbulent Fluid Flow, First Edition. Peter S. Bernard.

Companion website: www.wiley.com/go/Bernard/Turbulent_Fluid_Flow

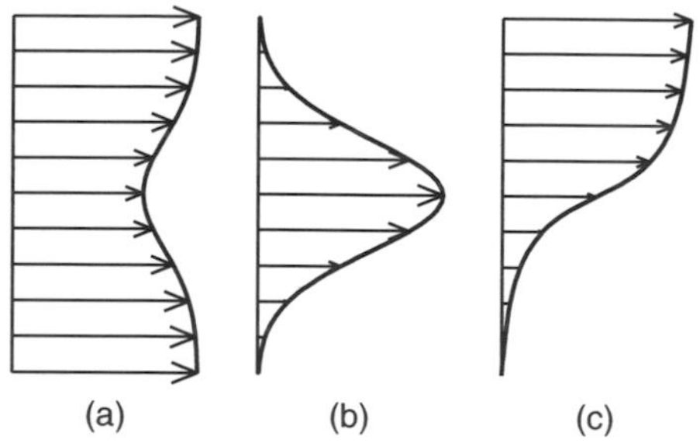

Figure 11.1 Basic characteristics of the mean flow in (a) wakes, (b) jets, and (c) mixing layers.

U_e is the velocity of the external flow. Jets without coflow may be analyzed similarly with the maximum velocity $\overline{U}_{max}$ acting as the velocity scale while for mixing layers the constant difference between the upper and lower flows $U_u - U_l$ is the appropriate velocity scale. For wakes and jets, a convenient length scale ℓ is the distance from the location of the minimum or maximum velocity, respectively, to a point within a fixed percentage of the free stream velocity. For mixing layers, ℓ may be taken to be one half the distance between where the mean velocity is within a fixed percentage of the outer velocities.

In all cases it will be seen that ℓ grows with downstream distance, yet in such a way that the ratio ℓ/x remains small. For turbulent wakes experiments show that this length ratio grows increasingly small as $x \to \infty$. On the other hand, ℓ/x becomes non-zero and constant with downstream distance for jets and mixing layers, with an asymptotic value $\lim_{x\to\infty} \ell/x \approx 0.06$. This is small enough so that the thin shear layer hypothesis is valid. When it is also the case that the rate of spread of the free shear flow is slow enough so that $d\ell/dx$ is small, then the flow may be considered to be approximately parallel.

A scaling analysis that is similar to that used in deriving the boundary layer equations [1, 2] justifies approximating the exact equation for $\overline{V}$ in Eq. (3.9) as a balance between convection and pressure forces the same as in Eq. (8.10) so that

$$\frac{\partial \overline{v^2}}{\partial y} = -\frac{1}{\rho}\frac{\partial \overline{P}}{\partial y}. \tag{11.1}$$

Integration of Eq. (11.1) across the thin shear layer gives

$$\frac{\overline{P}}{\rho} + \overline{v^2} = \frac{P_e}{\rho}, \tag{11.2}$$

where P_e is the pressure in the outer flow. Differentiating Eq. (11.2) with respect to x gives

$$\frac{1}{\rho}\frac{\partial \overline{P}}{\partial x} + \frac{\partial \overline{v^2}}{\partial x} = 0 \tag{11.3}$$

and substituting this into the steady streamwise mean momentum equation Eq. (3.9) yields

$$\overline{U}\frac{\partial \overline{U}}{\partial x} + \overline{V}\frac{\partial \overline{U}}{\partial y} + \frac{\partial}{\partial x}(\overline{u^2} - \overline{v^2}) + \frac{\partial \overline{uv}}{\partial y} = \nu\left(\frac{\partial^2 \overline{U}}{\partial x^2} + \frac{\partial^2 \overline{U}}{\partial y^2}\right). \tag{11.4}$$

Now applying the same thin layer scaling analysis as done in boundary layers to this relation gives the dominant balance in the streamwise direction as

$$\overline{U}\frac{\partial \overline{U}}{\partial x} + \overline{V}\frac{\partial \overline{U}}{\partial y} + \frac{\partial \overline{uv}}{\partial y} = 0. \tag{11.5}$$

This approximate relation is the basis for subsequent analyses of canonical thin free shear flows.

In the case of wakes and jets according to the mean continuity equation $(\overline{U} - U_e)(\partial\overline{U}/\partial x + \partial\overline{V}/\partial y) = 0$ and this may be added to the left-hand side of Eq. (11.5). Similarly, since U_e is constant, $\overline{U}\partial U_e/\partial x + \overline{V}\partial U_e/\partial y = 0$ and this relation may be subtracted from the left-hand side of Eq. (11.5). The result is the relation

$$\frac{\partial}{\partial x}[\overline{U}(\overline{U} - U_e)] + \frac{\partial}{\partial y}[\overline{V}(\overline{U} - U_e)] + \frac{\partial}{\partial y}\overline{uv} = 0. \tag{11.6}$$

For both jets and wakes $\overline{U} - U_e$ and $\overline{uv}$ go to zero when $|y|$ reaches the external potential flow on either side. Thus, in these two cases Eq. (11.6) can be integrated across the flow, yielding

$$\frac{d}{dx}\int_{-\infty}^{+\infty} \overline{U}(\overline{U} - U_e)dy = 0. \tag{11.7}$$

Thus, the total mean flux of momentum (per unit length in the spanwise direction) relative to the external flow, say M, and given by

$$M = \rho\int_{-\infty}^{+\infty} \overline{U}(\overline{U} - U_e)dy, \tag{11.8}$$

is constant as the wake and jet develop downstream. In the case of wake flow, M represents the mean streamwise momentum deficit, that is, the momentum lost to the external flow by the presence of the wake-producing body.

11.2 Turbulent Wake

One of the most studied of wake flows, and the one which will be considered here, is that of a circular cylinder with spanwise dimension large compared to the diameter [3]. Previously shown in Figure 1.7 is an example of the interesting vortical organization that can arise in such circumstances. Many of the structural aspects of this flow pass through several different regimes as the Reynolds number based on the far-field velocity and cylinder diameter increases. Starting at approximately $R_e = 47$ until $R_e = 180$ laminar 2D vortices of alternating sign shed in sequence from the top and bottom surface of the cylinder at a frequency f_s forming a von Kármán vortex street (also referred to as a Bénard–von Kármán vortex street). The non-dimensional shedding frequency, known as the Strouhal number, is measured to be approximately $St \equiv f_s d/U_e = 0.18$ and has some sensitivity to Reynolds number. The vortical wake is fully laminar in this regime and extends downstream approximately 75–100 diameters until the vortices decay exponentially in strength and ultimately disappear. The vortices grow in size with downstream distance, as does the width of the wake itself, by the entrainment of irrotational exterior fluid.

Beyond $R_e = 180$ spanwise waves are observed that signal the end of the 2D flow pattern and the onset of 3D behavior [4]. Relatively weak and small vortices that are not aligned in the spanwise direction appear that begin to distort the Kármán rollers. By $R_e = 360$ the wake tends to have arrived at a fully 3D state, with the shedding vortices accompanied by counter-rotating vortex pairs that appear in the region where the rollers are forming and stretch as they separate from the cylinder. The intensity of the streamwise-oriented vortices is observed to exceed that of the rollers themselves. Also observed, under some limited circumstances, are more exotic vortical patterns immediately behind the cylinder such as rhombus-shaped cells formed of oblique vortices from both sides of the cylinder. By $R_e = 540$ the near-wake pattern is strictly streamwise-oriented vortices interacting with the shedding rollers.

For larger Reynolds numbers, for example by $R_e = 1080$, the roller vortices form further downstream at approximately two diameters from the cylinder axis. Counter-rotating vortices form narrower filaments and continue to dominate the region prior to the rollers while subsequently wrapping around the rollers with varying complex 3D effects that somewhat distort the spanwise uniformity of the roller vortices. By $R_e = 5400$ the shear layers shedding off the top and bottom of the cylinder can be observed to undergo Kelvin–Helmholtz instability in which a series of co-rotating spanwise vortices develop over a short distance prior to the formation of the first roller vortex in the Kármán vortex street. In this regime the process of vortex shedding becomes better organized than at somewhat lower Reynolds numbers. The streamwise vortex pairs in this case quickly break up into a turbulent field wrapping around the rollers and having less of a spanwise influence.

An illustrative view of the cylinder wake at $R_e = 2200$ is shown in Figure 11.2. Figure 11.2a shows the development of the nearly 2D turbulent roller vortices at a few cylinder diameters downstream. These become increasingly disorganized beyond about 50 diameters downstream until they are no longer in evidence. Further downstream, in the fully turbulent wake, in Figure 11.2b, large-scale vortices are observed to reappear [3]. These are of considerably larger scale than the upstream Kármán vortices, are not very well organized, and appear to emerge in groups or packets agreeing with many other observations in the fully turbulent far wake [5, 6]. It is likely that the far wake vortices in this case are the product of an instability within the mean flow field [3]. The spectrum of the fully turbulent wake is broad-band and does not show concentrations of energy at particular frequencies such as that associated with the Kármán vortex street.

11.2.1 Self-Preserving Far Wake

Sufficiently far downstream in a wake flow it may be imagined that conditions conducive to a similarity solution for the mean velocity field are achieved. Specifically, the differences between the mean velocity profiles at two x locations should be attributable to

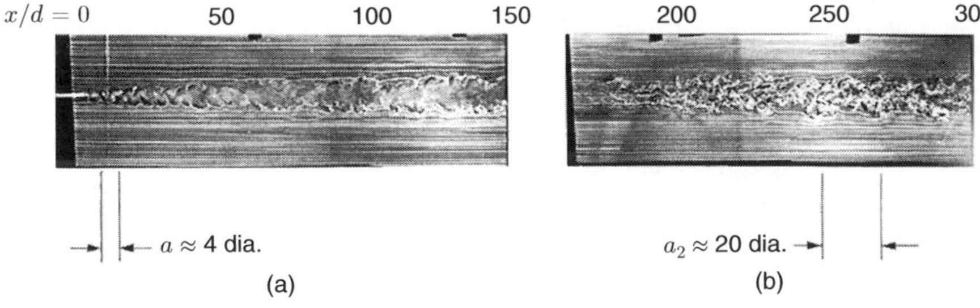

Figure 11.2 Circular cylinder wake at $R_e = 2200$; smoke wire at (a) $x/d = 1$ and (b) $x/d = 160$, [3]. Reprinted with permission of Cambridge University Press.

changes in scale only and not their functional form. Thus it makes sense to investigate whether a solution to Eq. (11.6) can be found in which the wake velocity is characterized by scales $\Delta U(x) = U_e - U_{min}(x)$ and $\ell = \ell(x)$, such that

$$U_e - \overline{U} = \Delta U\, f(\eta) \tag{11.9}$$

and

$$-\overline{uv} = (\Delta U)^2 g(\eta) \tag{11.10}$$

where

$$\eta \equiv \frac{y}{\ell} \tag{11.11}$$

is a similarity variable. Note that under the conditions of wake flow it is most reasonable to expect that the velocity defect, as in Eq. (11.9), obeys a similarity law rather than the mean velocity itself.

In 2D symmetric wakes, such as are of interest here, it follows from the definition of ΔU that

$$f(0) = 1 \tag{11.12}$$

while symmetry implies that

$$f'(0) = 0. \tag{11.13}$$

Moreover, anti-symmetry in the Reynolds shear stress implies that

$$g(0) = 0. \tag{11.14}$$

The goal now is to use the momentum equation (11.6) to explore the nature of the similarity solution, if it exists, corresponding to Eqs. (11.9) and (11.10), and the boundary conditions Eqs. (11.12)–(11.14).

From integration of the mean continuity equation it follows that at any fixed x position

$$\overline{V} = -\int_0^y \frac{\partial \overline{U}}{\partial x} dy, \tag{11.15}$$

since $\overline{V}$ is zero at the centerline $y = 0$. According to Eq. (11.9)

$$\frac{\partial \overline{U}}{\partial x} = -f\frac{d\Delta U}{dx} + \Delta U f' \frac{\eta}{\ell}\frac{d\ell}{dx}, \tag{11.16}$$

where the prime denotes differentiation with respect to η, and the relations $df/dx = (df/d\eta)(d\eta/d\ell)(d\ell/dx)$ and $d\eta/d\ell = -\eta/\ell$ have been used to obtain the second term on the right-hand side. Substituting this into Eq. (11.15) and converting the y integration into η integration gives

$$\overline{V} = \ell\frac{d\Delta U}{dx}G(\eta) - \Delta U\frac{d\ell}{dx}H(\eta) \tag{11.17}$$

where $G(\eta) \equiv \int_0^\eta f(\eta)d\eta$ and $H(\eta) \equiv \int_0^\eta f'(\eta)\,\eta\, d\eta$.

Noting that

$$\frac{\partial \overline{U}}{\partial y} = -\frac{\Delta U}{\ell}f' \tag{11.18}$$

and

$$\frac{\partial \overline{uv}}{\partial y} = -\frac{\Delta U^2}{\ell} g', \tag{11.19}$$

and using Eq. (11.17), (11.6) becomes, after dividing through by $\Delta U^2/\ell$,

$$-\alpha^* f + \beta^* \eta f' + \alpha^* \frac{\Delta U}{U_e}[-f'G + f^2] - \beta^* \frac{\Delta U}{U_e}[-f'H + \eta f f'] = g', \tag{11.20}$$

where

$$\alpha^* = \frac{U_e \ell}{(\Delta U)^2} \frac{d\Delta U}{dx} \tag{11.21}$$

and

$$\beta^* = \frac{U_e}{\Delta U} \frac{d\ell}{dx} \tag{11.22}$$

are dimensionless parameters. Equation (11.20) shows that a sufficient condition for a similarity solution to exist is that α^* and β^* be constant independent of x. Moreover, if attention is confined to the far wake, characterized by the condition that

$$\frac{\Delta U}{U_e} \to 0 \quad \text{as} \quad x \to \infty, \tag{11.23}$$

then the similarity form of the momentum equation is

$$-\alpha^* f + \beta^* \eta f' = g', \tag{11.24}$$

where α^* and β^* are constant.

Setting the expressions in Eqs. (11.21) and (11.22) to constant values leads to a coupled system of equations for ℓ and ΔU. Substituting $\Delta U/U_e$ from the second of these into the first and setting

$$n \equiv \frac{\alpha^*}{\beta^*} \tag{11.25}$$

gives

$$\frac{1}{\Delta U} \frac{d\Delta U}{dx} = n \frac{1}{\ell} \frac{d\ell}{dx}. \tag{11.26}$$

After integration this gives

$$\Delta U = C\ell^n, \tag{11.27}$$

where C is a constant. Substituting Eq. (11.27) into (11.22), defining $\alpha = (1-n)\beta^* C/U_e$, integrating, and introducing an integration constant, x_0, gives

$$\ell(x) = \alpha^m (x - x_0)^m \tag{11.28}$$

where

$$m = \frac{1}{1-n}. \tag{11.29}$$

Moreover, from Eq. (11.27) it then follows that

$$\Delta U(x) = C\alpha^{m-1}(x - x_0)^{m-1}. \tag{11.30}$$

It is convenient at this juncture to introduce the momentum thickness θ associated with wake flow that is similar to the same quantity given in Eq. (8.4) for boundary layers. In this case θ is defined as the thickness that a layer of fluid traveling at speed U_e must have in order to have the same defect momentum flux as occurs in the wake region. Thus, the balance is

$$-\rho U_e^2\theta = \rho \int_{-\infty}^{+\infty} \overline{U}(\overline{U} - U_e)dy \tag{11.31}$$

so that

$$\theta = \int_{-\infty}^{+\infty} \frac{\overline{U}}{U_e}\left(1 - \frac{\overline{U}}{U_e}\right) dy. \tag{11.32}$$

In view of Eqs. (11.8) and (11.31) it follows that θ is constant in wake flows.

Equation (11.32) provides a means of obtaining the value of the exponent m. Thus, substituting Eq. (11.9) into (11.32) gives

$$\frac{\Delta U}{U_e}\left[\int_{-\infty}^{+\infty} f(\eta)d\eta - \frac{\Delta U}{U_e}\int_{\infty}^{+\infty} f^2(\eta)d\eta\right] = \frac{\theta}{\ell}. \tag{11.33}$$

The second term on the left-hand side becomes small relative to the first in the far wake because $\Delta U/U_e \to 0$ so that Eq. (11.33) reduces to

$$\Delta U\ell = \frac{U_e\theta}{\int_{-\infty}^{+\infty} f(\eta)d\eta}. \tag{11.34}$$

The right-hand side of this equation is constant with the implication that the product of ΔU and ℓ is constant in the far wake. Using Eqs. (11.28) and (11.30) this means that $m + m - 1 = 0$ or $m - 1/2$. It is thus concluded that

$$\ell(x) = \alpha^{1/2}(x - x_0)^{1/2} \tag{11.35}$$

and

$$\Delta U = C\alpha^{-1/2}(x - x_0)^{-1/2}, \tag{11.36}$$

where x_0 is a virtual origin of the wake, and α and C are constant scales. It is evident from these results that the plane wake widens according to a 1/2 power law and the velocity defect decreases with an inverse 1/2 power law based on the distance downstream from the virtual origin of the wake.

Experiments show that the mean wake of a circular cylinder, for example, reaches a self-preserving state with the characteristics in Eq. (11.35) and (11.36) about 80–90 diameters downstream. It takes a considerably greater downstream distance for second and higher order properties of the velocity fluctuations to obtain a self-preserving state.

Equation (11.34), which is only valid in the far wake, also serves to establish an approximate relationship between θ and the drag, D, associated with the wake producing body. This requires performing a standard control volume analysis of streamwise momentum on a large rectangle exterior to the cylinder in question [7, 8]. In this, Eq. (3.9) is integrated over the control volume. The convective terms yield the difference in streamwise momentum flux across lines upstream and downstream of the body, while the viscous and pressure terms contribute to the drag force. The result is

$$D = \rho U_e \int_{-\infty}^{\infty} (U_e - \overline{U})dy \tag{11.37}$$

and, after changing the integration to η and using Eqs. (11.9) and (11.34), this gives

$$D = \rho U_e^2 \theta. \tag{11.38}$$

11.2.2 Mean Velocity

With knowledge of the x dependence of ℓ and ΔU it becomes possible to find the mean velocity field, $\overline{U}$, through a determination of $f(\eta)$ via Eq. (11.24) once a model for $g(\eta)$ is proposed. The traditional approach in this case is to assume the validity of a simple eddy viscosity model in which the eddy viscosity itself, ν_t, is taken to be constant. This latter assumption is reasonable in the central part of the wake where there is no intermittency. Closer to the outer part of the mean wake, where the flow is alternately turbulent and non-turbulent, the assumption may be less accurate.

By definition of ν_t,

$$\overline{uv} = -\nu_t \frac{\partial \overline{U}}{\partial y}, \tag{11.39}$$

and using Eqs. (11.10) and (11.18) it follows that

$$g = -\frac{f'}{R_t} \tag{11.40}$$

where $R_t = \ell \Delta U / \nu_t$ is a constant Reynolds number, as is evident from Eq. (11.34). Substituting Eq. (11.40) into (11.24) and noting from Eq. (11.25) that $\beta^* = -\alpha^*$, since according to Eq. (11.29) $n = -1$, a closed equation for $f(\eta)$ results in the form

$$f'' - R_t \alpha^* (\eta f' + f) = 0. \tag{11.41}$$

Integrating Eq. (11.41) with respect to η, applying the condition Eq. (11.13), and then integrating a second time and applying Eq. (11.12) gives

$$f(\eta) = e^{-\frac{U_e \alpha}{4\nu_t}\eta^2}, \tag{11.42}$$

where the definitions of R_t and α^* and Eqs. (11.35) and (11.36) have been used. It may be noticed that the constant, α, appearing in ℓ, ΔU, and f, only affects the scaling of distances and may be chosen arbitrarily. It is customary when considering the wake behind a cylinder of diameter d, to take

$$\alpha = d, \tag{11.43}$$

and with this choice it follows that

$$f(\eta) = e^{-R_d \eta^2/4} \tag{11.44}$$

is a Gaussian function and $R_d = dU_e/\nu_t$. Furthermore, with this choice of α, $\eta = y/\sqrt{d(x - x_0)}$.

Experimental measurements of

$$\frac{U_e - \overline{U}}{\Delta U} = f(\eta) \tag{11.45}$$

in the far wake of a circular cylinder at several cross-sections [9] are shown in Figure 11.3. For $\eta < 0.3$, Eq. (11.44) is seen to give a very good fit of the data when the empirically derived value $R_d = 61.04$ is used. As expected, the agreement is not as good in the

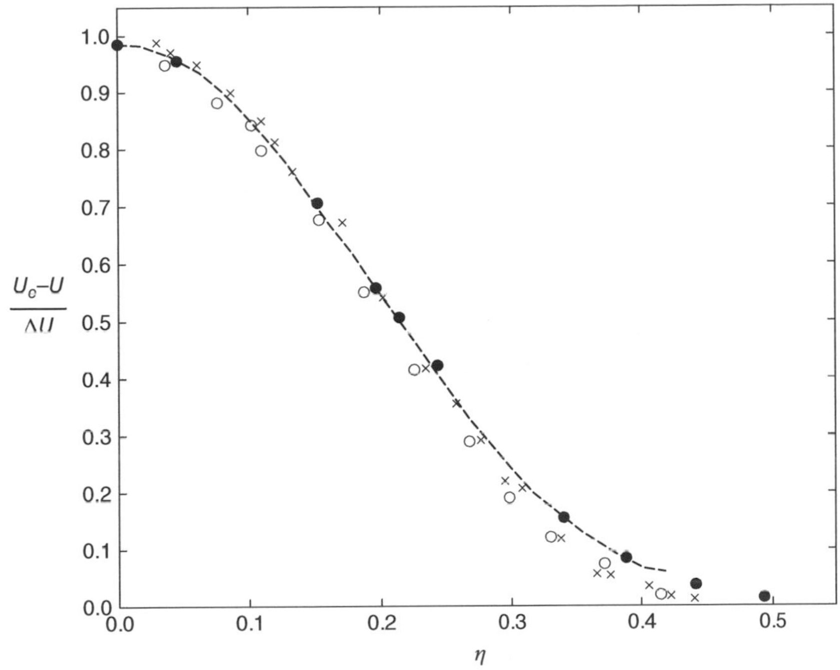

Figure 11.3 Comparison of the self-similar turbulent wake velocity profile of a cylinder with physical experiments at $R_d = 1360$: •, $x/d = 500$; +, $x/d = 650$; ○, $x/d = 800$; ×, $x/d = 950$; —, Eq. (11.44). Data from [9]. Reproduced from the *Australian Journal of Scientific Research* (Vol. A2, 1949), with permission of CSIRO Publishing.

outer part of the wake where the effect of intermittency becomes more prominent. With a more careful analysis, for example by including an intermittency factor $\gamma(\eta)$ in Eq. (11.39), better agreement of Eq. (11.41) with data may be achieved.

Carrying out the integration in Eq. (11.34) using Eq. (11.44) and noting that its left-hand side is C yields

$$C = \sqrt{\frac{R_d}{\pi}}\frac{U_e\theta}{2} = 2.204U_e\theta. \tag{11.46}$$

Now combining Eqs. (11.36), (11.43), and (11.46) it follows that

$$\frac{\Delta U(x)}{U_e} = 2.204\frac{\theta}{d}\sqrt{\frac{d}{x-x_0}}. \tag{11.47}$$

Introducing the drag coefficient

$$C_D \equiv \frac{D}{\frac{1}{2}\rho d U_e^2}, \tag{11.48}$$

and noting Eq. (11.38), it is found that

$$\frac{\theta}{d} = \frac{1}{2}C_D, \tag{11.49}$$

a relationship that also provides an alternative way of expressing $\Delta U/U_e$ in Eq. (11.47).

11.3 Turbulent Jet

Figure 1.4 showed a photograph of a turbulent axisymmetric round jet discharging into a motionless surrounding fluid. Near the nozzle mixing layers can be seen to form because of the difference in velocity between the approximately uniform velocity of the discharging flow and the stagnant fluid outside of the orifice. A similar presence of vortices developing between the juncture of the fast- and slow-moving outer flow is evident in the picture of a buoyant jet shown in Figure 1.5. In this way there is a degree of commonality between the early stages of jet flow and the mixing layers that will be considered below in the next section.

Near the jet orifice, the mixing layers formed from the fast- and slow-moving fluid grow outward into the surrounding fluid and inward toward the centerline until they meet. The mixing layers coincide with the shrinking of the region of essentially inviscid potential flow at the core of the jet. As Figures 1.4 and 1.5 show, the region of orderly vortex motion is short lived since a complete breakdown of the flow into turbulence is evident just a few diameters downstream of the orifice. The turbulent region that forms is filled with vortical motions of many scales. Far downstream it may be expected that the turbulent jet develops under the auspices of a similarity law with some of the same characteristics as the previous wake flow analysis, but with some significant differences as well. This is now considered for a plane jet, followed by a description of some of the statistical properties that have been measured in jets.

11.3.1 Self-Preserving Jet

At approximately 50 diameters downstream of the jet orifice, well beyond the point where the mixing layers formed at each side of the entering jet have met at the centerline, the jet flow reaches a fully developed state. In this, experiments show that the subsequent cross-stream distribution of the mean velocity has acquired the self-preserving form

$$\frac{\overline{U}}{\Delta U} = f(\eta), \tag{11.50}$$

where $\eta = y/\ell$ and $\Delta U = \overline{U}_{max}$ is the centerline velocity, as discussed previously. Both ΔU and ℓ change with downstream distance while Eq. (11.50) implies that the functional form of the mean velocity is fixed. The Reynolds stresses and higher order velocity statistics of the turbulent field reach a self-preserving state somewhat further downstream than the mean velocity.

As is done in developing the Blasius similarity solution to the laminar boundary layer [10], it is convenient to introduce a mean streamfunction $\overline{\Psi}(x, y)$ so that the $\overline{V}$ velocity component can be determined consistent with Eq. (11.50). A calculation using the stream function equation $\overline{U} = \partial\overline{\Psi}/\partial y$ shows that, in dimensional form

$$\overline{\Psi} = \ell \Delta U F(\eta), \tag{11.51}$$

where the similarity function $F(\eta)$ is related to f via

$$F'(\eta) = f(\eta). \tag{11.52}$$

Now using the relation $\overline{V} = -\partial\overline{\Psi}/\partial x$ it is seen that

$$\overline{V} = -\frac{d(\ell \Delta U)}{dx} F + \Delta U \frac{d\ell}{dx} \eta F'. \tag{11.53}$$

Substitution of these results into Eq. (11.5) gives

$$\frac{\ell}{\Delta U}\frac{d(\Delta U)}{dx}[F'^2 - FF''] - \frac{d\ell}{dx}FF'' = g', \tag{11.54}$$

where, as done in the case of the wake flow, it is assumed that

$$-\overline{uv} = (\Delta U)^2 g(\eta). \tag{11.55}$$

Equation (11.55) differs from (11.10) mainly due to the different velocity scalings. One way to achieve self-similarity in Eq. (11.54) is to require that

$$\frac{d\ell}{dx} = \alpha \tag{11.56}$$

and

$$\frac{\ell}{\Delta U}\frac{d(\Delta U)}{dx} = \beta, \tag{11.57}$$

where α and β are constants. It follows from Eq. (11.56) that

$$\ell = \alpha(x - x_0) \tag{11.58}$$

and from Eq. (11.57) that

$$\Delta U = C(x - x_0)^m \tag{11.59}$$

where C is a constant and the exponent, $m = \beta/\alpha$, is to be determined.

As shown previously, the momentum flux in the jet as expressed in Eq. (11.7) is constant. Changing the integration variable in Eq. (11.7) to η and using Eqs. (11.50) and (11.53) and the fact that $U_c = 0$ for a jet with no co-flow, it is derived that

$$\frac{d}{dx}\left(\ell \Delta U^2 \int_{-\infty}^{+\infty} F'^2\, d\eta\right) = 0. \tag{11.60}$$

After substituting for ℓ and ΔU from Eqs. (11.58) and (11.59), respectively, it is found that $1 + 2m = 0$ so that $m = -1/2$. Thus, in addition to Eq. (11.58) it follows that

$$\Delta U = C(x - x_0)^{-1/2}. \tag{11.61}$$

Thus, in the self-similar planar jet it appears that ℓ grows linearly with downstream distance while the mean jet centerline velocity decays according to $1/\sqrt{x - x_0}$. These results show that the Reynolds number, $\Delta U\ell/\nu$, increases as $(x - x_0)^{1/2}$ with x, so that the assumption that viscous terms may be neglected in Eq. (11.5) is increasingly well justified as the downstream distance increases.

11.3.2 Mean Velocity

As in the case of wake flow, the governing similarity equation, in this case Eq. (11.54), provides a means with which to determine the similarity form of the mean velocity field once an appropriate model of g' is proposed. Once again it is simplest to assume that $\overline{uv}$ can be modeled via a gradient law, in which case it may be shown that Eq. (11.39) also applies to the jet, and consequently

$$g' = \frac{1}{R_t}F''' \tag{11.62}$$

where

$$R_t = \frac{\Delta U \ell}{\nu_t} \tag{11.63}$$

must be constant if Eq. (11.54) is to yield a similarity law. If R_t is constant Eq. (11.63) implies that $\nu_t \sim \sqrt{x - x_0}$. Now, substituting Eqs. (11.56), (11.58), (11.61), and (11.62) into (11.54) yields an equation for $F(\eta)$ in the form

$$\frac{\alpha}{2}(FF'' + F'^2) + \frac{1}{R_t}F''' = 0. \tag{11.64}$$

Accompanying Eq. (11.64) is the boundary condition $F(0) = 0$ that forces the symmetry line to be a streamline, and the condition $F'(0) = 1$ from the definition of ΔU. A last condition is that $\lim_{\eta\to\infty} F'(\eta) = 0$ since the velocity is zero far from the jet centerline. Integrating Eq. (11.64) twice and applying the boundary conditions gives

$$F^2 + \frac{4}{\alpha R_t}(F' - 1) = 0, \tag{11.65}$$

which is an example of a Riccati equation [11] and is readily solved, yielding

$$F(\eta) = \frac{2}{\sqrt{\alpha R_t}} \tanh\left(\frac{\sqrt{\alpha R_t}}{2}\eta\right). \tag{11.66}$$

Taking a derivative of Eq. (11.66) and substituting into Eq. (11.50) gives

$$\overline{U} = \Delta U\left(1 - \tanh^2\left(\eta\frac{\sqrt{\alpha R_t}}{2}\right)\right), \tag{11.67}$$

for the mean similarity velocity field in a turbulent jet, a result first obtained by Görtler [12]. The similarity variable, η, depends on the arbitrary parameter α through ℓ. For convenience, it may be assumed that $\alpha = 4/R_t$ so that Eq. (11.67) becomes

$$\overline{U} = \Delta U(1 - \tanh^2(\eta)), \tag{11.68}$$

where $\eta = yR_t/(4(x - x_0))$. Equation (11.68) shows, via Eq. (11.61), that $\overline{U}$ ultimately depends on the two parameters R_t and C. The latter can be expressed in terms of the momentum flux, M, by substituting Eq. (11.68) into (11.8) and integrating. The result is

$$C = (3MR_t/16\rho)^{1/2}. \tag{11.69}$$

Finally, R_t is determined with reference to experimental data. Generally, it is found that a good fit occurs with $R_t = 25.7$. With these parameter values it may be observed from Eq. (11.68) that when $\eta = 1, y = \ell$ and $\overline{U}$ is approximately 42% of the centerline velocity.

Measurements of the mean and fluctuating properties of a plane jet with no co-flow [13] support the results of the similarity argument. For example, Figure 11.4 shows the development of the centerline velocity and the width of the jet, ℓ, with distance downstream. In this experiment, the jet has Reynolds number $Re_d = 3.4 \times 10^4$, d is the jet width at the exit plane, ℓ is measured from the jet centerline to where the mean velocity is $\Delta U/2$, and the velocity scale is the jet exit velocity. The approximate linear growth of $1/(\Delta U)^2$ for $x/d \geq 45$ confirms the validity of Eq. (11.61). Also evident is the linear growth of the normalized jet half-width, ℓ/d, for $x/d \geq 65$, confirming Eq. (11.58). The intercept of the two lines in the figure is taken to be the virtual origin, x_0.

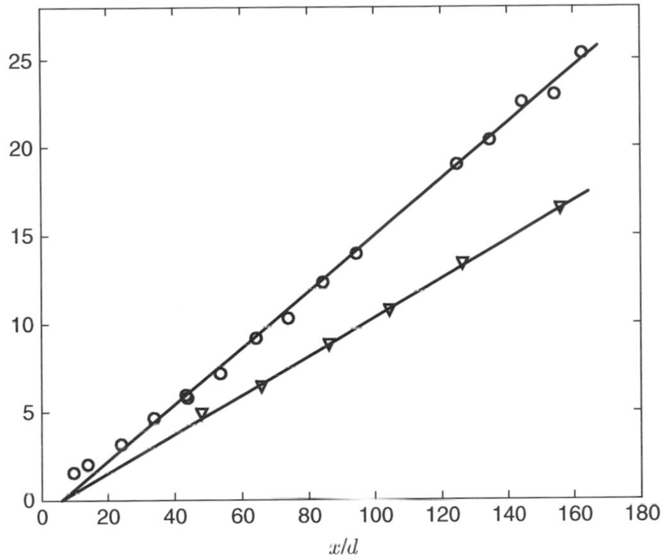

Figure 11.4 Centerline mean velocity and jet width development of a turbulent plane jet at $Re_d = 3.4 \times 10^4$. Data from [13]. ○, $1/(\Delta U)^2$; ▽, ℓ/d.

Figure 11.5 shows profiles of the mean jet velocity normalized by ΔU at six locations between $x/d = 47$ and 155 downstream of the exit plane of the nozzle. The cross-stream similarity variable is taken to be $\eta = y/(x - x_0)$, so that the choice $\alpha = 1$ is made in this instance. Also plotted is the predicted similarity form Eq. (11.68). The measured profiles clearly collapse on one another, displaying self-preservation, and the theoretical curve agrees well with the measured mean profile except near the exterior of the jet where the intermittency of the turbulence plays a role.

11.3.3 Reynolds Stresses

The growth of the root-mean-square streamwise velocity fluctuation, $u_{rms} \equiv \sqrt{\overline{u^2}}$, along the jet centerline, as determined from experiments and shown in Figure 11.6, is also linear for $x/d \geq 45$. The cross-stream distribution of the variances of the velocity fluctuations in the self-preserving region at $x/d = 101$ are shown in Figure 11.7. The streamwise variance is known to be approximately self-preserving for $x/d \geq 45$.

The variance of u is seen to have a peak more than twice that of v and about twice that of w. The variances of all three components drop to negligible values compared to ΔU for $\eta \geq 0.3$, which is located at about 2.5ℓ from the centerline of the jet. In this location the jet flow is mostly potential, since it is well outside the turbulent core of the jet. For $\eta \leq 0.15$ (within a cross-stream distance of approximately 1.3ℓ from the jet centerline), the flow is nearly always turbulent, though incursions of irrotational fluid occasionally penetrate nearly to the jet centerline.

The Reynolds shear stress, $\overline{uv}$, peaks at $\eta \approx 0.07$, which is at a cross-stream distance of approximately 0.6ℓ from the jet centerline, as seen in Figure 11.8. It is zero at the centerline due to symmetry of the mean flow. The figure also contains the predicted

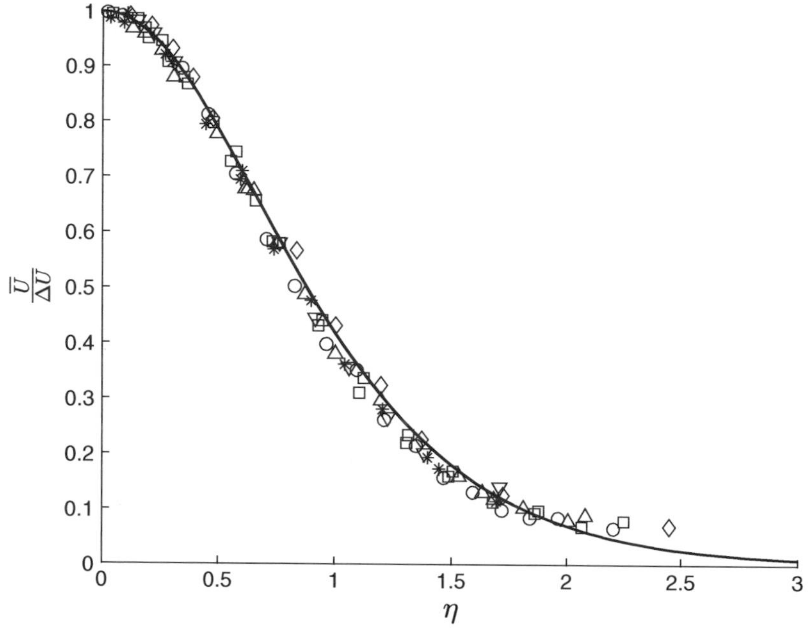

Figure 11.5 Mean streamwise velocity profiles of turbulent plane jet at $Re_d = 3.4 \times 10^4$ for ◇, $x/d = 47$; ○, 65; □, 85; ▽, 103; △, 125; ∗, 155; and, —, Eq. (11.68). Data from [13].

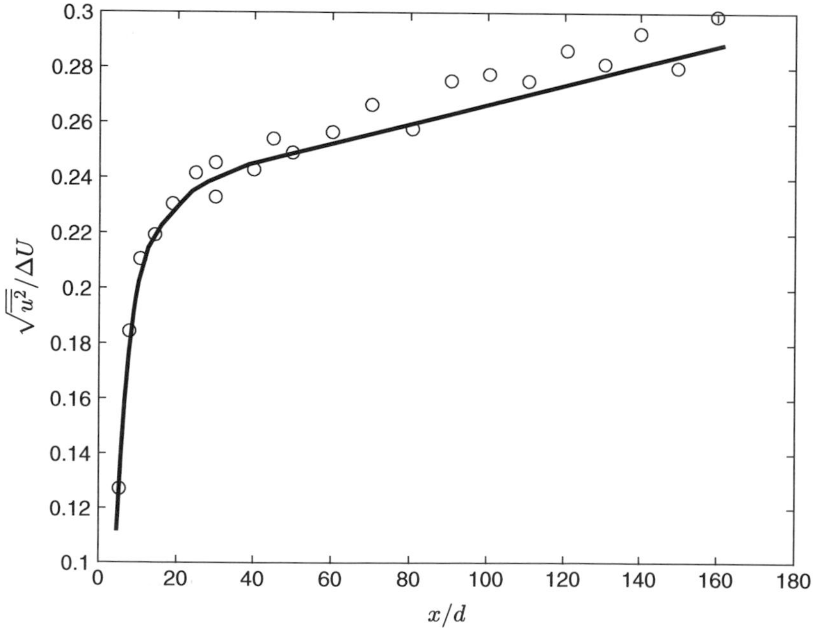

Figure 11.6 Growth of $\sqrt{\overline{u^2}}$ along the centerline of a turbulent plane jet at $Re_d = 3.4 \times 10^4$. ○, data from [13]; —, fit to the data.

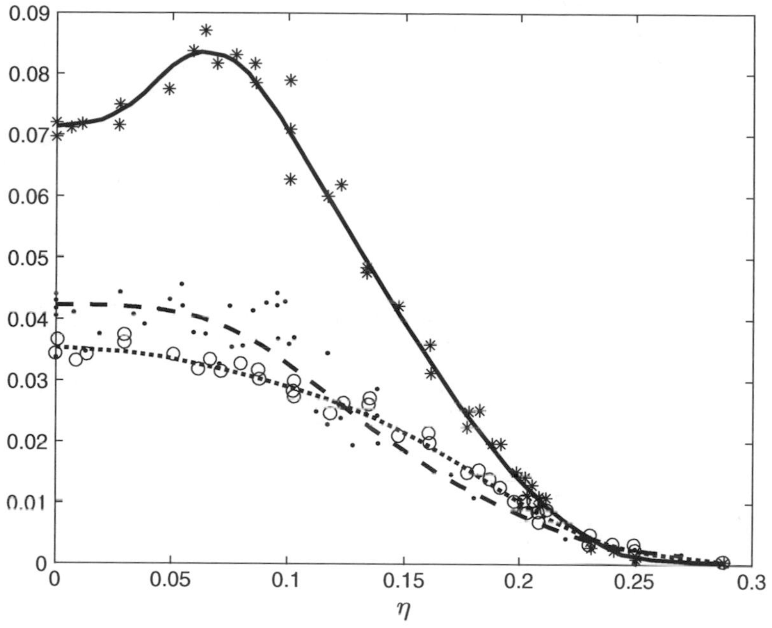

Figure 11.7 Velocity variances for turbulent plane jet at $Re_d = 3.4 \times 10^4$ and $x/d = 101$. Data from [13] with fitted curves: $*$ and —, $\overline{u^2}/(\Delta U)^2$; ○ and · · ·, $\overline{v^2}/(\Delta U)^2$; • and −−, $\overline{w^2}/(\Delta U)^2$.

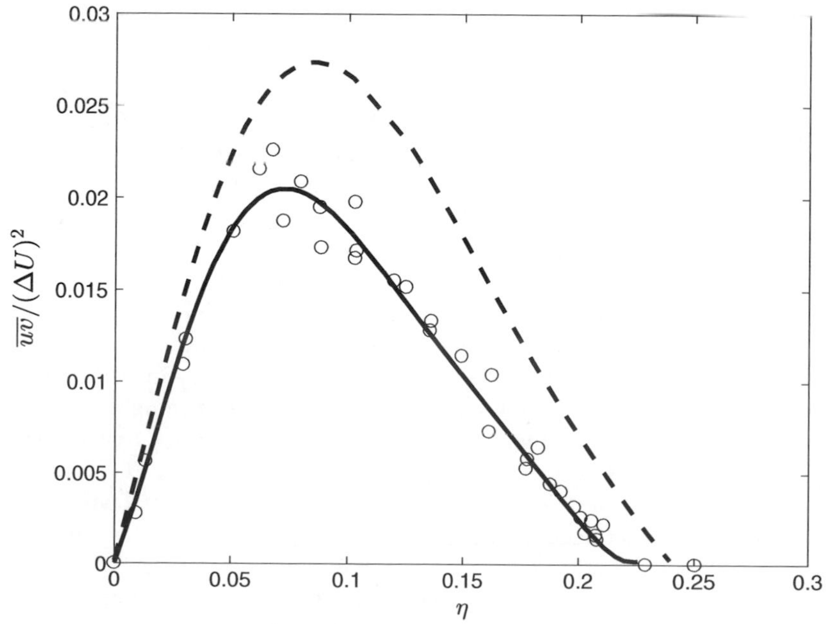

Figure 11.8 Reynolds shear stress distribution for turbulent plane jet at $Re_d = 3.4 \times 10^4$ and $x/d = 101$; ○, data from [13]; —, fit to data; −−, Eq. (11.70).

similarity solution for $\overline{uv}$ that is given by

$$g = -\sqrt{\frac{\alpha}{R_t}} \tanh\left(\eta\frac{\sqrt{\alpha R_t}}{2}\right)\left(1 - \tanh^2\left(\eta\frac{\sqrt{\alpha R_t}}{2}\right)\right), \tag{11.70}$$

a result that is obtained from Eq. (11.62). As in the previous figure, $\alpha = 4/R_t$ is assumed. It may be noted that with increasing distance from the jet centerline, the calculated values of $\overline{uv}$ systematically overestimate the measured values. Within about 0.7ℓ ($\eta \approx 0.08$) of the centerline the agreement is reasonably good. At larger distances the discrepancy in the figure may be partly due to measurement errors when the turbulence intensities become large and the mean velocity gets smaller, causing the assumptions used in justifying hot-wire measurements to become less valid.

11.4 Turbulent Mixing Layer

Turbulent mixing layers occur when two parallel fluid streams with different mean velocities are brought together at a sufficiently high Reynolds number. The previous sections showed how mixing layers appear as a natural component of wake and jet flows during their initial stages. They are also found in many common engineering applications involving two-phase flow and combustion, turbomachinery, and phenomena on the scale of the atmosphere as seen in Figure 1.3 and even in the interstellar medium within galaxies [14].

11.4.1 Structure of Mixing Layers

Figure 11.9 shows shadowgraph photos [15] of a turbulent mixing layer viewed from the side and above that is developing downstream of a thin plate separating an upper fluid stream of velocity U_u from a lower stream with velocity U_l for which the velocity ratio $U_l/U_u = 0.38$ and the Reynolds number based on the vorticity thickness defined as the length scale $\Delta U/(d\overline{U}/dy)_{max}$ is equal to 6500 at the location where the mixing layer becomes turbulent. The high shear in the interfacial region between the two streams gives rise to a Kelvin–Helmholtz-type instability resulting in the characteristic formation of spanwise vortices, referred to as "rollers", and seen in the lower part of the figure. The discovery of rollers in the original mixing layer visualizations of Brown and Roshko [16] was unexpected since the flow exhibited all the "random" characteristics of turbulence. The roller vortices grow as they convect downstream by ingesting irrotational fluid from the exterior flow. Though the vortices may often first appear under laminar flow conditions when the two fluid streams meet, they may quickly transition to turbulence by amplification of local small-scale perturbations. Turbulence is visible in Figure 11.9 in the irregular motions appearing within the vortices not far downstream of the splitter plate. A common occurrence in mixing layers is for the roller vortices to undergo pairing wherein two adjacent rollers merge into a single larger vortex [17]. The large-scale spanwise vortices that form and combine in the mixing layer may retain a high degree of coherency for considerable distances downstream. Vortex merger is a significant mechanism by which mixing layers grow in size with downstream distance. How far downstream roller vortices persist is not easily ascertained when the flow becomes fully turbulent, though there is evidence suggesting that observed linear

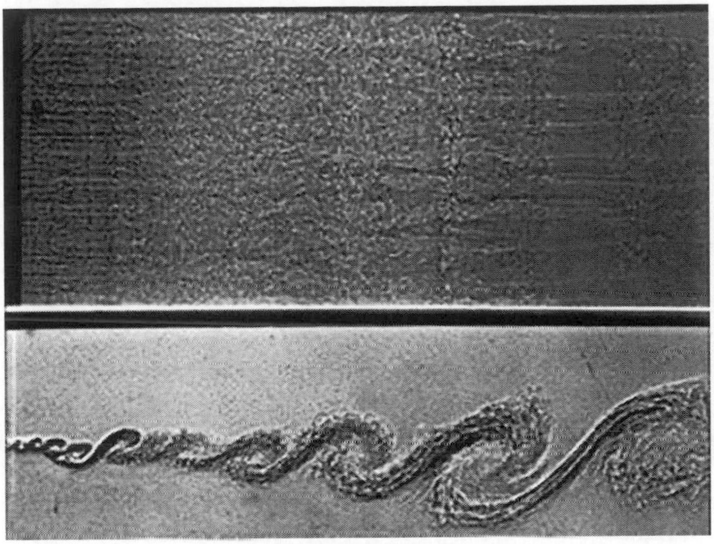

Figure 11.9 Growth of a turbulent mixing layer with x, y, and z denoting the streamwise, cross-stream, and spanwise coordinates, respectively. Top: viewed from above $(x - z)$ plane; bottom: viewed from the side $(x - y)$ plane. From [15].

growth in the mixing layer thickness coincides with linear growth of individual spanwise vortices within the self-similar turbulent flow region [18].

In addition to the roller vortices, investigations [15, 19] reveal the presence of additional coherent structures in the central plane of the mixing layer, most notably a system of relatively slender streamwise rib vortices oriented in the streamwise direction. They are evident within the top image in Figure 11.9. By examining the two parts of the figure, it appears that the secondary streamwise vortices appear on the thin layer of vorticity (referred to as the braid region) that extends between the adjacent roller vortices. The rib vortices wrap over the high-speed side of the downstream vortex and under the low-speed side of the upstream vortex. These have been observed [20] to occur in counter-rotating pairs during the early transition of the mixing layer to three-dimensionality soon after the appearance of the spanwise rollers. Moreover, the streamwise vortices appear first on the braids but then extend into the cores of the rollers. The downstream extent of the roller/rib structure appears to be limited and not evident in the downstream self-similar region [21]. There is also likely to be a special connection between how the vortical structure is first manifested in the mixing layer and the upstream perturbations existing at its initiation.

Roller/rib structures are shown in Figure 11.10 as computed in a spatially growing mixing layer simulated via a vortex filament scheme. Here, the initially spanwise oriented vortex filaments naturally agglomerate into rollers that merge into larger vortices. During the development of the mixing layer, braids appear containing streamwise rib vortices connecting the growing roller vortices. This process adds a three-dimensionality to the rollers that precedes their breakdown into turbulent flow.

A close-up view of the roller/rib structure is shown in Figure 11.11. The development of the rollers shows how they increasingly are affected by the action of the braids as they wind around the growing rollers, causing an uneven appearance in the spanwise

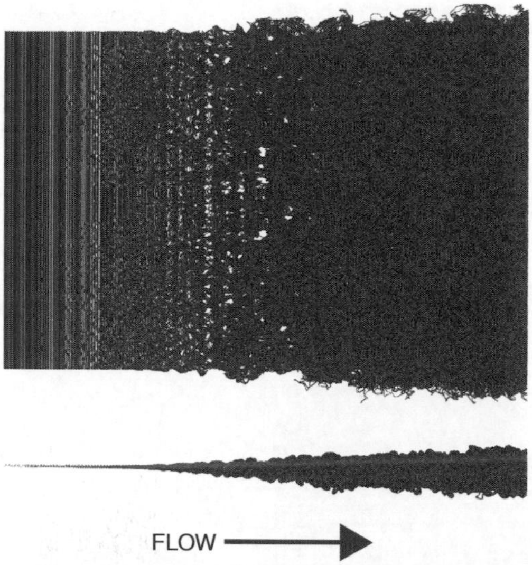

Figure 11.10 Overhead (top) and side (lower) views of a vortex filament computation of a mixing layer [22]. Used by permission of AIAA.

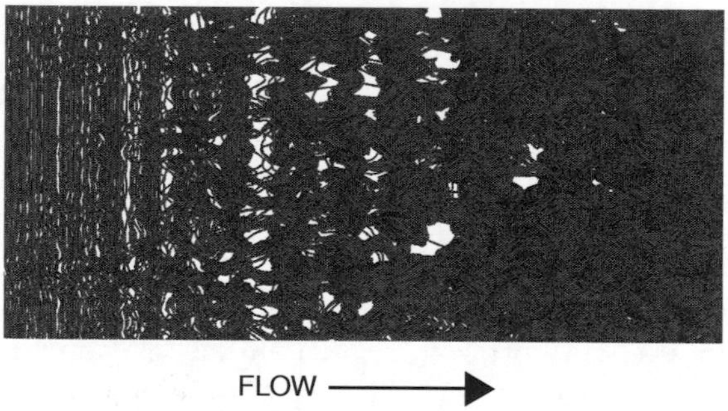

Figure 11.11 Detail of braid/roller vortices developing in a vortex filament simulation of a mixing layer [22]. Used by permission of AIAA.

direction. Out of this 3D structure the flow becomes fully turbulent. Besides the formation of roller/rib structures leading to the breakdown to turbulence, under some circumstances different vortical patterns can emerge in the mixing layer. Figures 11.12 and 11.13, respectively, show two other scenarios that have been observed in numerical calculations and physical experiments. In the first, Figure 11.12 shows how a chain-link fence pattern of vortices has developed out of the initial disturbances to the mixing layer. This kind of pattern has been found to occur when there is a degree of spanwise variation to the mean velocity field, a condition that is imposed in the numerical calculation in Figure 11.12 by not exactly enforcing spanwise periodicity. A third mode of structural development in the mixing layer is illustrated in Figure 11.13 wherein the normal roller development including braids occurs at an oblique angle. This behavior appears to be a consequence of having relatively strong perturbations to the upstream mixing layer. In the numerical study the perturbations came from a somewhat coarser implementation of the vortex filament scheme.

11.4.2 Self-Preserving Mixing Layer

The analysis of the mixing layer proceeds along the same lines as for the wake and jet. As in the case of those flows, the nearly parallel, thin flow form of the streamwise

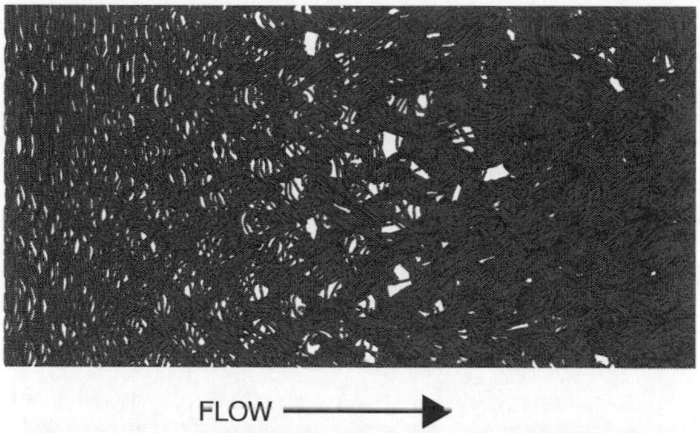

Figure 11.12 Detail of a chain-link fence vortex pattern developing in a vortex filament simulation of a mixing layer [22]. Used by permission of AIAA.

Figure 11.13 Overhead view of a mixing layer containing oblique roller/braid vortices.

momentum equation (11.5) is used to determine the similarity conditions and predict the resulting mean flow characteristics. For a mixing layer with U_l and U_u as the uniform mean velocities on the low- and high-speed sides, respectively, it is convenient to postulate that the self-preserving form of the mean streamfunction is:

$$\overline{\Psi} = \ell U_m F(\eta), \tag{11.71}$$

where $U_m = (U_u + U_l)/2$ and $\eta = y/\ell$ is the similarity variable. In this case

$$\overline{U} = U_m F'(\eta) \tag{11.72}$$

and, assuming that

$$\overline{uv} = -U_m^2 g(\eta), \tag{11.73}$$

it follows from Eq. (11.5) that after determining $\overline{V}$ from $\overline{\Psi}$ that

$$\frac{d\ell}{dx} FF'' + g' = 0. \tag{11.74}$$

As in the case of the jet, Eq. (11.74) implies that for a similarity solution to exist, ℓ should depend linearly on x, so that

$$\ell(x) = \alpha(x - x_0), \tag{11.75}$$

where α is a constant.

Assuming a constant eddy viscosity model as in Eq. (11.62) so as to get a closed system, Eq. (11.74) becomes

$$F''' + R_t \alpha F F'' = 0, \tag{11.76}$$

where

$$R_t = \frac{U_m \ell}{\nu_t}. \tag{11.77}$$

Since R_t must be constant to maintain similarity, it may be concluded that ν_t varies linearly with x to offset the x variation in ℓ, since U_m is constant.

The boundary conditions to be applied to Eq. (11.76) are

$$F'(\pm\infty) = 1 \pm \lambda \tag{11.78}$$

where $\lambda = \Delta U/(2U_m)$ and $\Delta U = U_u - U_l$. It may be noticed that Eq. (11.76) is identical in form to the classical Blasius boundary layer equation for a zero-pressure gradient boundary layer [10]. The only difference between that and the mixing layer is in the boundary conditions.

11.4.3 Mean Velocity

As in the case of boundary layer flow, Eq. (11.76) does not have a closed form solution. Nonetheless, a reasonably good approximate solution can be obtained when λ is small. This entails assuming F obeys a power series in the small parameter, λ, [12] as in

$$F(\eta) = \eta + \lambda F_1(\eta) + \lambda^2 F_2(\eta) + \dots \tag{11.79}$$

where the functions $F_1, F_2, \dots$ need to be determined. The leading term in Eq. (11.79) is chosen specially so that after applying Eq. (11.72) it follows that

$$\overline{U} = U_m + U_m \lambda F_1'(\eta) + U_m \lambda^2 F_2'(\eta) + \dots \tag{11.80}$$

Substituting Eq. (11.79) into (11.76) and collecting together terms of like powers of λ gives equations from which F_1, F_2, and so forth can be determined. The lowest order equation yields

$$F_1''' + \alpha R_t \eta F_1'' = 0 \tag{11.81}$$

which is a closed equation for F_1. One integration yields

$$F_1'' = Ce^{-\alpha R_t \eta^2/2}, \tag{11.82}$$

where C is a constant. The arbitrary constant, α, may be conveniently picked so that $\alpha R_t = 2$. In this case,

$$\eta = \frac{R_t}{2} \frac{y}{x - x_0} \tag{11.83}$$

and

$$F_1'' = Ce^{-\eta^2}. \tag{11.84}$$

If Eq. (11.79) is truncated at the level of F_1 so that $F = \eta + \lambda F_1$, then the boundary condition Eq. (11.78) implies that

$$F_1'(\pm\infty) = \pm 1. \tag{11.85}$$

Now integrating Eq. (11.84) and applying (11.85) it is found that

$$F_1' = erf\,\eta \tag{11.86}$$

where

$$erf\,\eta = \frac{2}{\sqrt{\pi}}\int_0^{\eta} e^{-\xi^2}\,d\xi \tag{11.87}$$

is the error function. Finally, it is seen from Eq. (11.80) that

$$\overline{U} \approx U_m\left(1 + \frac{\Delta U}{2U_m} erf\,\eta\right) \tag{11.88}$$

where higher order terms are neglected. There remains the parameter R_t that needs to be evaluated from experiment. Measurements show a dependence of R_t on U_u/U_l [23] and a typical value [24] is $R_t = 27$.

The momentum thickness θ defined for wakes and jets in Eq. (11.32) is not suitable for the mixing layer since the integral appearing in it is unbounded. An alternative definition that is appropriate to mixing layers is given by

$$\rho\theta(\Delta U)^2 = \rho\int_{-\infty}^{\infty}(U_u - \overline{U})(\overline{U} - U_l)dy \tag{11.89}$$

which is well defined since the integrand goes to zero for $y \to \pm\infty$. The right-hand side may be viewed as the momentum deficit (in relation to U_u) in the flux of momentum measured relative to U_l. θ, then, is the thickness of a layer one would need to account for this momentum defect if the fluid were uniformly traveling at speed U_u with its momentum flux measured relative to U_l.

Substituting Eq. (11.88) into (11.89) shows that

$$\frac{\theta}{\ell} = \frac{1}{4}\int_{-\infty}^{\infty}(1 - erf^2\eta)d\eta \tag{11.90}$$

or $\theta \sim \ell$. Thus, in view of Eq. (11.75), the rate of growth of θ with x is constant. It is empirically observed [23] that the growth rate parameter

$$r_\theta = \frac{U_m}{\Delta U}\frac{d\theta}{dx} \tag{11.91}$$

is a universal constant. In other words, this definition properly takes into account the dependence of $d\theta/dx$ on the velocities of the mixing streams.

Figure 11.14 shows a favorable comparison between Görtler's theoretical self-preserving mean profile in Eq. (11.88) and the results of a vortex filament simulation of a mixing layer [22]. In this, θ varied linearly with downstream distance in agreement with Eqs. (11.75) and (11.90). Equally good verification of Eq. (11.88) is found from physical experiments [25, 26].

11.4.4 Reynolds Stresses

The normal Reynolds stresses scaled by $(\Delta U)^2$ as measured in physical experiments [25, 26] and computed in a temporally varying DNS [21] are compared in Figure 11.15. The cross-stream similarity variable in this figure is taken to be $\eta = y/\theta$. The maxima of the velocity fluctuation variances are seen to occur at $\eta = 0$, where the mean velocity

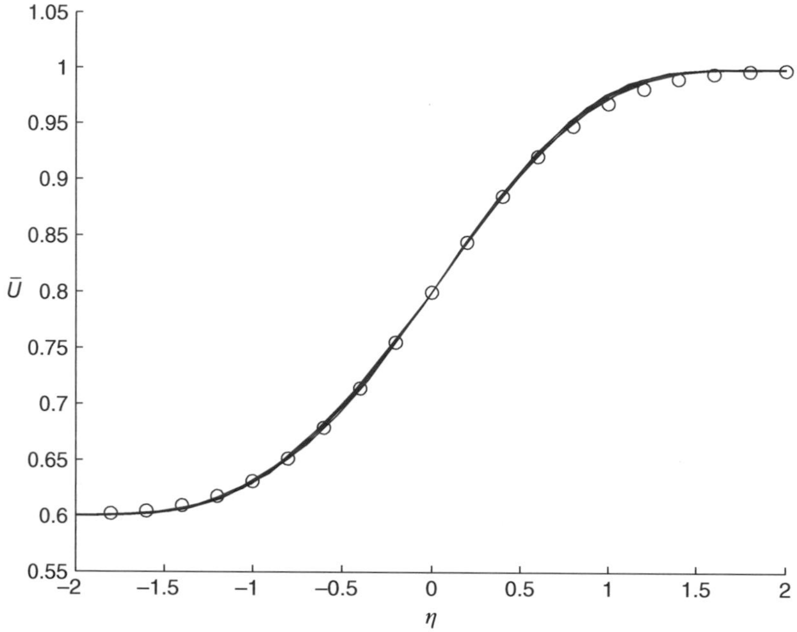

Figure 11.14 Mean streamwise velocity distributions in a self-preserving two-stream mixing layer. Solid lines are results from seven locations across a vortex filament simulation of a mixing layer ([22]). o, Eq. (11.88). Used by permission of AIAA.

profile has its inflection point and maximum gradient. The streamwise variance is largest and the cross-stream variance the smallest. It should be noted that earlier investigations [27–29] show considerable variation in the peak values, a fact that is often attributed to differences in initial conditions of the mixing layer experiments. In general, the experimental values and the DNS curves in Figure (11.15) are in good agreement and show that self-preservation has been reached.

Figure 11.16 shows the distribution of the Reynolds shear stress for the same conditions as in Figure 11.15. The experimental results agree well with the DNS prediction, suggesting that $\overline{uv}$ exhibits self-preservation. Analysis of the eddy viscosity implied by the measured and computed Reynolds shear stress confirms that it is reasonable to assume that it is constant through the center portion of the mixing layer, between the limits of $\eta = \pm 3$. Outside of this region the model is less well justified.

11.5 Chapter Summary

Our consideration of wakes, jets, and mixing layers in this chapter describes some of the distinctive properties of turbulence free of the direct influence of boundaries. Despite the different circumstances out of which these free shear flows develop, there is a significant degree of commonality between them in regard to vortical structures and the fact that they develop self-similar regions at sufficiently far downstream distances from their sources.

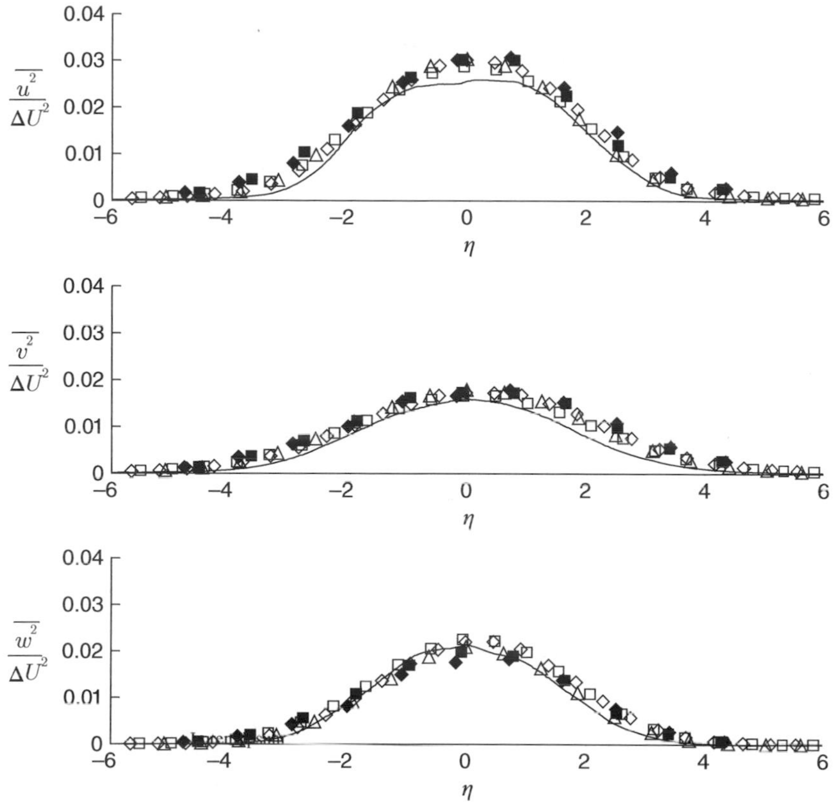

Figure 11.15 Normal Reynolds stresses in a two-stream mixing layer. Here, $\eta = y/\theta$. Experiment [25]: ◆, $R_\theta = 1{,}792$; ■, $R_\theta = 2{,}483$. Experiment [26]: □, $R_\theta \approx 3{,}820$; ◇, $R_\theta \approx 4{,}380$; △, $R_\theta \approx 6{,}365$. DNS [21]: —, averaged for $R_\theta = 1{,}500$–$2{,}000$.

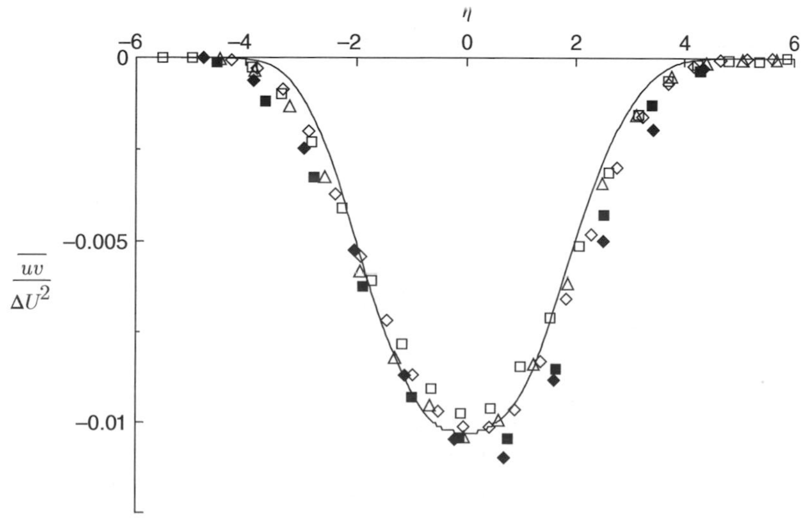

Figure 11.16 Reynolds shear stress distributions in a two-stream mixing layer. Here, $\eta = y/\theta$. Symbols as in Figure 11.15.

The formation of spanwise vortical structures out of shearing is a common occurrence. Accompanying such roller-type vortices is a network of streamwise-oriented vortices that develop and influence the breakdown to turbulent flow. Coincident with the growth of vortex structures and turbulence is a tendency for free shear flows to grow laterally by absorption of potential flow and the natural proclivity of turbulence produced in the central core of the free shear layers to transport to the outer edges of the flow though the mechanism of turbulent transport.

References

1 Schlichting, H. (1968) *Boundary Layer Theory*, McGraw-Hill Book Co., New York, 6th edn.
2 Townsend, A.A. (1976) *The Structure of Turbulent Shear Flows*, Cambridge University Press, Cambridge.
3 Cimbala, J., Nagib, H., and Roshko, A. (1988) Large structure in the far wakes of two-dimensional bluff bodies. *J. Fluid Mech.*, 190, 265–298.
4 Scarano, F. and Poelma, C. (2009) Three-dimensional vorticity patterns of cylinder wakes. *Exp. Fluids*, 47, 69–83.
5 Mumford, J.C. (1982) The structure of the large eddies in fully-developed turbulent shear flows. 1. the plane jet. *J. Fluid Mech.*, 118, 241–268.
6 Townsend, A.A. (1979) Flow patterns of large eddies in a wake and in a boundary layer. *J. Fluid Mech.*, 95, 515–537.
7 Hinze, J.O. (1975) *Turbulence*, McGraw-Hill, New York.
8 Takahashi, T.T. (1997) On the decomposition of drag components from wake flow measurements. *AIAA Paper*, 97-0717.
9 Townsend, A.A. (1949) The fully developed turbulent wake of a circular cylinder. *Aust. J. Sci. Res.*, A2, 451–468.
10 Bernard, P.S. (2015) *Fluid Dynamics*, Cambridge University Press, Cambridge.
11 Bender, C.M. and Orszag, S.A. (1978) *Advanced Mathematical Methods for Scientists and Engineers*, McGraw-Hill, New York.
12 Görtler, H. (1942) Berechnung von aufgaben der frein turbulenz auf grund eines neuen nahrungsansatzes. *ZAMM*, 22, 244–254.
13 Heskestad, G. (1965) Hot-wire measurements in a plane turbulent jet. *Trans. ASME, Ser. E, J. Appl. Mech.*, 32, 721–734.
14 Begelman, M.C. and Fabian, A.C. (1990) Turbulent mixing layers in the interstellar and intracluster medium. *Mon. Notices Roy. Astro. Soc.*, 244, 26P–29P.
15 Konrad, J.H. (1977) *An experimental investigation of mixing in two-dimensional turbulent shear flows with applications to diffusion-limited chemical reactions*, Ph.D. thesis, California Institute of Technology.
16 Brown, G.L. and Roshko, A. (1974) On density effects and large structure in turbulent mixing layers. *J. Fluid Mech.*, 64, 775–816.
17 Winant, C.D. and Browand, F.K. (1974) Vortex pairing: the mechanism of turbulent mixing-layer growth at moderate Reynolds number. *J. Fluid Mech.*, 63, 237–255.
18 McMullan, W.A., Gao, S., and Coats, C.M. (2015) Organized large structure in the post-transition mixing layer. Part 2. Large-eddy simulation. *J. Fluid Mech.*, 762, 302–343.

19 Breidenthal, R. (1982) Structure in turbulent mixing layers and wakes using a chemical reaction. *J. Fluid Mech.*, 109, 1–24.
20 Lasheras, J.C., Cho, J.S., and Maxworthy, T. (1986) On the origin and evolution of streamwise vortical structures in a plane, free shear-layer. *J. Fluid Mech.*, 172, 231–258.
21 Rogers, M.M. and Moser, R.D. (1994) Direct simulation of a self-similar turbulent mixing layer. *Phys. Fluids*, 6, 903–923.
22 Bernard, P.S. (2008) Grid-free simulation of the spatially growing turbulent mixing layer. *AIAA J.*, 46, 1725–1737.
23 Yule, A.J. (1972) Spreading of turbulent mixing layers. *AIAA J.*, 10, 686–687.
24 Reichart, H. (1942) Gesetzmä sigkeiten der freien turbulenz, *Tech. Rep. 414*, VDI-Forschungsheft.
25 Loucks, R. (1998) *An experimental examination of the velocity and vorticity fields in a plane mixing layer*, Ph.D. thesis, University of Maryland.
26 Bell, J.H. and Mehta, R.D. (1990) Development of a two-stream mixing layer from tripped and untripped boundary layers. *AIAA J.*, 28, 2034–2042.
27 Spencer, B.W. (1970) *Statistical Investigation of Turbulent Velocity and Pressure Fields in a Two-Stream Mixing Layer*, Ph.D. thesis, University of Illinois.
28 Yule, A.J. (1971) Two-dimensional self-preserving turbulent mixing layers at different free stream velocity ratios, *Tech. Rep. 3683*, Aero. Res. Counc. Rep. & Mem.
29 Oster, D. and Wygnanski, I. (1982) The forced mixing layer between parallel streams. *J. Fluid Mech.*, 123, 91–130.

Problems

11.1 Defining $\eta_{1/2}$ as the value of η in a wake flow where the velocity defect is half its maximum, that is $f(\eta_{1/2}) = 0.5$, show that $\eta_{1/2} = 0.213$.

11.2 Use a symbolic program (such as *syms* in MATLAB with commands *int, diff, simplify, eval*) to compute $\overline{V}$ and the mean vorticity $\overline{\Omega}$ from Eqs. (11.15), (11.44), (11.45), and (11.47) in the self-similar region of a cylinder wake flow. Scale velocities by U_e lengths by d and assign constant values to R_d and C_D, and set $x_0 = 0$. Make contour plots of $\overline{U}, \overline{V}$ and $\overline{\Omega}$ in the region in scaled coordinates $10 \leq x \leq 500$, $-10 \leq y \leq 10$. Make spanwise line plots of the mean velocities and vorticity at some x locations.

11.3 Show that the scaled eddy viscosity $\nu_t/(dU_e)$ depends on $(\theta/d)^{2/3}$ in the self-similar region of a wake flow.

11.4 Use a symbolic program (such as *syms* in MATLAB) to compute $\overline{U}, \overline{V}$, and the mean vorticity $\overline{\Omega}$ beginning from Eq. (11.51) for the mean stream function in the self-similar region of a jet flow. Scale velocities by $C/\sqrt{d}$ where d is an arbitrary length scale such as the diameter of the jet orifice. Set $R_t = 25.7$ and take $x_0 = 0$. Make contour plots of $\overline{U}, \overline{V}$ and $\overline{\Omega}$ in the region in scaled coordinates

$10 \le x \le 500$, $-100 \le y \le 100$. Make spanwise line plots of the mean velocities and vorticity at some x locations.

11.5 For a planar jet flow, derive an expression for the eddy viscosity scaled by the width of the entrance region d and the velocity scale $\sqrt{M/(d\rho)}$.

11.6 Determine $\overline{uv}$ in the self-similar region of a mixing layer flow that corresponds to Eq. (11.88).

11.7 Use a symbolic program (such as *syms* in MATLAB) to compute $\overline{U}, \overline{V}$, and the mean vorticity $\overline{\Omega}$ beginning from Eq. (11.71) for the mean stream function in the self-similar region of a mixing layer. Use the approximation in Eq. (11.79) through the term F_1. Scale velocities by U_m, take $\Delta U = 2$, and scale lengths by an arbitrary length scale d. Set $R_t = 27$ and take $x_0 = 0$. Make contour plots of $\overline{U}, \overline{V}$ and $\overline{\Omega}$ in the region in scaled coordinates $10 \le x \le 500$, $-100 \le y \le 100$. Make spanwise line plots of the mean velocities and vorticity at some x locations.

11.8 Find an expression for the eddy viscosity scaled by U_m and an arbitrary length d in a mixing layer.

12

Calculation of Ground Vehicle Flows

This chapter provides an examination of the practical capabilities of several of the RANS and LES turbulence prediction schemes that were discussed previously. A selection of specific results is presented for the flow past automobiles and trucks that illustrate the successes and failures of flow prediction as it exists currently. In general, the account given here is focussed on those methods that are commonly employed in the field of ground vehicle aerodynamics. It can be expected that other related models may typically give comparable results to these and may even produce superior results in some settings. This fits in with the general fact that no one model can be identified that is superior to all others.

The examples considered include both automobile and truck flows. The results presented are intended to illustrate the kinds of data that are often acquired computationally and the kind of experimental data they are compared to. The results give an indication of the quantitative and qualitative accuracy with which the essential physical aspects of ground vehicle flows are captured. Inaccuracies will be seen to be in some cases fundamental as, for example, when essential phenomena are missed entirely in the calculations. One definite trend that emerges is that LES schemes employing many mesh points tend to more faithfully capture the main properties of ground vehicle flows than do solutions of the RANS equations. Depending on the situation, the substantial extra cost of LES may or may not be justified.

For calculations of complex fluid flows such as are encountered in ground vehicle aerodynamics, numerical considerations such as grid resolution are likely to have a noticeable effect on results. Ideally, predictions of mean flow quantities through RANS modeling should be fully converged and hence grid independent since the solutions are fully deterministic. In practice, even with very large meshes it is possible that convergence isn't complete, though it may be sufficient for engineering purposes. In the case of LES computations grid dependence tends to be built into the calculation of the random flow fields so grid effects are expected. In such cases, the grid parameters become additional variables that might be adjusted besides those in the subgrid models.

12.1 Ahmed Body

Validation of turbulence models in the realm of ground vehicle aerodynamics centers on a few specific configurations that have received extensive study via physical experiments. Perhaps the most widely used geometry for assessing the capabilities of

Turbulent Fluid Flow, First Edition. Peter S. Bernard.

Companion website: www.wiley.com/go/Bernard/Turbulent_Fluid_Flow

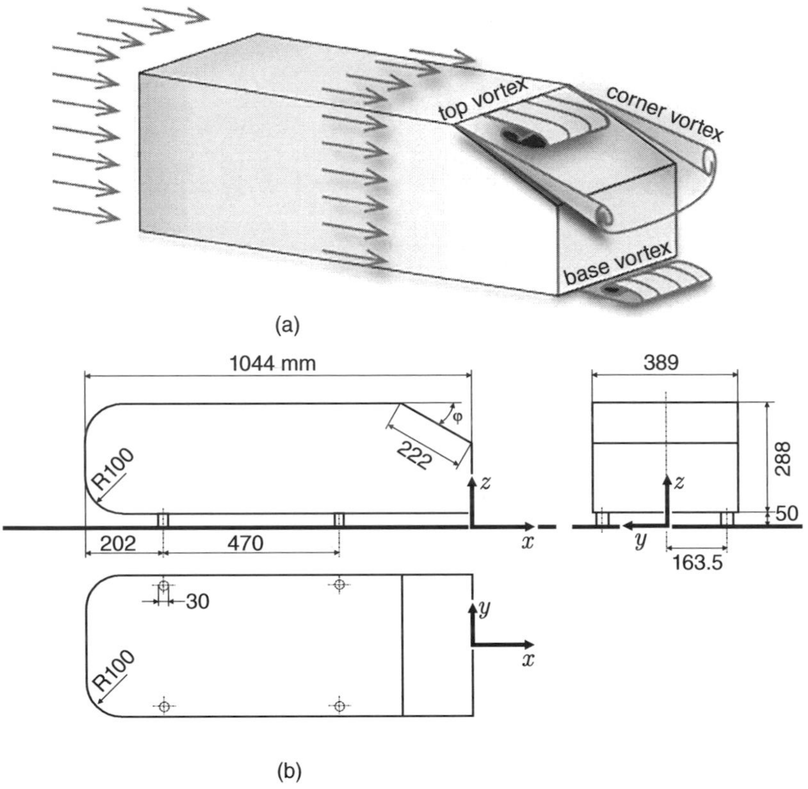

Figure 12.1 Ahmed body. From [4]. Used with permission of Elsevier. (a) Some of the principal flow features.; (b) Body dimensions from three orientations.

models is the Ahmed body [1, 2] that is shown in Figure 12.1. The shape is reminiscent of a van with rounded front edges, which in this case prevents flow separation in the front, and a slanted rear window reminiscent of a hatchback design. The model is held above a stationary ground plane on four pillars allowing flow underneath that affects the wake. Though much simplified compared to a real automobile shape, the Ahmed body nonetheless contains some of the phenomena that have been most challenging for successful turbulent flow prediction [3]. In particular, the flow in the wake of the Ahmed body and the net pressure force this generates on the surface are sensitive functions of the precise angle formed by the rear window. For the flow at $R_e = 7.68 \times 10^5$ that is often studied, a fundamental change in the character of the flow occurs as the rear slant angle (ϕ in Figure 12.1b) is lowered from 35° to 25°. In the former case, the flow separates at the top edge of the rear slanted window and remains separated traveling into the wake. In contrast, for angle 25° as illustrated in Figure 12.1a, the flow separates as in the 35° case but reattaches on the slanted surface. This behavior is a result of the influence of strong counter-rotating vortices produced on the edges of the slanted surface as also shown in the figure. The vortices supply high-speed momentum toward the center of the slanted surface, resulting in reattachment of the separated flow, as also seen in the figure. Capturing the physics that is responsible for this change in flow pattern is a major challenge for turbulence models and one of great importance to

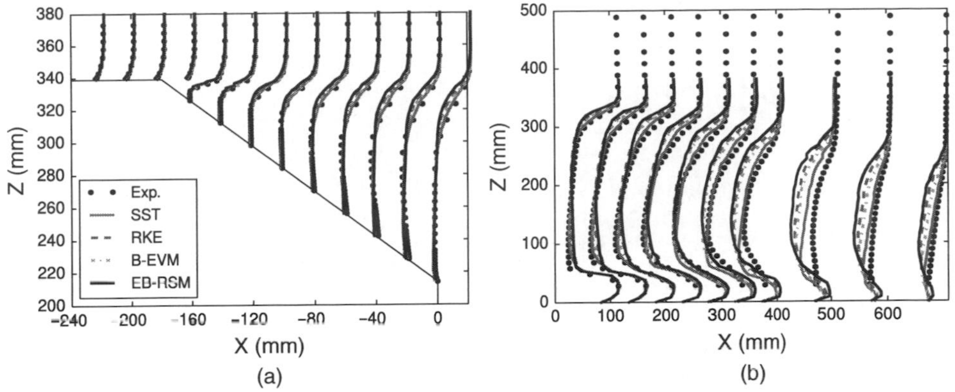

Figure 12.2 Prediction of the mean velocity for the Ahmed body with rear slant angle 35°. (a) Locations over the rear slant and (b) locations in the wake. From [4]. Used with permission of Elsevier.

ground vehicle design, since the induced drag on the vehicle is substantially higher for the flow with the 25° slant than it is with the 35° slant.

To be most useful in engineering work, model predictions of the flow past the Ahmed body should be sensitive to the changes in flow structure when the rear slant angle changes from 35° to 25°. In fact, computations show that this basic feature cannot be reliably reproduced by RANS models as well as many LES and hybrid LES/RANS schemes. For many of the approaches the 35° case can be somewhat accurately portrayed while the 25° case is the more challenging. Figure 12.2 shows the result of model calculations of the mean velocity along the rear slant and into the wake of the Ahmed body for the 35° case. The particular models considered here are the SST model discussed in Section 9.2.4, the realizable variant of the $K-\epsilon$ closure [5] (RKE) that was discussed in Section 9.2.2, and formulations of the elliptic relaxation or blending model [6] both as an eddy viscosity model (B-EVM) and as a Reynolds stress model (EB-RSM) [7]. The latter models include an equation for the wall-normal Reynolds stress as discussed in Section 9.2.2.4.

Figure 12.2 shows that the models have a good capability for capturing the mean velocity trends over the slant surface. In the wake of the body, the relaxation of the mean flow back to the free-stream values is best captured by the SST model and appears to be most problematical for the EB-RSM model. Note the jetting of the fluid from beneath the Ahmed body that enters into the wake. In contrast to the modeling capabilities on display in Figure 12.2, predictions of the mean velocity in the 25° rear slant case are much less successful, as shown in Figure 12.3. Over the rear slanted surface, the various models behave in very much the same way and fail to capture the reattachment of the flow near the midpoint of the slanted surface. The velocity profiles actually maintain the same character they had in the 35° case, suggesting that they are not sensitive to the physics of flow separation in this case. In the wake the models under-predict the mean velocity field and fail to predict its qualitative behavior. Among the models, SST recovers most quickly toward the experimental values in the far wake. The plots of kinetic energy in Figure 12.3c corresponding to Figure 12.3a show it to be significantly under-predicted in the beginning of the separated shear layers. Failing to provide sufficient magnitude of turbulent kinetic energy K reduces the effective turbulent

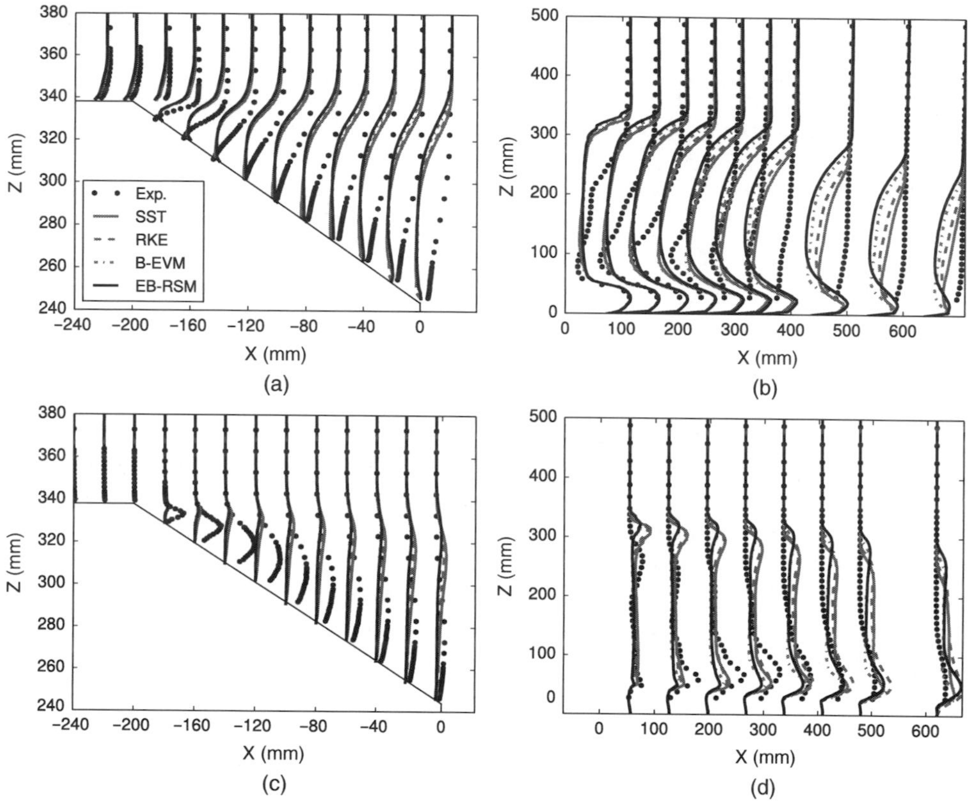

Figure 12.3 Prediction of mean velocity (a and b) and kinetic energy (c and d) for the Ahmed body with rear slant angle 25°. (a) and (c) are over the rear slant and (b) and (d) are in the wake. From [4]. Used with permission of Elsevier.

transport in the separated flow region so that the latter extends further downstream than occurs in the comparable experiment.

While Figure 12.1 gives an accurate portrayal of the average behavior of the flow over the rear slant on the Ahmed body, the transient flow tends to be considerably more complicated. For example, during some random time intervals the flow may separate from the rear of the roof and not reattach, and at any given time there is likely to be many convecting vortical structures along the slanting surface. To the extent that turbulence models need to account for such effects, LES is in a better position than RANS models to provide such details of the Ahmed body flow. A number of studies have applied LES schemes to the solution of the Ahmed body flow field. Figure 12.4 shows the effectiveness of DES methods, such as were previously discussed in Section 10.4.1, in modeling the 25° Ahmed body rear surface. The IDDES model utilizing the SA and SST models is compared in Figure 12.4a while Figure 12.4b contrasts the DDES vs. IDDES models adapted from the SA model. For both figures it is evident that there are only small differences between the various specific implementations of DES. The prediction of mean velocity over the rear slanted surface is somewhat better than the RANS predictions shown in Figure 12.3, though there is still room for improvement.

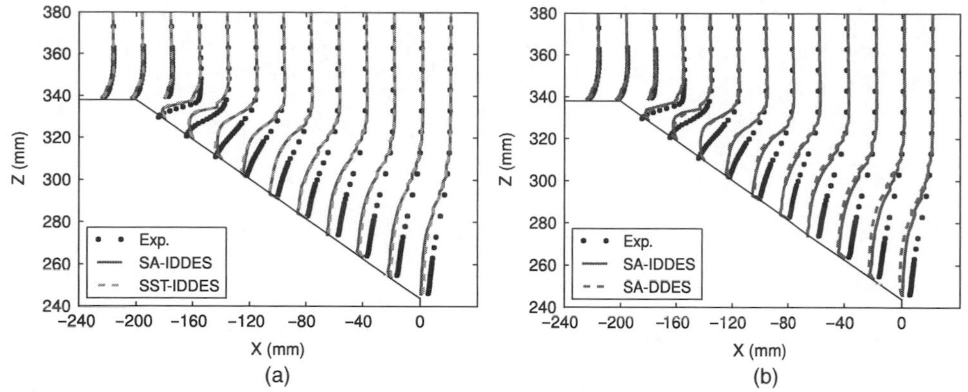

Figure 12.4 Prediction of mean velocity for the Ahmed body with rear slant angle 25°. (a) Comparison of SA-IDDES and SST-IDDES with experiment. (b) Comparison of SA-DDES and SA-IDDES with experiment. From [4]. Used with permission of Elsevier.

One reason the DES results show some advantage over RANS is that they tend to predict considerably more kinetic energy than the RANS models. However, the continued failure of the DES calculations to capture the correct size of the separation region (see also [8]) reflects a need for even higher levels of K at the inception of the separated region than are typically computed. Downstream of the body as the wake flow develops the SST-IDDES model gives excellent prediction of the mean velocity field. It should also be noted that ad hoc embedding of synthetic turbulence just upstream of the rear slant is beneficial in raising the kinetic energy in the beginning shear layer, leading to an improved prediction of the 25° separated flow region. Techniques such as this, however, are not readily applied in more general flow geometries.

Another study that has focused on evaluating LES schemes in simulating the 25° rear slant angle flow field includes the WALE model considered in Section 10.3.2 and a variational multiscale model (VMS) that utilizes a comparatively coarse mesh. In the latter, the mathematical form of the WALE model is applied to the small-scale part of the filtered velocity field [9]. Figure 12.5 compares the computed streamwise velocity to physical measurements along the slant back for these LES methods. The WALE model is

Figure 12.5 Prediction of mean velocity for the Ahmed body with rear slant angle 25°. From [9]. Used with permission of Elsevier.

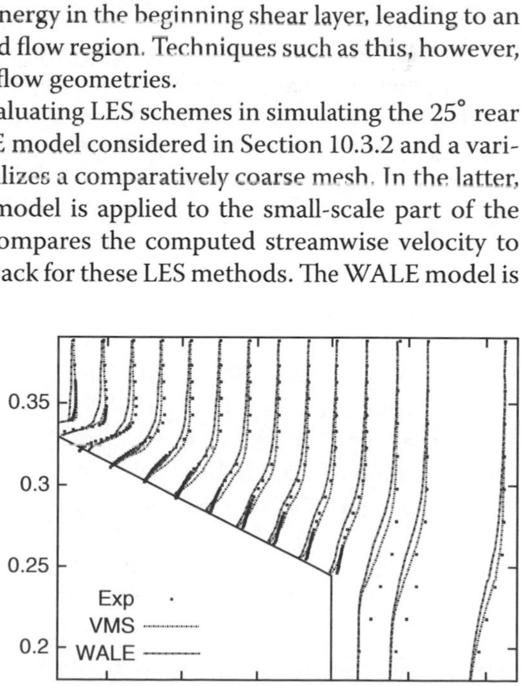

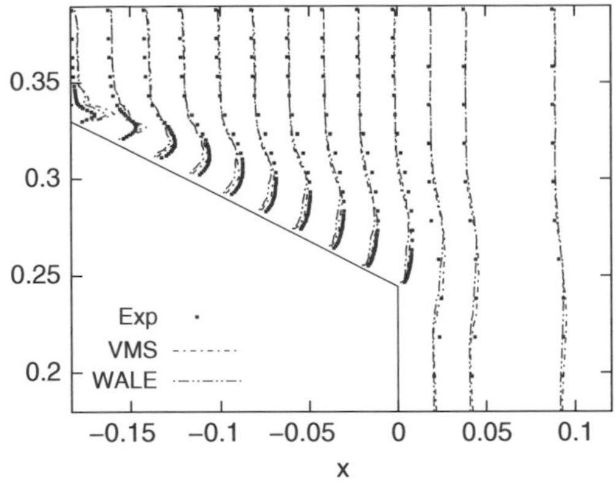

Figure 12.6 Prediction of u_{rms} for the Ahmed body with rear slant angle 25°. From [9]. Used with permission of Elsevier.

somewhat successful in capturing the trend in mean velocity and does so better than the VMS model. There is a clear improvement in accuracy over the RANS and DES methods displayed in Figures 12.3 and 12.4. The plots of u_{rms} shown in Figure 12.6 help explain the better results since it is seen that the turbulent energy in the initial stages of the separated shear layer is much better accounted for than in the previously considered methods. In the wake, the results for both models shown in Figure 12.5 are initially inaccurate but improve significantly with downstream distance, finally agreeing well with the physical experiments. In the case of u_{rms} the VMS model gives better results than the WALE model. The drag coefficient for the Ahmed body in these calculations is measured in physical experiments to be 0.285 while the VMS/WALE model predicts 0.290 and the traditional WALE model has drag equal to 0.304.

In another study [10] four LES models applied to the Ahmed body flow were directly compared. These include the traditional Smagorinsky subgrid scale model incorporating a near-wall model (LES-NWM) based on the use of wall functions as mentioned in Section 10.3.1, the dynamic Smagorinsky model with the near-wall flow resolved (LES-NWR) as discussed in Section 10.3.4, the K–ω SST variant of DES (DES-SST), and a high-order spectral vanishing viscosity (LES-SVV) method that uses spectral approximations to stabilize the calculations. The latter approach inserts an artificial viscosity into the high-frequency range of the spectral approximation so as to model dissipation of the subgrid scales. Each of these calculations were made on relatively large meshes, though none provided DNS resolution near the boundary. Mean velocity predictions for these cases are shown in Figure 12.7, where it is seen that the dynamic Smagorinsky model predicts a fully attached boundary layer over the slant and poor agreement with the experiment. Such flow behavior is seen in experiments for rear slant angles less than 12.5°. Accompanying this result is a significant under-prediction of the turbulent kinetic energy.

The calculation with DES-SST shows that the flow has separated in the form of a shear layer over the entire length of the slant surface. The kinetic energy in this case is under-predicted at the start and rises later. These DES results are consistent with those observed in reference to Figure 12.4. The Smagorinsky model with wall functions also predicts separation over the length of the slant with somewhat better accuracy than DES in the first part of the region and equivalent accuracy in the lower half. Accompanying

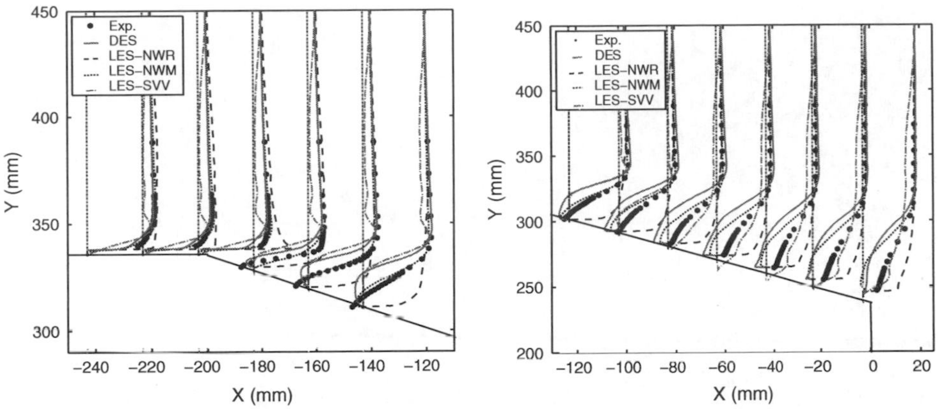

Figure 12.7 Prediction of mean velocity for the Ahmed body with rear slant angle 25°. From [10]. Used with permission of Elsevier.

this is a somewhat better prediction of the turbulent kinetic energy near the leading edge than the DES model, but less accuracy downstream.

The calculation with the SVV approach fails to capture the upstream boundary layer over the roof where its slope at the wall is too small and the mean velocity is over-predicted, creating a jet-like profile. These defects reflect the legacy of upstream facets of the flow field such as separation at the forebody and diversion of the mean flow from the center plane. Like the DES calculation the shear layer formed by the separating fluid is much further from the wall than in experiments so initially the trend in mean velocity is underpredicted. On the lower half of the slant surface, the flow reattaches and has better agreement with the experiments than the other models.

It is instructive to observe the way in which vortical structures appear on the Ahmed body shape as computed in the different simulations. Figure 12.8 shows isosurfaces of λ_2 as discussed in Section 8.5.3 for a calculation with the SST IDDES model of the 25° base slant angle flow. A prominent part of the calculation is the appearance of vortices on either side of the slanted rear surface, though this feature is not sufficiently accurate to coincide with reattachment of the boundary layer on the slanted surface. More success in the later regard was achieved by the VMS/WALE model simulation [9], whose prediction of the vortical features is shown in Figure 12.9 for the 25° slanted rear surface. In this figure the quantity Q as discussed in Section 8.5.3 is used to demarcate rotational regions. Reattachment in a mean sense is along the slanted surface, as was noted in Figure 12.7.

Averaged streamlines for four LES calculations of the flow along the rear centerline of the Ahmed body are compared to experimental measurements in Figure 12.10. Both the DES calculation and Smagorinksy LES with wall functions show separation from the front of the base slant that does not reattach to the body, in other words the same condition as seen in the 35° base slant angle case. The better resolved dynamic Smagorinksy model fails to separate from the rear slant surface, a result that agrees with the observation that side vortices such as those seen in Figures 12.8 and 12.9 fail to form out of the well-resolved near-wall flow in this case. The SVV method is seen to predict

Figure 12.8 Isosurfaces of λ_2 for the Ahmed body with 25° rear slant computed from the SST-IDDES simulation. From [4]. Used with permission of Elsevier.

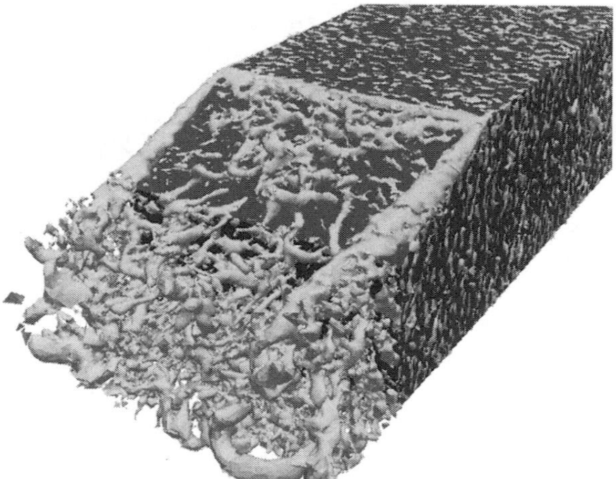

Figure 12.9 Isosurfaces of Q for the Ahmed body with 25° rear slant computed from the VMW/WALE simulation. From [9]. Used with permission of Elsevier.

flow reattachment along the surface that is qualitatively in agreement with experiments, though it has some idiosyncratic 3D features that are not apparent in the figure. Beyond the rear surface of the body all of the simulations contain mean counter-rotating recirculating regions that form from separation over the rear edge of the vehicle. It is evident that the SVV approach yields a vortex pattern that is closest to that seen in experiments. For the LES-NWM and DES-SST results, the separated region over the slanted surface is cojoined to the comparatively large recirculation zone formed on the top part of the rear surface, a behavior seen as well for the 35° base slant angle case.

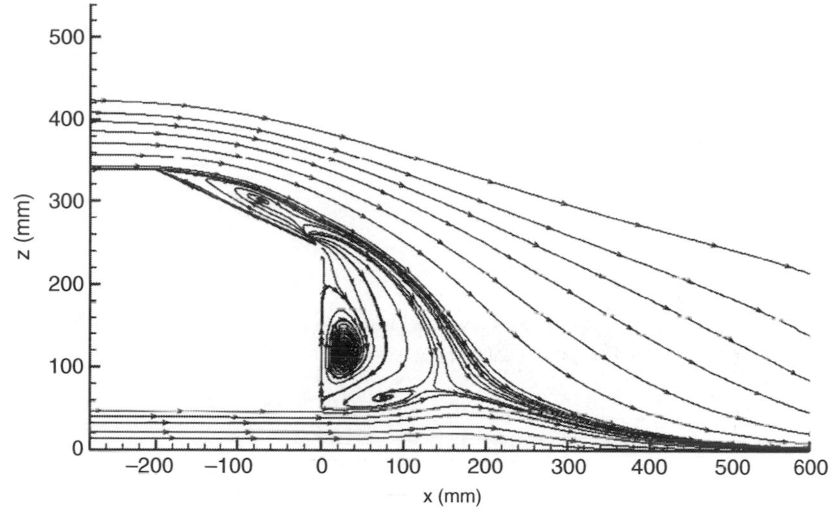

(a)

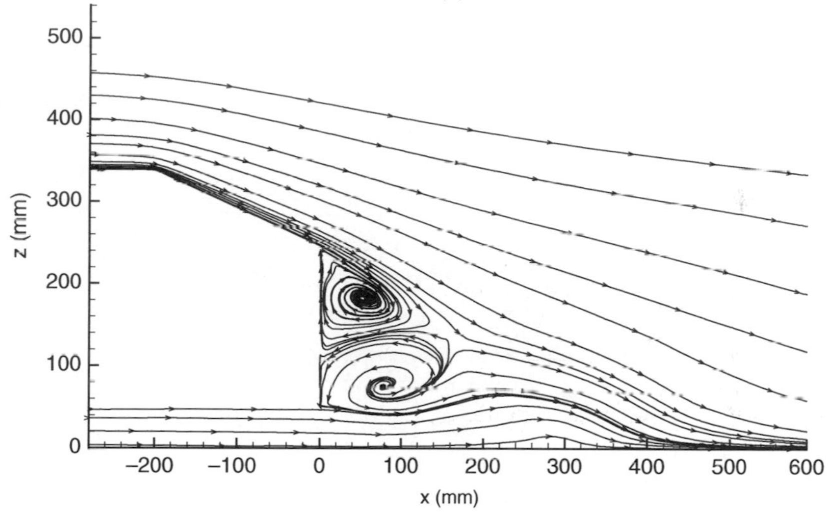

(b)

Figure 12.10 Mean streamlines on the central plane of the Ahmed body. From [10]. Used with permission of Elsevier. (a) DES-SST; (b) LES-NWR; (c) LES-SVV; (d) LES-NWM; (e) Experiments [11]

12.2 Realistic Automotive Shapes

A generic vehicle shape that is somewhat more realistic than the Ahmed body and for which non-proprietary data is available for assisting in turbulence model development is the drivAer models shown in Figure 12.11. As seen in the picture, the model comes in three variations depending on how the rear surfaces are rendered. For the fastback model, the rear separation over the top surface must be at the very rear of the body. In contrast, for the estate model, there is a rear slanted window similar to that of the Ahmed body that can be expected to provide a significant challenge to flow prediction.

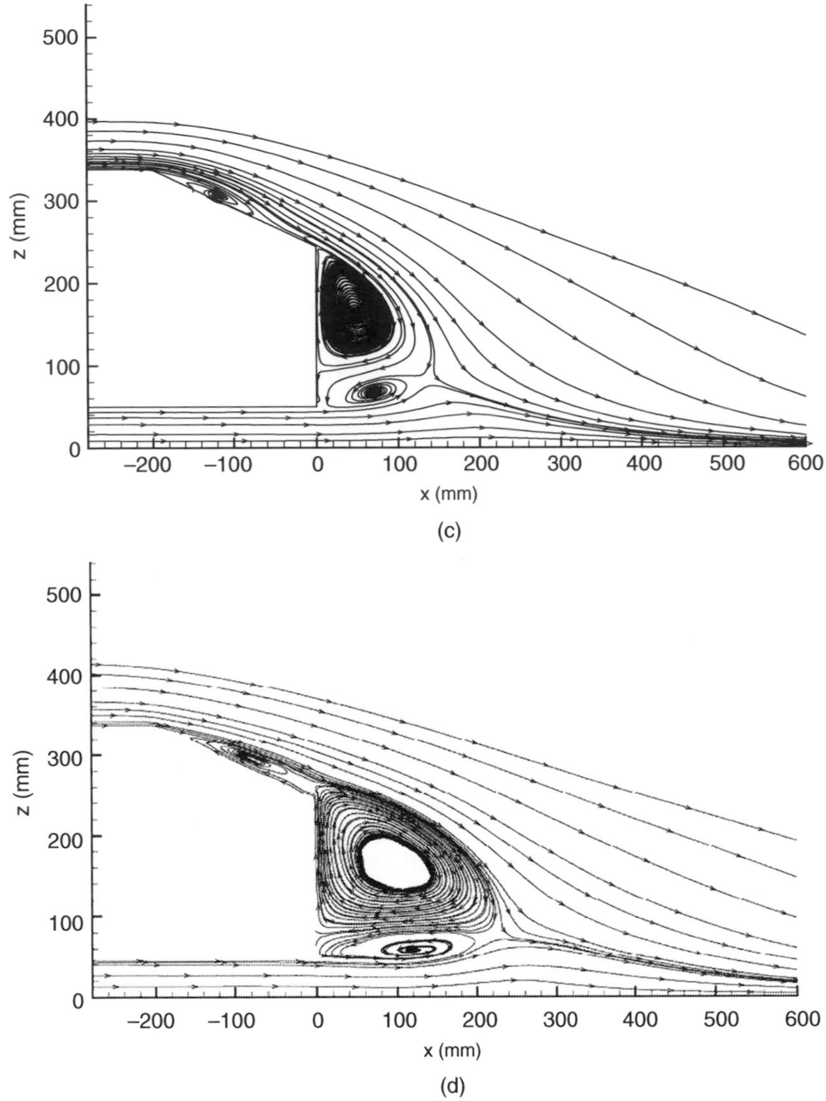

Figure 12.10 (*Continued*)

Mean pressure data obtained over the centerline of the drivAer geometries is compared to calculations in Figure 12.12 using the SST RANS and IDDES models. The effect of grid resolution is tested by showing results with IDDES for coarse and fine meshing. While the pressure predictions capture the general trends in the experimental data, a range of significant differences are apparent, such as at the rear of the body where the calculations tend to under-predict the pressure. Ultimately, it is the accumulated effect of pressure over the entire surface that is the main concern in predicting forces. Some idea of the overall accuracy of the method is given in Table 12.1 showing drag and lift coefficients for the drivAer fastback and estate models determined in experiments and computed using several RANS approaches and hybrid LES/RANS schemes in the form

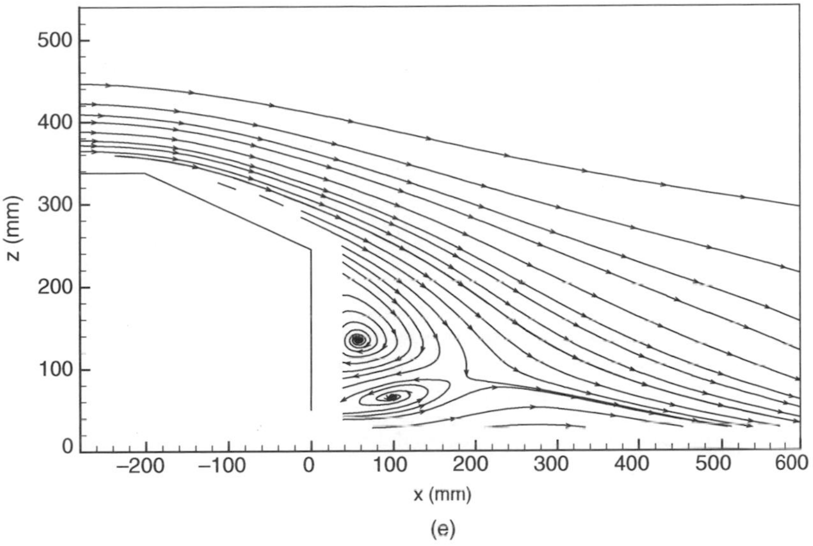

Figure 12.10 (*Continued*)

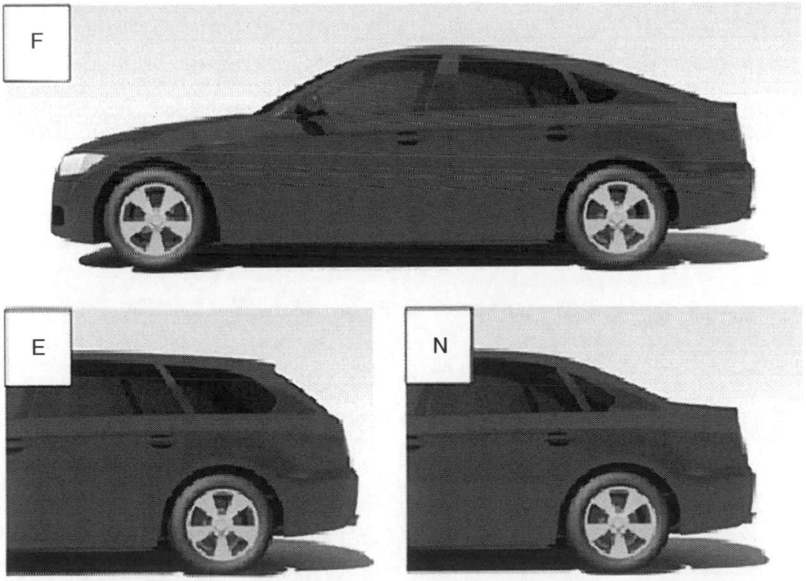

Figure 12.11 DrivAer models. F, fastback; E, estate; N, notchback. From [4]. Used with permission of Elsevier.

of DES. On the whole, the greater physical realism of hybrid LES/RANS methods pays off in superior predictions of drag and especially the lift force. For both models the RANS calculations tend to significantly over-predict the lift forces compared to DES. Some of the RANS models predict a positive lift force for the fastback model where experiments show that it is, in fact, negative. Capturing the correct trend in such cases is an important goal of many engineering studies.

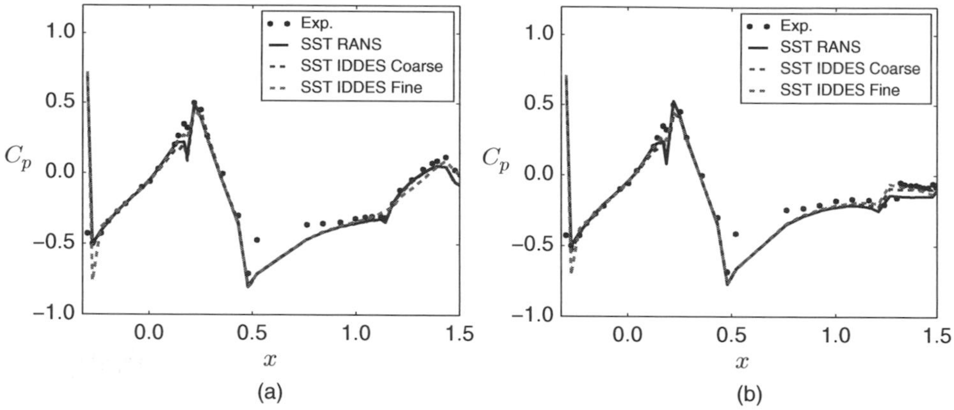

Figure 12.12 Mean pressure coefficient on the top center of the vehicles. From [4]. Used with permission of Elsevier. (a) Fast back; (b) Estate

Table 12.1 Drag coefficient for the estate and fastback configurations using RANS and DES models. From[4]

	Fastback C_D	Fastback C_L	Estate C_D	Estate C_L
Experiment	0.294	−0.12	0.261	0.01
Spalart–Allmaras (SA)	0.280	0.054	0.260	0.136
Realizable$K$–ϵ (RKE)	0.260	−0.026	0.244	0.085
K–ω SST	0.275	0.0436	0.260	0.124
K–ϵ B-EVM	0.253	0.007	0.2435	0.116
EB-RSM	0.256	−0.029	0.2482	0.075
SST IDDES (coarse)	0.310	−0.096	0.268	0.011
SST IDDES (fine)	0.307	−0.131	0.2615	0.024
SA IDDES (fine)	0.313	−0.136	–	–
SA DDES (fine)	0.307	−0.13	–	–

The significant differences between RANS and LES/RANS predictions is visible in the structure of the wake flow, as shown in Figures 12.13 and 12.14 for the drivAer fastback and estate models, respectively. The RANS models tend to predict relatively large recirculation regions that differ both qualitatively and quantitatively from that computed using the LES/RANS models. Since the latter results are associated with superior drag and lift predictions, one expects that these wake predictions are likely to be closest to the actual physical behavior.

Since the nature of the wake flow is often important in its own right, in addition to the actual drag and lift values, this provides some incentive for seeking LES/RANS solutions, despite the limitations to their accuracy in predicting separation and pressure forces on flows such as that of the estate model. From the point of view of practicality, however, it should be noted that the LES/RANS calculations require approximately 17

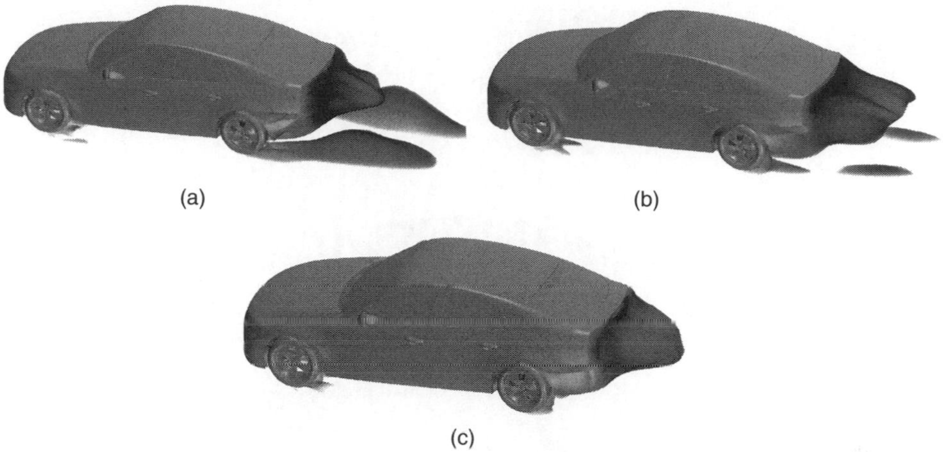

Figure 12.13 Fastback wake structure indicated by the region of reverse flow ($U < 0$). From [4]. Used with permission of Elsevier. (a) SST-RANS; (b) EB-RSM; (c) SST-IDDES

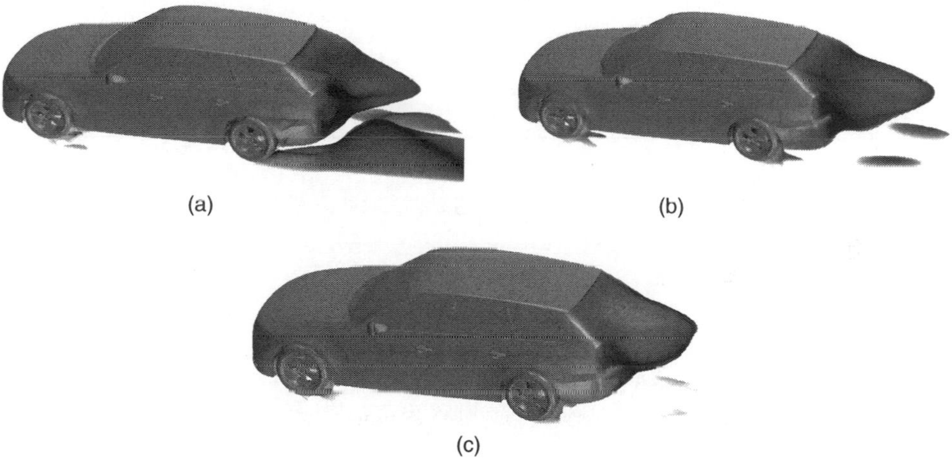

Figure 12.14 Estate wake structure indicated by the region of reverse flow ($U < 0$). From [4]. Used with permission of Elsevier. (a) SST-RANS; (b) EB-RSM; (c) SST-IDDES

times the computational cost of the RANS studies. This can be an important consideration in whether or not the LES/RANS models are ultimately useful for a particular application.

Another generic car shape that is often studied is the Asmo vehicle shown in Figure 12.15. In this case for flow at $R_e = 2.7 \times 10^6$ experiments show that the pressure drag = 0.153, which is exactly matched by the VMS/WALE calculation on a mesh with 2.8M elements [9]. For the same calculation the WALE model itself had the pressure drag as 0.162. A detailed look at the mean pressure predictions on the back centerline of the Asmo vehicle is shown in Figure 12.16 for the VMS/WALE and WALE models and in Figure 12.17 for a LES calculation using the Smagorinsky model and a RANS calculation with the $K\!-\!\epsilon$ closure [12]. The latter computations were done for both fine and coarse meshes. Among these results it is seen that the WALE model gives a good

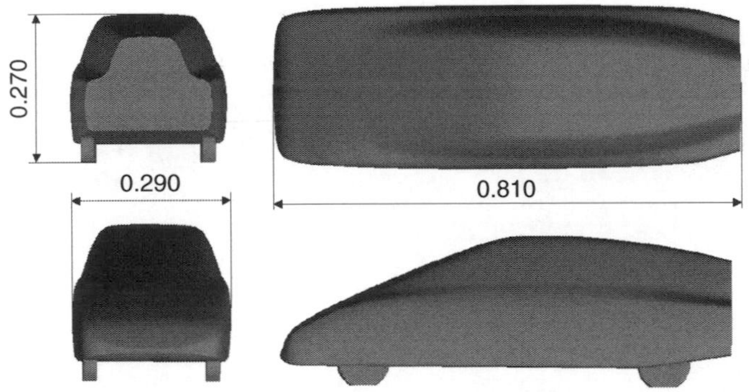

Figure 12.15 Asmo vehicle. From [9]. Used with permission of Elsevier.

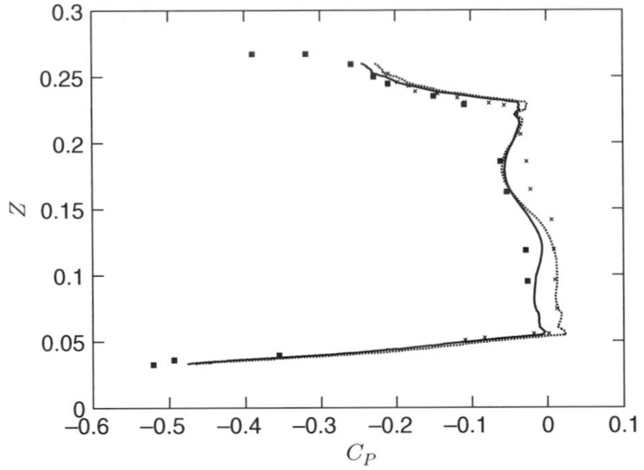

Figure 12.16 Pressure on back surface of the Asmo vehicle. · · ·, VMS/WALE; —, WALE; ■, data from Daimler-Benz; ×, data from Volvo. From [9]. Used with permission of Elsevier.

description of pressure on the back surface of the car, though it is not consistently so for other surfaces such as the front and underbody. The Smagorinsky model shown in Figure 12.17 gives a superior representation of the rear pressure compared to the RANS model. This should translate into significant improvements in the quality of drag predictions.

The value of developing large meshes in LES that enable better prediction of flow in the near-wall region is evidenced in a study of the flow past a small utility vehicle with variation in the slant of the roof [13] with values 0.5°, −2.5°, −7.5°, and −12.5° as shown in Figure 12.18. Plotted in these figures are contours of velocity magnitude surrounding the van. The flow is computed using the Smagorinsky model at Reynolds number $R_e = 2.68 \times 10^6$ based on the vehicle height. The mesh in these calculations has 2 billion elements, with extra care taken in supplying a fine near-wall mesh. The latter is found to be of significant importance in improving the accuracy of the prediction of surface forces compared to coarser meshes. The mesh density is found to have a significant effect on wake structure. For example, Figure 12.18 shows that in all cases there is essentially

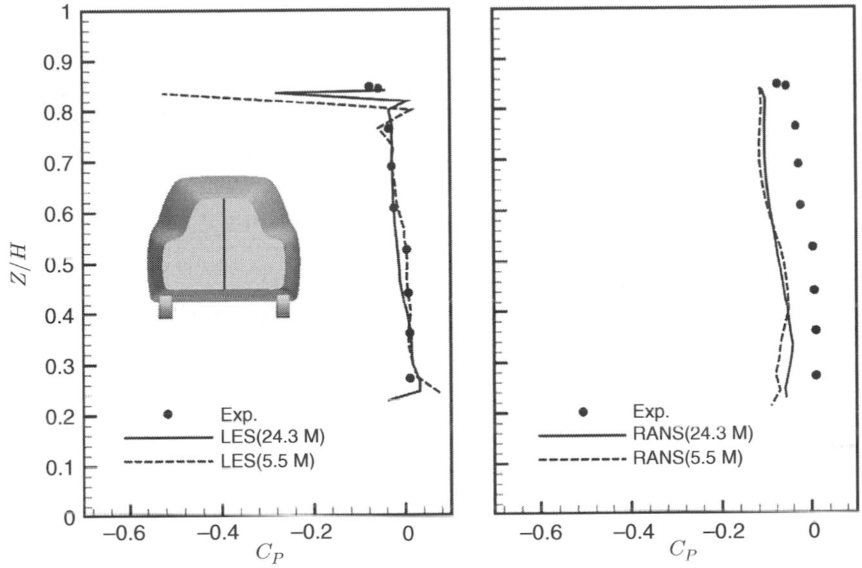

Figure 12.17 Pressure on flat back surface of the Asmo vehicle. Left, the Smagorinsky LES model; right, the K–ϵ closure. From [12]. Used with permission of Elsevier.

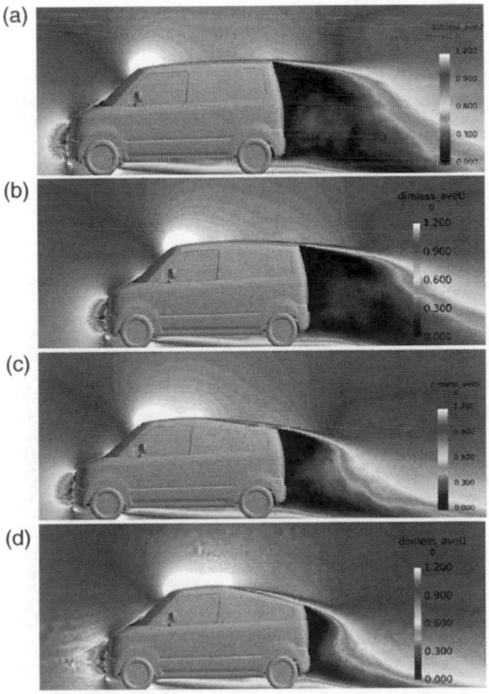

Figure 12.18 Contours of velocity magnitude on the central plane computed with the Smagorinsky LES scheme for flow past a van with variable roof angle. (a) 0.5°; (b) −2.5°; (c) −7.5°; (d) −12.5°. From [13]. Used with permission of SAE International.

a single region of low velocity in the wake of the van that angles toward the ground at increasing distances from the vehicle. However, the region of low velocity becomes smaller as the slant of the roof increases. Similar calculations with a coarser mesh at the wall yield a qualitatively different wake structure with two distinct regions of low velocity: one centered in the upper wake region and the other in the lower region. Experimental measurement of the drag coefficient gives the values 1.0, 0.978, 0.988, and 0.966, respectively, for the slant angles 0.5°, −2.5°, −7.5°, and −12.5°. With the fine mesh, the predicted drag coefficients in these cases are 0.977, 0.943, 0.969, and 0.977, respectively, showing that the error varies between 1.1% and 3.5%. The errors are substantially greater with the coarser mesh.

12.3 Truck Flows

The economic and ecological benefits of reducing aerodynamic drag on trucks has spurred considerable research into developing low-drag modifications such as enclosing the gap between the tractor and trailer via an aerodynamic fairing, adding appendages to the rear of the truck that streamline the wake flow, and skirts beneath the sides of the trailer. For the most part such developments have been designed and tested via physical experiments, though an increasing role of simulations can be expected as the capabilities of CFD and turbulent flow modeling mature. A primary source of data for evaluating turbulent flow predictions for heavy trucks is the generic conventional model (GCM) shown in Figure 12.19 whose flow characteristics have been measured in wind tunnels [14]. Data is available for surface pressure distributions, drag forces, and some velocity measurements.

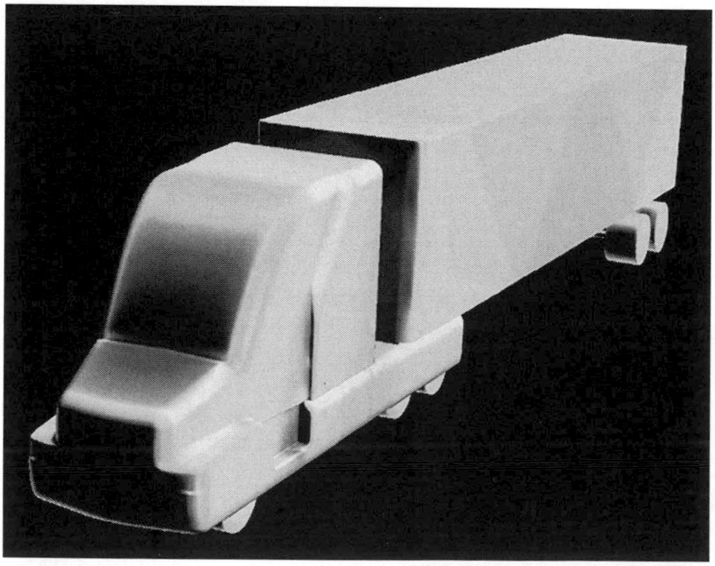

Figure 12.19 The generic conventional model used in testing turbulence prediction schemes. Used with permission of Elsevier.

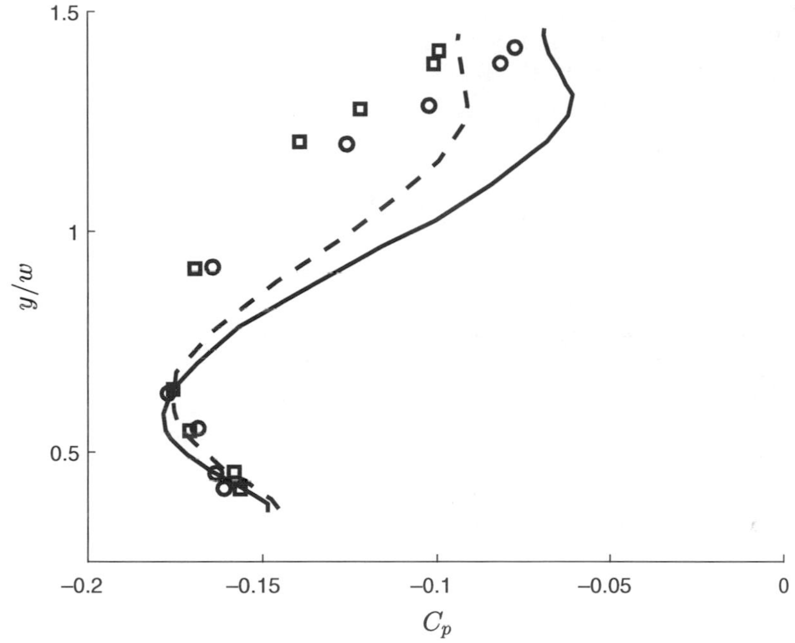

Figure 12.20 Mean pressure over the rear surface of the trailer. —, computed on centerline $z = 0$; −−, computed at $z = 0.25$; □, experiment on centerline; ○, experiment at $z = 0.25$. From [15]. Used with permission of Elsevier.

Simulations of the GCM flow [15] have been accomplished with the scale adaptive simulation (SAS) hybrid LES/URANS scheme described in Section 10.4.6. In this the Reynolds number $R_e = 1.15 \times 10^6$ scaled by the trailer width and the grid contains $\approx 65M$ elements. While much of the measured pressure distribution can be reproduced by the simulations, and predictions of overall drag are considered good, nonetheless a detailed look at the pressure field reveals some significant differences in important parts of the flow. For example, Figure 12.20 compares the predicted pressure vs. physical experiments across the rear of the trailer. Evidently, the modeling provides only some general qualitative agreement in the pressure distribution with some substantial quantitative errors.

Simulations provide a potentially useful means of studying the stability of ground-vehicle flows in the face of side winds and transient changes to the local wind distribution as when two trucks are passing on the road. An example of the potential that such technology may have on the engineering of stable and drag efficient trucks is shown in Figure 12.21 for the case of a 10° yaw with respect to the wind. Here, the instantaneous vorticity magnitude is shown for a plane at the approximate mid-height of the trailer. Accompanying pressure values show similar accuracy as in the zero-yaw case. A dearth of physical measurements for the velocity field in the depicted circumstances prevents more rigorous testing of predictions of separated regions. Nonetheless, such calculations can have value in revealing trends and in otherwise exposing areas of the flow field worthy of more extensive examination.

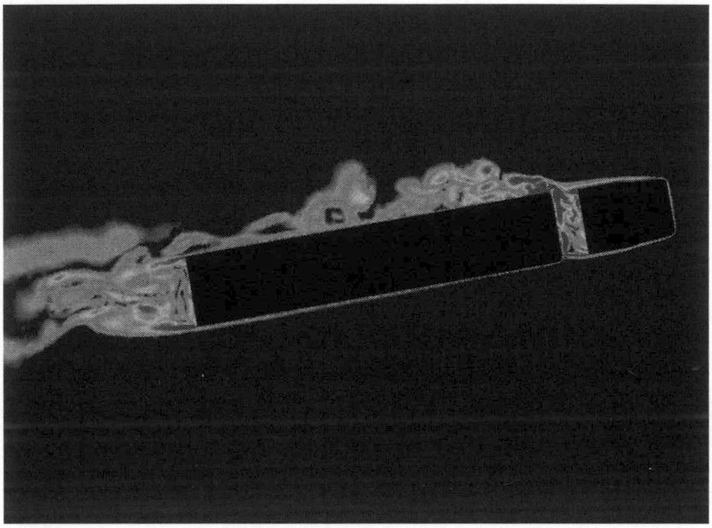

Figure 12.21 Instantaneous contours of vorticity magnitude on a plane approximately half-way through the truck at a 10° yaw with respect to the wind. From [15]. Used by permission of Elsevier.

12.4 Chapter Summary

The prediction of several important aspects of ground vehicle flows considered here generally reveals the superiority of simulation methodologies such as LES and LES/RANS in comparison to RANS modeling. It is also the case that there is considerable need for further development and testing of the simulation methods. Thus, while some qualitative and quantitative success can be had in reproducing fundamental features such as the modes of flow separation, and in predicting pressure, velocities, and energy, there are also numerous facets of the computed flow fields that disagree with experiment. One promising direction that can be taken to improve predictions in all aspects of ground vehicle flow fields lies in the usage of grids with enhanced resolution, particularly in the near-wall flow field. Larger meshes reduce the reliance on modeling so that even calculations with simplified subgrid models, such as the Smagorinsky model, are seen to achieve some favorable results in test problems.

Beyond applications to test configurations such as the Ahmed, drivAer, Asmo, and GCM bodies for which detailed experimental data is available for comparisons, new studies are venturing into the calculation of turbulent flow around complex automotive geometries, including realistic underbody details, engine compartment flows, and other details of actual ground vehicles [12]. The results from simpler shapes suggest that provision must be made for retaining high near-wall mesh resolution even as additional geometrical features are accommodated. This requires the use of very large meshes and simulation approaches that can provide for such calculations in terms of mesh generation, storage, parallel efficiency, and the practicality of data post-processing. Exactly what role improvements in sub-grid modeling will play in such future work remains to be seen, particularly in regard to whether or not more computationally demanding subgrid models can deliver sufficient improvements in accuracy so that they are cost effective.

References

1 Ahmed, S.R., Ramm, G., and Faltin, G. (1984) Some salient features of the time-averaged ground vehicle wake, *Paper 840300*, SAE, Warrendale, PA.

2 Choi, H., Lee, J., and Park, H. (2014) Aerodynamics of heavy vehicles. *Ann. Rev. Fluid Mech.*, 46, 441–468.

3 Zhang, B.F., Zhou, Y., and To, S. (2015) Unsteady flow structures around a high-drag Ahmed body. *J. Fluid Mech.*, 777, 291–326.

4 Ashton, N., West, A., Lardeau, S., and Revell, A. (2016) Assessment of RANS and DES methods for realistic automotive models. *Computers and Fluids*, 128, 1–15.

5 Shih, T.H., Liou, W.W., Shabbir, A., Yang, Z., and Zhu, J. (1995) A new $K-\epsilon$ eddy viscosity model for high Reynolds number turbulent flows. *Computers and Fluids*, 24, 227–238.

6 Billard, F. and Laurence, D.A. (2012) A robust $K-\epsilon-\overline{v^2}/K$ elliptic blending turbulence model applied to near-wall, separated and buoyant flows. *Int. J. Heat Fluid Flow*, 33, 45–58.

7 Lardeau, S. and Manceau, R. (2014) Computations of complex flow configurations using a modified elliptic-blending Reynolds-stress model, in *ETMM10: 10th International ERCOFTAC Symposium on Engineering*.

8 Guilmineau, E., Deng, G., and Wackers, J. (2011) Numerical simulation with a DES approach for automotive flows. *J. Fluids Struct.*, 27, 807–816.

9 Aljure, D.E., Lehmkuhl, O., Rodríguez, I., and Oliva, A. (2014) Flow and turbulent structures around simplified car models. *Computers and Fluids*, 96, 122–135.

10 Serre, E., Minguez, M., Asquetti, R., Guilmineau, E., Deng, G.B., Kornhaas, M., Schäfer, M., Frählich, J., Hinterberger, C., and Rodi, W. (2013) On simulating the turbulent flow around the Ahmed body: A French–German collaborative evaluation of LES and DES. *Computers and Fluids*, 78, 10–23.

11 Lienhart, H., Stoots, C., and Becker, S. (2000) Flow and turbulence structures in the wake of a simplified car model (Ahmed body), in *DGLR fach symposium der AG STAB*, Stuttgart University.

12 Tsubokura, M., Kobayashi, T., Nakashima, T., Nouzawa, T., Nakamura, T., Zhang, H., Onishi, K., and Oshima, N. (2009) Computational visualization of unsteady flow around vehicles using high performance computing. *Computers and Fluids*, 38, 981–990.

13 Tsubokura, M., Kerr, A., Onishi, K., and Hashizume, Y. (2014) Vehicle aerodynamics simulation for the next generation on the K computer: Part 1 development of the framework for fully unstructured grids using up to 10 billion numerical elements. *SAE Int. J. Passeng. Cars – Mech. Syst.*, 7, 1106–1118.

14 Storms, B., Satran, D., Heineck, J., and Walker, S. (2006) A summary of the experimental results for a generic tractor-trailer in the Ames Research Center 7- by 10-foot and 12-foot wind tunnels., *TM 2006-213489*, NASA.

15 Hyams, D.G., Sreenivas, K., Pankajakshan, R., Nichols, D.S., Briley, W.R., and Whitfield, D.L. (2011) Computational simulation of model and full scale Class 8 trucks with drag reduction devices. *Computers and Fluids*, 41, 27–40.

Author Index

Turbulent Fluid Flow, First Edition. Peter S. Bernard.

Companion website: www.wiley.com/go/Bernard/Turbulent_Fluid_Flow

Subject Index

Turbulent Fluid Flow, First Edition. Peter S. Bernard.

Companion website: www.wiley.com/go/Bernard/Turbulent_Fluid_Flow